Progress in

PHYSICAL ORGANIC CHEMISTRY

VOLUME 6

Progress in

PHYSICAL ORGANIC CHEMISTRY

VOLUME 6

Editors

ANDREW STREITWIESER, JR., *Department of Chemistry*
University of California, Berkeley, California

ROBERT W. TAFT, *Department of Chemistry*
University of California, Irvine, California

1968

INTERSCIENCE PUBLISHERS

a division of John Wiley & Sons · New York · London · Sydney · Toronto

PRINTED IN THE UNITED STATES OF AMERICA

Introduction to the Series

Physical organic chemistry is a relatively modern field with deep roots in chemistry. The subject is concerned with investigations of organic chemistry by quantitative and mathematical methods. The wedding of physical and organic chemistry has provided a remarkable source of inspiration for both of these classical areas of chemical endeavor. Further, the potential for new developments resulting from this union appears to be still greater. A closening of ties with all aspects of molecular structure and spectroscopy is clearly anticipated. The field provides the proving ground for the development of basic tools for investigations in the areas of molecular biology and biophysics. The subject has an inherent association with phenomena in the condensed phase and thereby with the theories of this state of matter.

The chief directions of the field are: (a) the effects of structure and environment on reaction rates and equilibria; (b) mechanism of reactions; and (c) applications of statistical and quantum mechanics to organic compounds and reactions. Taken broadly, of course, much of chemistry lies within these confines. The dominant theme that characterizes this field is the emphasis on interpretation and understanding which permits the effective practice of organic chemistry. The field gains its momentum from the application of basic theories and methods of physical chemistry to the broad areas of knowledge of organic reactions and organic structural theory. The nearly inexhaustible diversity of organic structures permits detailed and systematic investigations which have no peer. The reactions of complex natural products have contributed to the development of theories of physical organic chemistry, and, in turn, these theories have ultimately provided great aid in the elucidation of structures of natural products.

Fundamental advances are offered by the knowledge of energy states and their electronic distributions in organic compounds and the relationship of these to reaction mechanisms. The development, for example, of even an empirical and approximate general scheme for the estimation of activation energies would indeed be most notable.

The complexity of even the simplest organic compounds in terms of physical theory well endows the field of physical organic chemistry with the frustrations of approximations. The quantitative correlations employed in this field vary from purely empirical operational formulations to the approach of applying physical principles to a workable model. The most common procedures have involved the application of approximate theories to approximate models. Critical assessment of the scope and limitations of these approximate applications of theory leads to further development and understanding.

Although he may wish to be a disclaimer, the physical organic chemist attempts to compensate his lack of physical rigor by the vigor of his efforts. There has indeed been recently a great outpouring of work in this field. We believe that a forum for exchange of views and for critical and authoritative reviews of topics is an essential need of this field. It is our hope that the projected periodical series of volumes under this title will help serve this need. The general organization and character of the scholarly presentations of our series will correspond to that of the several prototypes, e.g., *Advances in Enzymology*, *Advances in Chemical Physics*, and *Progress in Inorganic Chemistry*.

We have encouraged the authors to review topics in a style that is not only somewhat more speculative in character but which is also more detailed than presentations normally found in textbooks. Appropriate to this quantitative aspect of organic chemistry, authors have also been encouraged in the citation of numerical data. It is intended that these volumes will find wide use among graduate students as well as practicing organic chemists who are not necessarily expert in the field of these special topics. Aside from these rather obvious considerations, the emphasis in each chapter is the personal ideas of the author. We wish to express our gratitude to the authors for the excellence of their individual presentations.

We greatly welcome comments and suggestions on any aspect of these volumes.

ANDREW STREITWIESER, JR.
ROBERT W. TAFT

Contents

Barriers to Internal Rotation about Single Bonds

JOHN P. LOWE*†

Department of Chemistry, The Johns Hopkins University, Baltimore, Maryland

CONTENTS

I. INTRODUCTION

The existence of energy barriers hindering internal rotation about single bonds was first recognized in the 1930's, when it was found that the observed thermodynamic properties of ethane did not agree with those properties calculated from statistical mechanics under the assumption of free internal rotation (1). The problem of understanding and measuring such barriers has attracted the attention of many scientists, and the development of theories and experimental techniques over the

* Supported by a Public Health Service postdoctoral fellowship (5-F2-GM-24, 839-02) from the National Institute of General Medical Sciences, 1964–1966.

† Present address: Department of Chemistry, The Pennsylvania State University, University Park, Pennsylvania 16802.

1

years is a remarkable testament to the importance of the phenomenon. The increasing importance of questions of conformation and the energetics of conformational change in macromolecules has stimulated an upsurge of interest in the process of hindered internal rotation about single bonds, making apparent a need for an up-to-date compilation of information on this subject.

This review has three major aims: to describe methods used to measure barriers, to present a table of barriers so far measured, and to describe the major theoretical approaches in the light of our present knowledge. The description of each experimental method includes a brief qualitative discussion of the principles involved, an indication of the kinds of molecules to which it is limited, and an evaluation of the accuracy in the barrier information produced by the method.

An effort has been made to make the tabulation of data as complete as possible, but there probably are omissions. No attempt has been made to include all references to work which has been superseded by more accurate measurements. An attempt has been made to include references to all theoretical papers written on the subject of barriers to internal rotation.

The organization of this article is very similar to that used by E. B. Wilson in his excellent 1959 review of the same subject (2).

II. EXPERIMENTAL METHODS

A. Thermodynamic Method

The partition functions for translation, overall rotation, and vibration of a molecule may be evaluated from the molecular mass, moments of inertia, and constants describing normal vibrational modes. The usual sources for such data are infrared and Raman spectra. If the frequencies associated with the torsional motion are unknown, the vibrational partition function is evaluated assuming completely free internal rotation. The resulting theoretical values for such thermo-dynamic properties as free energy, entropy, and heat capacity will differ from the experimental values by an amount which depends on the nature of the barrier to internal rotation. Tables are available which relate the discrepancy between observed and theoretical values for thermodynamic observables to the height of a barrier of simple periodic form (3).

Many molecules can, in principle, be studied in this way, since it is not necessary that the molecule have special properties such as a permanent dipole moment. There are, however, several practical limitations which restrict the applicability of the method. One of these is that the moments of inertia and the frequencies of the normal vibrational modes of the molecule must be known. For most molecules, this information is not yet available. Another requirement is that the form of the barrier should be fairly well characterized by one or two numbers, since in practice the method does not produce enough information to evaluate more than this. Symmetry often gives a barrier such a simple form. For example, methyl alcohol has a barrier which is quite well characterized by V_3, the difference between the energy of a peak and that of a valley, in the potential energy as a function of torsional angle. This is because the symmetry of the methyl group requires that the barrier have the form

$$W(\alpha) = (V_3/2)(1 - \cos 3\alpha) + (V_6/2)(1 - \cos 6\alpha) + \cdots \qquad (1)$$

where α is the angle of internal rotation. Experimental measurements have shown that, for methyl alcohol and other molecules possessing methyl rotors, the higher terms of eq. (1) are quite small compared to the first term (unless V_3 is equal to zero by symmetry as it is in nitromethane), so the value of V_3 suffices to accurately characterize the barrier. A molecule such as CH_2Cl-CH_2Cl, on the other hand, may possess a barrier which has two different peak heights and two different valley depths, requiring three numbers to characterize it. Sometimes thermodynamic measurements are used to get results on such a barrier, but it is usually necessary to assume a simplified form (e.g., all peak heights the same) and determine an "average" height.

The accuracy of this method varies widely. In some cases, where all information used is very accurate, a very reliable barrier value may be obtained. The barrier in ethane has been fixed within 5% on the basis of thermodynamic measurements, however this is exceptional. Many values have been quoted which are based upon assumed or erroneously assigned vibrational frequencies. Interactions between molecules, such as hydrogen bonds, can produce difficulties in absolute entropy determinations unless degrees of randomness at very low temperatures are completely understood (4). The same interactions may cause serious errors to occur in vapor-phase heat-capacity measurements unless corrections are made for nonideality. When one compares barrier values found by thermodynamic methods with those found by more accurate methods, it becomes apparent that the former are often no more than indicative.

B. Sound Absorption Method

For a gas or liquid comprised of simple molecules that take up or release energy only from such modes as translation, overall rotation, and nontorsional vibration, the velocity of sound waves is independent of their frequency over a wide range. This is because the factors determining sound velocity, namely the density, the isothermal elasticity, and the heat capacities at constant pressure and volume, are all independent of the sound-wave frequency. Indeed, measurement of sound-wave velocities in such systems provides an accurate way of measuring heat capacities. However, when a low-frequency vibrational mode, such as a torsional mode, is present, the velocity of sound waves becomes frequency dependent, being higher at higher frequencies, and the apparent heat capacity measured from the velocity is lower at higher frequencies. The reason for this behavior is that the low-frequency torsional mode, while able to maintain equilibrium with the alternate compressions and rarefactions at lower frequencies, becomes inoperative at higher frequencies. Loss of this mode of energy storage lowers the heat capacity and, in a sense, increases the rigidity of the molecule, thereby increasing the velocity of the sound waves. If the molecule being studied has only one stable rotameric form, the amount of sound absorption and dispersion will be quite small. To measure the total heat capacity of a substance then, it is necessary to use frequencies sufficiently low to activate all the modes of the molecules. This method has provided data for barrier determinations by the thermodynamic method (5).

In contrast with this, a pronounced increase in the *absorption* of sound may occur in the presence of molecules with different rotameric forms of *different* energy (394). An equilibrium usually exists between such rotamers, and this equilibrium tends to adjust to the local changes in temperature and pressure produced by the passage of the sound waves through the medium. As before, the equilibrium will keep pace with low-frequency fluctuations, but will become inoperative at high frequencies. This shift in equilibrium will result in a marked absorption of sound energy at the lower frequencies, relative to the higher frequencies. Measurements of the sound velocity or absorption may be used to arrive at heat-capacity values, and these will again be smaller for higher frequencies. By taking the difference between heat capacities at very low and very high frequencies, one can evaluate the energy difference between stable rotameric forms. In addition, the inflection point on the curve of frequency versus velocity may be used to obtain the relaxation time for the process, which leads to a value for the rate constant. By studying sound propagation at several temperatures, it is possible to set the temperature dependence of the rate constant, which gives a barrier height through the Arrhenius relationship.

For barriers of several kilocalories per mole, the absorption of sound method may be reliable to about 10–20%. Since only a small number of molecules has been studied by this method, it is difficult to estimate the accuracy.

C. Dipole Moment Method

If the two parts of the molecule which rotate with respect to each other *each* have a dipole moment component perpendicular to the torsional axis (e.g., H_2O_2), then the molecular dipole moment will vary with torsional angle. For a given barrier, the temperature dependence of the distribution of molecules over different torsional angles is known in principle, and hence the average molecular dipole moment is accessible through bond-moment considerations. Comparison of the calculated and experimental dipole moments over a range of temperature provides a test for an assumed barrier (7).

Few reliable barrier values have been found by this method. The major difficulty is that the dipole-moment requirement mentioned above prohibits either part of the molecule from having a nontrivial rotational symmetry axis coincident with the torsional axis. In other words, the barrier shape will necessarily not be simple, and the method is not reliable enough for complicated barriers. There are, however, cases in which the dipole-moment method provides information about the nature of the stable rotameric form. For example, the sizeable dipole moment of H_2O_2 indicates that the stable rotameric form is not *trans*.

D. Microwave Methods

During the past fifteen years, a large number of barriers to internal rotation have been accurately determined by analysis of microwave spectra. Absorption in the microwave region occurs when molecules containing permanent dipole moments undergo changes in their

rotational quantum states only. However, the energy levels for rotation
of the molecule as a whole are influenced by the state of the internal
rotor, and various theoretical methods for obtaining barrier values
from observed spectra are available (8,448,449,472).

Figure 1 shows the energy levels of the hindered methyl rotor in
CH_3CHO. The barrier has threefold symmetry since the molecule is
unchanged by a methyl rotation of $2\pi/3$ radians with respect to the
carbonyl frame. In the limit of infinitely high barriers, the solutions
would be the equally spaced levels characteristic of a harmonic oscillator, but they would be triply degenerate since there are three physically
indistinguishable ways for the ordered methyl protons to be placed in
the potential wells. For a finite barrier, the torsional levels are unequally
spaced. Also, because the rotor can tunnel through the barrier, the
levels are not triply degenerate, but exhibit splitting which increases as
the torsional levels approach the top of the barrier.

The Schrödinger equation has been solved for the rotation of a
methyl-like rotor (a symmetric top) in the potential given by eq. (1).
The torsional levels are split by tunneling into a single A-type level and
a doubly degenerate E-type level. (The letters refer to symmetry characteristics of the solutions and correspond to periodicity for rotations of
$2\pi/3$ or 2π radians, respectively.)

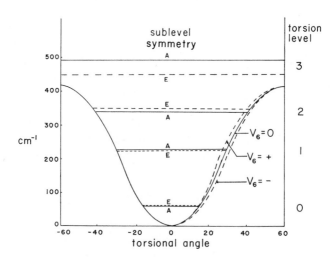

Fig. 1. Torsional levels and sublevels in acetaldehyde (ref. 104).

1. The Intensity Method

Microwave spectra have been interpreted to give information on both kinds of energy difference shown in Figure 1. The energy differences between torsional levels are not directly observable in the microwave region since the energies for such transitions typically correspond to frequencies in the far-infrared. But the pure rotational spectrum of a molecule will be slightly different for different torsional states, due to slight changes in the motion-averaged moments of inertia of the molecule. The observed spectrum for an ensemble of molecules will consist then of intense lines accompanied by weaker satellites corresponding to rotational transitions of molecules in the ground and higher torsional states, respectively. Correlation of satellites with appropriate ground-state lines is helped by the fact that similar Stark effects are shown by lines corresponding to the same change in rotational quantum numbers. By comparing the intensity of the main line with that of the strongest satellite, one finds the ratio between the number of molecules in the ground state and the first excited torsional level. The Boltzmann relation then can be used to obtain the energy difference between levels. This energy difference determines V_3 of eq. (1) if higher terms are neglected.

Many barrier values which have been found by measurements of microwave intensities are not very accurate. Early equipment was designed for measurement of frequencies and was ill suited for intensity measurements. Difficulties also occur because of the tunneling-induced splitting of the torsional levels. In some cases, this leads to splitting in the lines of the observed spectrum. Since the amount of splitting differs for different torsional levels, simple comparisons of maximum intensities become inadequate, and it is necessary to measure the areas under the peaks. Verdier and Wilson (9) and Esbitt and Wilson (443) have described apparatus which is capable of giving more reproducible intensity measurements. They have also shown how to correct for complications due to splitting, and have shown that, with these improvements, the intensity method gives barrier values which are more accurate.

Although the intensity method has been applied primarily to molecules with symmetric-top rotors, it is applicable in principle to any molecule with a dipole moment. However, the fact that intensity ratios can be found for only a few torsional levels at best has so far made it impractical to apply this method except when the barrier is characterized by symmetry to within one or two constants.

2. *The Splittings Method*

The total angular momentum for a molecule is the resultant of the angular momenta of the rotor and the remainder, or framework, of the molecule. Hence there is a cross-product term in the Hamiltonian corresponding to the coupling of these momenta. For a symmetric-top molecule such as CH_3CF_3, the coupling term is the same for A- or E-type symmetry of the torsional substate. But if the framework is asymmetric, as in CH_3CHO, the coupling term differs for rotors in different sublevels of a torsional state. This leads to splitting of the lines observed in the microwave spectrum. The amount of splitting observed, then, is related to the splitting between the torsional sublevels, and this in turn is related to the height of the barrier through which tunneling occurs.

Since the moments of inertia for a molecule with a symmetric-top rotor are independent of the torsional angle of the top, the theory for such molecules is less complicated than that for molecules in which neither rotor nor framework is symmetric. Further simplification results if the framework has a plane of symmetry containing the torsional axis. Most barriers determined by the splittings method are for molecules with both of these characteristics, although theory has been developed for molecules in which both the rotor and the framework are asymmetric (10). Studies have also been made of the spectra of molecules, such as acetone, containing two symmetric-top rotors. The indications are that coupling between methyl groups is quite weak in most cases (11–13), so that analyses based on the assumption that the methyl groups are independent may still give fairly accurate barrier values.

In molecules having very high barriers, the splitting of the ground torsional state becomes too small to be detectable in the rotational spectrum. However, it is sometimes possible to do a splittings analysis of lines arising from excited torsional states. In cases where independent barrier measurements are made from splittings in different torsional levels of the same molecule, very good agreement has been found (14). This gives an indication of the precision of the method and also demonstrates that terms higher than V_3 in eq. (1) are comparatively small since, if they were large, their neglect would lead to discrepancies in the various calculated values for V_3.

Molecules with low barriers to internal rotation can be accurately treated by the splittings method, whereas other methods are unsuitable. The solutions for completely free rotation are taken as a starting point, and the effects of the barrier are treated as a perturbation. By such an approach, an upper limit for the barrier in CH_3—$C\equiv C$—SiH_3 of 3 cal/mole has been determined (15).

Most barriers have been determined under the assumption that the two parts of the molecule are rigid and do not change with change of torsional state or angle. However, it has already been pointed out that the microwave intensity method *depends* on the rotational constants being slightly different in different torsional states. It also seems clear that the forces which favor one rotamer over another should cause slight changes in average structure on rotation. This has been observed for distinct stable rotamers like *trans*- and *gauche*-3-fluoropropene, and it is undoubtedly true for, say, staggered and eclipsed ethane too. Finally, it is possible for nontorsional vibrational modes to interact with both the torsional motion and overall molecular rotation. This provides an indirect way for internal rotation to influence the microwave spectrum. Thus, while the rigid molecule theory predicts that symmetric-top molecules cannot exhibit splitting of their microwave spectra from splittings of their torsional levels, the theory of vibration–internal rotation interactions indicates that this indirect coupling can produce splitting in the spectra of such molecules (16,17). This predicted splitting has been observed (18). Fortunately, it is found that barriers measured using theories which account for effects of nonrigidity are in agreement with those measured using the simpler, rigid-molecule approximation (478).

Microwave spectra are easily analyzed for the moments of inertia of a molecule. By analyzing the spectra of enough different isotopically substituted molecules, it is possible to make a complete determination of the bond lengths and angles. Accurate structures are necessary in determining the barrier since the moments of inertia about the torsional axis of the methyl group must be separated from the moments for the molecule as a whole for the purposes of calculation. When structures are carefully determined, the stable conformation is generally found too. For example, CH_2DCHO will have one-third of its deuterium atoms eclipsed with either O or H, depending on whether the stable form features a methyl group staggered or eclipsed with respect to H. The moments of inertia will differ for the two cases, and comparison of the predicted microwave spectra with the observed spectrum suffices to determine the preferred conformation. Usually, a complete structure determination is not made, and educated guesses are made instead. This is the source of a large part of the uncertainty in most microwave splittings barrier measurements. Whereas the precision of the splittings method is high enough to give barriers to within about 2%, the uncertainties in structure, as well as the approximations in theoretical treatment of the data, often raise this to about 5%.

E. Infrared Method

The separation between torsional levels for rotor groups is often such that transition energies correspond to the far-infrared region of the spectrum. Until recent years, this region has been inaccessible, and torsional frequencies have been assigned from bands where they occur in combination with other vibrational transitions. Once a torsional frequency has been obtained in this way it is simple to estimate a barrier height by assuming a simple mathematical form for the barrier.

Infrared spectra have been used to measure conformational energy differences also. Different conformations will have different spectra, and intensities of corresponding transitions in the two forms provide a measure of the relative abundance of these forms. By determining the intensity ratio and using the Boltzmann equation, one may get the conformational energy difference.

Most of the work on barriers done by IR methods has been of the type just described. Comparison of results of more accurate methods indicates that such IR work commonly has a 10 or 15% error.

When the separation between torsional levels is small enough, several excited levels will be appreciably populated at normal temperatures. Hence it is possible, in principle, to observe various transitions $0 \rightarrow 1$, $1 \rightarrow 2$, etc. in the far-infrared spectrum of some molecules (see Fig. 1). For such transitions to be IR active, it is necessary that there be a component of the dipole moment of the molecule perpendicular to the axis about which the torsional motion occurs.

Recently, a number of molecules have been carefully studied and analyzed using far-infrared techniques. Fately and Miller (19) have observed a number of transitions between torsional levels in several molecules and have sought to obtain the value of V_6, the second term in the cosine series expression for the barrier. Whereas V_3 is a measure of barrier height, V_6 pertains to barrier shape. A positive value of V_6 will produce a sharper minimum and a broader maximum in the potential barrier curve, while a negative V_6 will produce the opposite effect. This is shown schematically in Figure 1. The spacing between torsional levels is influenced by such factors as the height and shape of the barrier (characterized by V_3 and V_6 if higher terms are assumed to be equal to zero), change in molecular structure, and change in vibration –internal rotation interaction. If one ignores these complications and calculates V_3 independently for each of the various observed transitions, the values differ slightly, and often show a monotonic increase or decrease with increasing torsional level. Fately and Miller (19) have shown that the "drift" in value of V_3 is too large to be attributed entirely to changes in structure. They have computed values for V_6 which would be appropriate to correct the drift if the molecules were rigid, and have obtained both positive and negative values for this quantity in different molecules. So far, however, no one has succeeded in separating and elucidating all of the factors which, in concert, determine the observed transition frequencies, and all V_6 values quoted must be considered to be approximate (except for those cases where $V_3 = 0$).

Accurate barrier determinations from infrared measurements require accurate structural data at one point in the straightforward solution of the pertinent equations, however, such data usually are not available. It is possible to circumvent this difficulty by combining information from the infrared and microwave spectra of a single molecular species, as Fately and Miller (19) have done, following a suggestion by Lide. This removes the necessity for laborious preparation and spectral analyses of isotopically substituted molecules. This marriage of methods probably will prove to be a fruitful source of new information on the detailed nature of the barrier to internal rotation.

The fact that several torsional transitions may be observed makes it possible to characterize barriers which have complicated forms. For example, the barrier in H_2O_2 has recently been expressed as a three-term Fourier series on the basis of infrared spectral studies (20). In this case, there is some uncertainty in the structural parameters used (21).

Because the far-infrared spectrum for torsional transitions is complicated by superimposed lines due to rotational transitions, it is difficult to locate absorption peaks with precision. This problem is aggravated by the fact that the torsional levels are generally split by tunneling through the barrier. Thus, a transition from a practically unsplit ground level to a slightly split first excited level will result in two slightly separated absorption energies. To identify and assign the corresponding peaks with accuracy is difficult in the presence of rotational bands. Nevertheless, it appears that, in favorable cases, the far-infrared method is extremely accurate. When coupled with microwave data, the error is probably about 1%.

F. Raman Method

The Raman method (7) has usually been used along with the IR method since the two spectra give similar kinds of information. Because the selection rules differ, however, Raman spectra may supply additional data on frequencies which can serve to put the entire assignment on a firmer basis. Also, the torsional frequency itself, when active, may be readily observed in the Raman spectrum even though it is an inaccessible part of the IR spectrum.

It is possible to measure conformational energy differences by studying the intensity ratios in exactly the same way as described for IR spectra.

Since most Raman spectra are taken on liquids or solutions, it is difficult to compare results with gas-phase work. It is well known that conformational energy differences change markedly between liquid and gas states for molecules whose dipole moments vary with change in torsional angle.

The accuracy of barriers determined from Raman spectra is usually about 10–20%.

G. Nuclear Magnetic Resonance Method

Nuclear magnetic resonance spectra may be interpreted to give information both on energy differences between conformers and on energy barriers (22).

Conformational energy differences can be determined by measuring the relative populations of various conformers at a single temperature. Use of the Boltzmann equation then leads to the energy differences between the conformers. In order to measure the relative populations of various conformers by NMR, it is necessary that each conformer have a different NMR spectrum, that the energy differences between conformers be on the order of 1 kcal/mole or less (so that all conformers are present in observable quantities), and that the barrier between conformers be higher than about 5 kcal/mole (so that exchange between them is slow at temperatures accessible in present-day NMR spectroscopy). The NMR spectrum of a mixture of conformers satisfying these criteria is the sum of the spectra for the component conformers. Once the various peaks have been assigned, the ratio of populations of the conformers is found by comparing areas under appropriate peaks. Recently, coupling constants have also been used to obtain conformational analyses (431).

Determination of barrier heights requires comparison of NMR spectra at several temperatures. As the temperature of a mixture of conformers is increased, the rate of interconversion increases. If the barrier is not too high, interconversion may become so rapid that a single nucleus moves through several local environments within the time required for the (microscopic) NMR measurement. The resulting spectrum often becomes simpler since the separate chemical shift peaks for a nucleus in different conformations merge into a single peak representative of the averaged environments. Thus, as a sample of N,N-dimethylformamide is warmed, the two sharp peaks in the NMR spectrum corresponding to the *cis* and *trans* methyl groups slowly collapse while a single peak develops between them. Several techniques (88) are available for determining the rate at which interchange occurs from the appearance of an intermediate-temperature spectrum. After the rate of interchange at several temperatures has been found, it is possible to estimate the free energy of activation from reaction-rate theory. Subtraction of the entropy of activation gives the barrier height.

Several determinations of the barrier to internal rotation about the nitrogen–carbonyl single bond in dimethyl formamide have been made by the NMR method. The resulting values range from 7 to 13 kcal/mole, indicating that results from this method must be accepted with caution. Although it appears that systematic errors (23) may invalidate comparisons between barriers measured by different techniques on different instruments, comparisons of barriers done by the same method on the same apparatus ought to be meaningful.

Molecules which have been studied by this method fall into two groups. The first group comprises the amides, amines, and nitrites, which exhibit fairly high barriers presumably due to partial double-bond character in the bond about which torsion occurs. These molecules generally have simple barriers, characterized by one number—the barrier height. The second group comprises the polyhalosubstituted ethanes, many of which have high barriers presumably due to end-to-end steric repulsions. Because these molecules have barriers featuring several different maxima and minima, they are not amenable to treatment by most other methods. Unfortunately, because many of these molecules have dipole moments which are a strong function of conformation, the conformational energy differences determined by NMR in the liquid state or in a polar solvent may not apply for isolated molecules. Recent work (432) suggests that conformational energy differences are also temperature dependent in such cases.

H. Electron Spin Resonance Method

When a system containing an unpaired electron is placed in a magnetic field, the magnetic moment of the electron can assume one of two orientations with respect to the field. As in NMR, an observable absorption of energy will occur when the system is subjected to electromagnetic radiation of energy equal to the energy difference between these two states. In the absence of any further sources of magnetic field, the spectrum for an ensemble of such systems would show a single resonance. However, when a system contains nuclei with nonzero nuclear spin, coupling occurs between the magnetic moment of the odd electron and the magnetic moments of these nuclei. This leads to hyperfine structure in the resonance spectrum. Since the coupling constant between a nucleus and an electron depends upon their relative orientations, it is possible to analyze ESR spectra for information on radical conformations and barriers. Thus the equivalence of the methyl proton couplings with the odd $p-\pi$ electron in the $CH_3C(COOH)_2$ radical indicates that the methyl group is undergoing free rotation (24). Where rotation is restricted, nuclei will occupy a weighted range of positions. The observed coupling constants will reflect this situation, and they can be employed to find the stable conformation and barrier height (25).

A more common approach for estimating barrier heights from ESR spectra is through the determination of the temperature dependence of line broadening. This is analogous to the method described in the foregoing section on NMR.

The ESR method is important because it yields information on barriers for radicals. The method has been applied in irradiated crystals and also in solution, and results obtained with it so far can only be considered approximate.

I. Electron Diffraction Method

Electron-diffraction patterns produced by molecules in the gas phase may be analyzed to give a radial distribution function which contains information on relative interatomic distances in the molecules.

For a molecule like acetaldehyde, some of these distances (e.g., the distance between methyl carbon and carbonyl oxygen) are independent of conformation, whereas others (e.g., those between methyl hydrogens and carbonyl oxygen), are not. Similarly, the radial distribution function for *gauche*-propionaldehyde will differ from that for *cis*-propionaldehyde. It is possible to theoretically derive the distribution function appropriate for any of these forms or a mixture of them. By comparing these with the experimentally determined distribution function, one can make an estimate of the relative amounts of various conformers present and, using the Boltzmann relation, obtain the relative energies of the minima in the potential curve.

The radial distribution function contains peaks rather than lines, and the peak widths may be utilized to give information on the nature of the vibrational motions of the molecule, including torsional vibrations. If each of two atoms in a molecule is very tightly bound, the distance between them will not vary much, and the peak corresponding to the internuclear distance will be relatively sharp. If one assumes various barrier heights and takes classical distributions of molecules with regard to torsional angle, one can get theoretical distribution functions. Comparison of these with the observed distribution function leads to a rough estimate of barrier height (26).

The uncertainties in population of different conformational states are around 10–15%, which leads to greater uncertainties in ratios of populations. Hence the electron-diffraction method can give reliable information on the nature of the stable conformations (i.e., locate the torsional angles corresponding to some potential minima), but the numerical estimates of relative energies of minima and also of barrier heights are quite approximate. It appears, however (430), that greater accuracy is forthcoming.

J. Neutron Scattering Method

Very recent experiments (27) indicate that information on barrier heights can be obtained from the inelastic scattering of low energy, or "cold," neutrons ($E_n \leq 0.005$ eV) from solids. The procedure is to allow a chopped monoenergetic beam of low-energy neutrons to impinge upon a thin sample of the solid, and to analyze the energy spectrum of the emergent neutrons by measuring their time-of-flight to a detector. Inelastically scattered neutrons are produced when they acquire energy released as a molecule drops from one vibrational quantum state to another. Thus, a plot of energy of scattered neutrons versus frequency shows peaks which correspond to molecular energy transitions. Rush and Taylor (27) have found that hexamethyl benzene (below 110°K) has such a peak corresponding to a neutron energy gain of 137 cm^{-1}. By assuming this to correspond to the $1 \rightarrow 0$ transition in the methyl torsional mode and assuming a simple barrier shape, they compute $V_3 = 1.07$ kcal/mole.

In practice, the impinging neutron beam is not really mono-energetic. Also, the time-of-flight analysis involves an error due to the finite time for each burst of neutrons, and there are problems of peak identification and determination of the proper form of the barrier. It is reasonable to expect that some of these difficulties will be overcome in the future and that the method will be fairly useful, at least for solids.

III. COMMENTS ON EXPERIMENTAL RESULTS

A comprehensive table of experimental barrier data is given in Section V of this review. There are some striking regularities and irregularities in these experimental barriers to internal rotation. The roughly 3:2:1 ratio of barrier heights in CH_3CH_3, CH_3NH_2, and CH_3OH is a relationship that was noticed early, and it is satisfied by virtually all semiempirical theories which attribute the barrier to the H atom or the X—H bond. This ratio is not repeated for the somewhat analogous series CH_3—SiH_3, CH_3PH_2, CH_3SH. Indeed, although CH_3SiH_3 has a barrier which is about half that in CH_3CH_3, CH_3PH_2 and CH_3SH have barriers equal to and greater than CH_3NH_2 and CH_3OH, respectively.

The effects on the barrier of halosubstitution in ethane and similar molecules is usually small unless substitution by Cl or Br occurs on *both* ends, in which case the barrier rises markedly. Also, a barrier decrease has been noted when substitution occurs in the *cis* position of propylene. These effects have generally been ascribed to steric hindrance, and so far molecules in which this complication is thought to occur have been avoided in theoretical treatments. Fluorosubstitution in ethane shows an interesting irregularity. When the first fluorine is substituted, the barrier rises by about 400 cal/mole. Addition of a second (*gem*) fluorine causes the barrier to drop by about 100 cal/mole. In methyl silane, the effect of fluorosubstitution is monotonic but still irregular. Substitution of one and then another fluorine on silicon causes the barrier to drop first by about 100 cal/mole and then by another 300 cal/mole. This behavior suggests that mere additivity of group effects will not suffice to explain the successive effects upon the barrier of substitution and that interactions between substituted groups may be important.

In "ethanelike" molecules [i.e., of the form $(X_1X_2X_3)$ C—D $(Y_1Y_2Y_3)$ where atoms X_1, X_2, and X_3 are each separately bonded to C and similarly Y_1, Y_2, and Y_3 are separately bonded to D, and where Y_1 and/or Y_2 may be nonbonded electron pairs] all conformational studies indicate the stable form to be a staggered one. In molecules containing a double bond adjacent to the threefold axis of rotation (e.g., acetaldehyde) the stable conformation, when determined, almost always (416) has a rotor hydrogen eclipsing the double bond. Thus the ethanelike molecules have barriers which are in phase with internuclear repulsion energies, whereas the acetaldehydelike molecules are often exactly out of phase with these energies.

The effects of halogen substitution in the secondary position in propylene are almost the same as those in ethane. The effects of similar substitutions in acetaldehydes are much smaller, and in the case of fluorine substitution, the barrier is actually lowered slightly. Substitution in the *trans* position of propylene hardly affects the barrier.

The values for V_6 are always of the order of tens of calories or less, regardless of the value of V_3.

Barriers for molecules with a onefold rotor on a given frame (e.g., formic acid) are generally much higher than barriers for a threefold rotor on the same frame (acetaldehyde).

The large amount of experimental information available today on barriers to internal rotation provides simultaneously a challenge to chemists and a rigorous test for their theories.

IV. THEORIES OF THE BARRIER

Ever since the phenomenon of hindered rotation about single bonds was discovered, efforts have been made to find a conceptually simple, quantitatively useful theory of the source of the barrier. These attempts range from empirical to *ab initio* and have focused on the axial single bond (28–30,417) the nonaxial bonding electrons, or the nonaxial substituents as the entities providing the interactions responsible for the barrier. The gradual accumulation of reliable experimental barrier values has served to rule out many of these approaches. Also, recent theoretical calculations have tended to shift the emphasis toward theories based on more intimate knowledge of the detailed changes in electron energies and distributions. Since the earlier work has been reviewed elsewhere (1,31) it will be touched on only briefly here. Recent developments will be discussed more fully.

Purely empirical methods, eschewing any consideration of the physical causes for a barrier, have sought merely to relate barrier height to some characteristic molecular distance (32,33,87), such as l, the length of the torsional axis. For example, it has been pointed out (15) that the barriers for CH_3SiH_3, CH_3GeH_3, CH_3SnH_3, and CH_3—C≡ C—SiH_3 are related through a factor of l^{-7}. However, such a relation poorly fits certain other molecules of the same type and fails completely for molecules with different types of framework such as acetaldehyde. There seems to be no reason to hope that a single simple empirical relationship will be capable of correlating the large amount of experimental data now in existence.

Various sources have been postulated for the forces causing the barriers. These postulates have been made the bases for semiempirical methods for barrier evaluation. Most of these methods may be distinguished as attributing the barrier to repulsions between nonaxial bonds (e.g., C—H bonds repel C—F bonds in CH_3CF_3) or else to forces between atoms on opposite ends of the rotor axis which are not directly bonded to each other (e.g., H---F interactions in CH_3—CF_3). These two basic approaches will be discussed separately.

Several investigators have made calculations of electrostatic bond–bond repulsions in ethane and related molecules. Lassettre and Dean (35), following an earlier suggestion by Kistiakowsky, Lacher, and Ransom (36), assumed neutral C—H bonds, each containing two electrons and two positive charges, one at the hydrogen nucleus and one on carbon. The electron distribution was taken to be given by a wavefunction made up of an sp^3 hybrid on carbon and a $1s$ orbital on hydrogen. The variational parameters were fixed by requiring that the calculated bond dipole moment be either $+0.4$ or -0.4 D, which were guesses for the C—H bond dipole moment. Lassettre and Dean computed the electrostatic repulsion between such bonds at opposite ends of ethane, treating the bond charge distributions in terms of a multipole expansion arbitrarily truncated after the quadrupole term. This led to an estimate which was too small.

Oosterhoff (37) modified this approach by assuming that the C—H bonds could be ionic. He expanded the C—H overlap charge in terms of multipoles up to the octupole and did not attempt to fit the experimental value for the C—H bond moment. He found good agreement with the observed barrier only if the C—H bond was very ionic.

Tang (38) and Luft (39) made some extensions on this method, but there appears to be little active interest in it today. In addition to the problem of unreasonableness of parameters, there is no unique way to extend the method to other than a fairly limited class of molecules. Also, it is now generally realized (40,82) that the kinetic energy change is fully as great as the total energy change, and that efforts to attribute the barrier entirely to potential energy changes have neglected this contribution.

The other theory of bond–bond interactions is related to early valence bond calculations of Gorin, Walter, and Eyring (28), but it was Pauling (41) who more recently revived the basic idea and showed how it could be extended to a large number of molecules. As long as the tetrahedral orbitals on carbon in ethane are taken to be made up of s- and p-type orbitals only, there can be no barrier-creating interaction between them. Pauling suggested that inclusion of small amounts of d- and f-orbital character in these hybrids would produce the correct behavior. He emphasized only the exchange interactions between the tetrahedral orbitals and ignored other interactions which might be equally important, such as the interaction between a hydrogen on one end of ethane and the tetrahedral orbitals on the other end. Earlier calculations (28) had in fact indicated that inclusion of these other effects may lead to a prediction that eclipsed ethane is more stable than staggered ethane. Hence the basis for Pauling's approach is not rigorous. Nevertheless, he was able to correlate and explain a good deal of the barrier data available in 1958. The proper conformations for molecules like acetaldehyde were predicted if the double bond was assumed to act as though it were two bent tetrahedral single bonds. Also the small changes in barrier height with substitution in ethanes were consistent with the theory. In this picture, the roughly 3:2:1 ratio of barriers in ethane, methyl amine, and methyl alcohol, is reasonable because nonbonding electrons would not be expected to contain d or f character. However, the method is not capable of giving more than a rough estimate of barrier heights (42). Moreover, the assumptions underlying this approach are put in serious doubt by recent, fairly accurate molecular orbital (SCF—LCAO—MO) calculations (43–46,453), which demonstrate that reasonable barriers are computed even when no d or f character is allowed in the bond orbitals. Also, Bartell and Guillory (26) have found that cyclopropyl carboxaldehyde is stable in a conformation which the Pauling method predicts to be unstable. Recent experimental determinations (19) of V_6, the second term in the barrier expansion for a threefold rotor, indicate that this quantity may be positive or negative, whereas the Pauling theory is consistent with only negative values.

Dale (417) has recently proposed a very similar explanation for rotational barriers, He suggests that the bonds interact indirectly through trigonal polarization of the axial bond. However, this should again require some d or f character in the nonaxial bonds. This makes it difficult to rationalize Dale's hypothesis that lone pairs of electrons play the same role in barrier production as do bonding electrons.

A variety of methods based upon interactions between nonbonded atoms have been proposed for computing barriers to internal rotation. These methods have recently been reviewed by Cignitto and Allen (31). It is well known that nonbonded atoms attract each other at large interatomic distances and repel each other at very short distances. It seems reasonable that such forces exist between nonbonded atoms in the same molecule as well as between those in different molecules. The problem is to find the correct interaction energy as a function of distance for a given molecule. Several different approaches have been tried.

Theoretical calculations (31,47,48) have been carried out using Mulliken's (49) "magic formula" which employs overlap integrals, exchange integrals, and ionization potentials. The values for some of these quantities depend upon assumptions made about orbitals describing the regions around the hydrogen, so the calculations are indicative, not rigorous. The barriers predicted in this way for ethane, methyl silane, and methyl germane are, respectively, about one-half, one-third, and one-third of the observed barrier values (31). Valence-bond calculations (50–52) have been made which relate the interaction to the potential curves for the lowest $^1\Sigma$ and $^3\Sigma$ states of H_2. Several theoretical calculations of these potential curves have been made, and curves based upon spectroscopic data also exist. Hence the predicted barrier in ethane varies from $+0.42$ to -0.07 kcal/mole, depending on how one chooses the potential curves (31).

Since experiments on molecular scattering, viscosity, etc. provide data on intermolecular forces, one can try to transfer this information to problems involving intramolecular forces. There are some unresolved difficulties which prevent this procedure from being so meaningful as it might at first seem, however. First, what molecules have an interaction between themselves that is most like the end-to-end interaction within ethane molecules? The interactions are quite different between H_2 molecules and between CH_4 molecules, so that choice of the source of the interaction potential is very important (31). Even more difficult is the problem of the orientation of the interacting groups. Whereas intermolecular forces are generally expressed only in terms of intermolecular distance, it seems clear that a meaningful calculation of the interaction between two C—H hydrogens must be a function of relative bond angle as well as distance.

Numerous other calculations have been made, principally of the barrier in ethane, by considering various combinations of the factors discussed above. The usual procedure has been to construct a potential curve which is repulsive at short distances and attractive at greater distances. The attractive part has generally been taken to have the r^{-6} dependence corresponding to the instantaneous dipole–dipole interaction portion of the van der Waals force. The repulsive part has been taken to be; the $^3\Sigma$ potential curve for H_2 (53), this same $^3\Sigma$ curve with an angularly dependent factor (54), the negative of the $^1\Sigma$ curve for H_2 (55), and the function br^{-12} (56,57). Some of these methods have been extended to many molecules (92).

There are serious deficiencies in the barriers predicted by such methods when they are compared with experimental results (58,482). First of all, when parametrized to give the correct barrier for ethane, they generally predict barriers for CH_3SiH_3, CH_3GeG_3, GeH_3GeH_3, etc. which are too low. Also, they predict that the stable rotamer in some molecules (e.g., propylene) is that which is known from experiment to be unstable. [Hence, comparisons (32,55) of predicted barrier values with experimental values are not a sufficient test for a method. It is necessary to check that the predicted stable form agrees with experiment.] Thus, while it seems fairly certain that nonbonded interactions are responsible for a large part of the barrier in molecules such as CH_2BrCH_2Br, it appears that the major cause of the barrier in molecules such as ethane lies elsewhere.

Attempts have been made to predict barriers using molecular orbital formalism (34,95) together with empirical parametrization (59, 60,86). Hoffmann (59) has used an extended Hückel method which employs a limited basis set for sigma as well as pi electrons and which computes two-center energy integrals from orbital overlaps. Hence, molecular energies are dependent upon conformation, not just topology as in simple Hückel theory. The single parameter is adjusted to give the proper barrier value for ethane. Predictions of stable conformers for other molecules are quite successful (466). Related calculations have been made by Pople and Santry (90) and Pople and Segal (91). Their computed barriers for ethane, methylamine, and methyl alcohol are somewhat lower than the observed values, but the observed 3:2:1 ratio in barrier heights is duplicated (91). Analysis of the calculations indicates that the barrier is produced by antibonding between *cis* hydrogens brought about by partial multiple bond character in the central bond. Another suggestion (60) has been to adjust the Coulomb integral, α, with conformation, since α depends on distance-dependent two-center and penetration integrals. Using *n*-butane and cyclohexane barriers for parametrization, barriers were predicted for a number of equatorial–axial transitions in cyclic molecules. Within this limited class, agreement with the scanty experimental results has been good. It does not appear that this approach can be easily extended to more than those few classes of molecules where a great deal of similarity between members exists.

Until quite recently, *ab initio* calculations have been of the valence bond type, and have been either rather incomplete treatments (28,50, 61,62,95) of a molecule or else more complete calculations on some fragment of a molecule (63–66). Computational limitations prevented consideration of all 492 VB structures in ethane and these approximate calculations were unable to determine unequivocally the source of the barrier. Increased computer capabilities have now made it possible for *ab initio* molecular orbital calculations to be carried out on molecules of the complexity of ethane (43–46,67,453). These have so far mostly been limited-basis, self-consistent, LCAO–MO calculations, but calculations involving the use of extended basis sets (e.g., $2p$ orbitals on hydrogens, $3d$ orbitals on carbons) will appear soon (473). The molecules for which barriers have been computed by this method so far include ethane (43–46a,93,453), methylamine (46b,453), methyl alcohol (46a, 453), hydrazine (94,46c,453), hydroxylamine (46c,453), and hydrogen peroxide (46a,46b,67,453,461). The agreement between calculated and observed barriers for ethane and methyl alcohol is quite good, but the computed barrier in the hydrogen peroxide calculations (46a,b,453) is about twice as high as experiments indicate, and the computed minimum is at 180° (*trans*) whereas the experimental minimum is around 110°. For hydrazine, the computed barrier is also about two times too large.

This discrepancy in quality of barrier prediction is not as great as it first appears. Calculation of a barrier for a molecule such as hydrogen peroxide is a much more rigorous test for a method than is a calculation for the barrier of a methyl rotor. This is because the symmetry of the methyl group may force a great deal of error cancellation whereas the errors will still be highly evident in the case of hydrogen peroxide. The reason for this is fairly easily seen. Suppose we consider the methyl alcohol molecule as an example. The experimental barrier in methyl alcohol may be thought of as being the sum of three barriers, one corresponding to each of the methyl hydrogens. That is, as each methyl hydrogen revolves about the axis of internal rotation, it experiences forces which may be expressed in terms of an energy curve. So we might imagine a curve for such a methyl proton as a function of angle from 0 to 180°. When methyl alcohol is in the eclipsed conformation, the appropriate energy will be found by taking one methyl hydrogen to have the energy at 0° on the curve and two more to have the energy at 120°. When methyl alcohol is staggered, the appropriate energies occur at 60, 60, and 180°. Hence the barrier for methyl alcohol would be found by taking

$$\Delta W = W(0) + 2W(120) - 2W(60) - W(180) \qquad (2)$$

Shown in Figure 2 are two curves for which the quantities ΔW are identical. Clearly these curves are very different. One resembles the barrier curve for H_2O_2, whereas the other has a barrier height almost

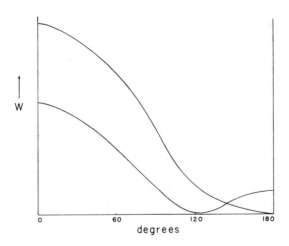

Fig. 2. Hypothetical curves with identical threefold components.

twice as great and has its minimum at 180°. (Construction of such curves is trivial. Starting with a given curve, simply make sure that the net modification at the 0 and 120° positions is equaled by the net modification at 60 and 180°, remembering that changes at 60 and 120° have double weight.) This shows that it is possible for calculations of threefold barriers to appear to be more accurate than they really are. Therefore, it is wise to make calculations for molecules of lower symmetry when possible.

Unfortunately, computations on molecules with lower symmetry meet a new difficulty. It appears that the structural changes which accompany internal rotation are much greater in molecules such as hydrogen peroxide or nitrous acid (68,69) than in methyl alcohol or ethane. Since the usual procedure is to pick a structure and not allow it to vary except for torsional angle, calculated barriers may contain an error from this source. Simple estimates based on force constants indicate that a symmetric deformation of the O—O—H angle by 2° would raise the molecular energy of H_2O_2 by about 300 cal/mole, while a symmetric stretch of the O—H bonds by 0.02 Å would account for about 400 cal/mole. These estimates neglect interaction terms, but they do suggest that, if such deformations *actually* occur, then *ab initio* calculations which ignore them may be in error by several hundred calories. It seems reasonable then, that for small effects like the *trans* barrier in H_2O_2 (1.1 kcal/mole), molecular deformation could be important. Ideally, minimization of the energy with respect to bond angles and distances should be carried out at each torsional angle, but this would require more computer time than anyone has so far been able to provide.

The problem of structural change with internal rotation has been examined in connection with certain semiempirical methods by Sovers and Karplus (70). However, because these effects are at least partially included in semiempirical methods by the parametrization processes, it is possible to make fairly good barrier predictions without explicitly accounting for them, at least for groups of similar molecules.

The errors in the calculated barriers for hydrazine and hydrogen peroxide seem too large (3–7 kcal/mole) to be due solely to deformation effects. A likely source of error is neglected polarization effects in the nonbonding electron pairs. These molecules differ from the methyl tops in that such pairs occur on *both* ends of the molecule. Interaction between these pairs seems likely; only calculation can tell whether such interactions are sufficient to remove the errors in computed barriers.

Another type of *ab initio* calculation has been reported by Hoyland (467). He has constructed energy-optimized bond functions for methane from a Gaussian basis set of functions. Transfer of these bond functions to ethane without modification (except to maintain orthogonality) gives a barrier of 2.82 kcal/mole compared to the experimental value of 2.88 kcal/mole. Transfer to propane gives an average barrier (both methyls rotate) of 4.0 kcal/mole/methyl group, compared to the rough experimental value of 3.4 kcal/mole. It is not yet clear how far such a simple procedure can be extended before breaking down. This calculation is much easier than the SCF calculations mentioned above.

Even though the *ab initio* MO calculations *could* be missing features which influence the barrier and still give good results, it seems more likely that they do account for the essential processes leading to a barrier and that the poorer results for hydrogen peroxide are due to some exceptional cause such as changing polarization of nonbonded electrons and/or structural inadequacies. Assuming this to be the case, what can be gleaned from these calculations regarding the nature of the barrier? Since only *s*- and *p*-type orbitals were used on carbon in ethane, it is apparent that the *d* and *f* orbitals proposed by Pauling (41) are unnecessary. Theoretical inclusion of van der Waals forces would require mixing higher harmonics on the hydrogen atoms, so van der Waals attraction (71,72) is apparently not important for barrier production in these molecules. In addition, effects due to correlation of electronic motion (73) do not appear to be responsible for the computed barrier values. While it is easy enough to show that the barrier is not due to these special effects, it is more difficult to sort out of the mass of information a simple explanation of what the barrier *is* due to. There is no unique way to approach this problem. One may go about obtaining models and physical arguments for the barrier from wavefunctions in several ways. The results may vary in usefulness and may emphasize completely different aspects of the total process.

Pitzer (44) transformed the MO wavefunctions for staggered and eclipsed ethane into localized orbitals and examined the energy changes with rotation associated with C—H bonds, the C—C bond, the hydrogen atoms, etc. He was unable to find any simple physical picture for the barrier source consistent with all his data. Very slight changes in electron distribution are seen to occur upon rotation. It appears from these localized orbitals that the electrons in the C—H bonds drift out towards the protons when the eclipsed form is obtained from the staggered. Comparisons of overlap populations for the two rotamers suggests that changes in bonding and antibonding character between hydrogen atoms may be a major factor. This would tend to justify the Mulliken "magic formula" approach mentioned earlier.

A quite different approach to the barrier problem has been taken by Wyatt and Parr (74), who have analyzed the Pitzer-Lipscomb wavefunctions (43) by using the integral Hellmann-Feynman theorem (75,76). Upon multiplying the real wavefunctions for staggered (ψ_s) and eclipsed (ψ_e) ethane together, integrating over all spin coordinates and the space coordinates of all electrons but one, and renormalizing, a one-electron spinless transition density, ρ_{es}, is obtained:

$$\rho_{es}(1) = \frac{\int \psi_s(1, 2, \ldots, n)\psi_e(1, 2, \ldots, n) \, d\tau(2)\cdots d\tau(n) \, d\omega(1)}{\int \psi_s(1, 2, \ldots, n)\psi_e(1, 2, \ldots, n) \, d\tau(1) \, d\tau(2)\cdots d\tau(n)} \tag{3}$$

This density is merely an overlap function between the wavefunctions. It may have negative as well as positive regions, and does not correspond to any actual intermediate state for the process of rotation. The integral Hellmann-Feynman formula states that the total energy for the change from the eclipsed state to the staggered one is equal to the difference in attractions of the eclipsed and staggered nuclear frames for the transition density plus ΔV_{nn}, the change in internuclear repulsion:

$$\Delta W = \Delta V_{nn} + \int \rho_{es}(1) \, \Delta V_{ne}(1) \, dv(1) \tag{4}$$

As a trivial example of the use of the integral Hellmann-Feynman theorem, one can consider the translation of a hydrogen atom in field-free space. The transition density for this process is simply an overlap density (renormalized) symmetrically disposed between the before and after positions of the nucleus. The internuclear repulsion term is zero in each case and the attractions between the transition density and the protons in the before and after positions are identical, leading to an energy of zero for the translation.

For the process of internal rotation, the transition density may appear quite different depending on how one chooses to visualize the rotation. Twisting each end by 30° in opposite directions will produce a different transition density than rotation of one end by 60° while the other end is fixed. In Figure 3 is a sketch of what one might conceive for the transition density for internal rotation in methyl alcohol if the hydroxyl group is fixed while the methyl group rotates 60°. The transition density on the fixed end is pictured to be much like the electron

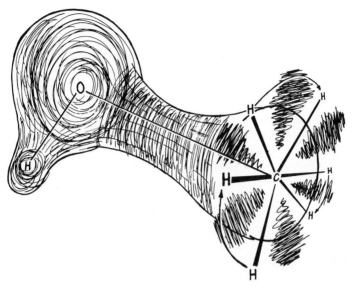

Fig. 3. The transition density for internal rotation in methyl alcohol. The methyl group, on the right, has rotated 60°.

density in staggered or eclipsed methyl alcohol since internal rotation presumably has little effect on the electron distribution (i.e., $\psi_s\psi_e \approx \psi_s\psi_s \approx \psi_e\psi_e$ on the fixed end). On the rotating end, the transition density is pictured as being mostly symmetrically disposed between the bonds in the two forms analogously to the transition density for the translation of hydrogen discussed above. If the C—H bonds were absolutely unchanged by rotation, this portion of the transition density would be sixfold symmetric (77,447). The barrier energy is computed by taking the difference in attractions between such a transition density and the staggered and eclipsed methyl protons and adding to this the change in repulsion between the hydroxyl proton and the staggered and eclipsed methyl protons.

Wyatt and Parr (74) found that the barrier in ethane computed in this way is 2.4 kcal/mole compared to the 3.3 kcal/mole obtained by taking the difference between the computed total energies of staggered and eclipsed ethanes. From an examination of the detailed nature of the transition density, it was possible to attribute certain contributions-to-the-barrier to certain regions of the molecule. (Of course, this can produce statements about the source of the barrier that are meaningful only within the framework of the method.) For ethane it was found that the regions around the protons provide most of the threefold component of the transition density. In the case where one end of the molecule is held fixed while the other end is rotated 60°, -3.8 kcal/mole comes from the interaction of the rotated protons and the transition density on the fixed end of the molecule, while $+1.4$ kcal/mole comes from the interaction of the rotated protons with transition density on the rotated end. (ΔV_{nn} supplies $+4.8$ kcal/mole of energy.) The deviations from sixfold symmetry of the transition density on the moving end are slight, but occur so close to the rotated protons that they contribute substantial energy (according to the Pitzer-Lipscomb wavefunctions).

Lowe and Parr (78,79) have extended this approach in a semi-empirical way to other molecules. By neglecting the effect of transition density on the rotated end, they arrive at a simple electrostatic model which, by virtue of its connection with the integral Hellmann-Feynman theorem, does not neglect kinetic energy changes. Implicit here is the assumption that the energy contributions which are neglected are at least roughly proportional to the total energy change. For cases in which the rotating group is always the same or very similar, this may be true. The semiempirical method requires making assumptions about the changes in transition density which occur as one goes from molecule to molecule. Thus, the problem of rationalizing or computing barriers to internal rotation resolves itself into such questions as: How does the the electron distribution in the O—H end of CH_3OH compare with that in the S—H part of CH_3SH? or What electronic redistributions occur when a fluorine is substituted in ethane? Questions such as these may be discussed in terms familiar to the chemist, such as electronegativity, induction, and resonance. Furthermore, since the electronic changes of interest in the molecule occur in regions distant from the moving protons, only fairly gross effects are important.

It is still too early to tell whether the assumptions behind the method are sufficiently reliable to justify its general use, but some success has already been achieved in predicting barrier values, and some previously unexplained relationships between barriers for different molecules are consistent with this approach. Lowe and Parr consider the trend in the relationship between the barriers in ethanelike molecules and the change in internuclear repulsion (40,80),

$$\Delta W = f \cdot \Delta V_{nn} \qquad (5)$$

From the standpoint of the electrostatic model, the barrier is due to the repulsion change, ΔV_{nn}, plus a change in the attraction between the threefold part of the transition density on the fixed end of the molecule and the moving protons. Hence the value of f is related to the amount of threefold transition density present. Since the threefold component is related to the occupation of the $1s$ orbitals on the hydrogens, it seems reasonable to determine whether the value of f is correlated with the electronegativity of the atom on the fixed end to which the hydrogens are bonded. The results of such a comparison are shown in Figure 4. For molecules containing elements in the first and second rows of the periodic table, such a correlation exists. The high f value for CH_3OH appears as a result of the hydroxyl proton being less effectively screened by electrons (in a threefold way) than a proton in ethane. [The three-foldedness of the transition density is important since most other symmetries (e.g., cylindrical symmetry) can have no effect on the barrier.] In molecules containing one Ge or Sn atom, d and f orbitals should mix into the bonds and increase the threefold shielding component on the fixed end, leading to low f values. In digermane, the bonds at *both* ends may have increased threefold character, leading to

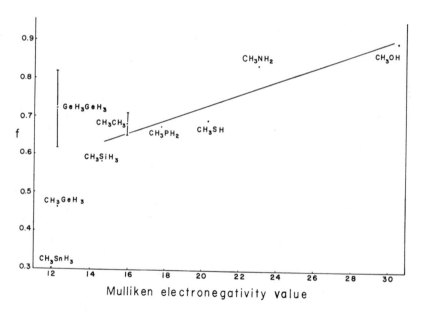

Fig. 4. Effective charge f on fixed protons versus electronegativity of heavy atom on fixed end.

an additional mechanism for barrier production and hence an increased value for f.

On the basis of such arguments, it has been predicted that CH_3SeH would have an f value of 0.5–0.55 (81). Using approximate structures, this leads to a barrier value of 900–1200 cal/mole. Use of the exact structure gives 990 ± 47 cal/mole, compared to the new experimental value of 1010 cal/mole (89). This illustrates an important limitation in the method—accurate structures are required for accurate predictions. Since accurate structures are often determined in the same experiment that gives the barrier (as was the case above), it is normal to find oneself predicting barriers on the basis of approximate structures.

The effects of fluorosubstitution in ethane and methyl silane have been shown (78), through use of the electrostatic model, to be consistent with inductive transfers of electronic charge. Substitution of the first fluorine in ethane is postulated to produce a fixed-end transition density, which, compared to that in ethane, is somewhat rarified about the protons and somewhat concentrated about the fluorine. In other words, the rotating protons "see" two unusually positive hydrogens and one unusually negative fluorine, leading to an increased barrier. Upon substitution of the second fluorine, the rotating protons see one unusually positive hydrogen and two somewhat negative fluorines producing a lower barrier, consistent with what is observed. A similar argument applies to the fluorinated methyl silanes.

Molecules containing a double bond have also been treated by this method (79). It has been shown that such molecules would be expected to have transition densities which would tend to attract the methyl protons into the correct stable conformation. In acetaldehyde the important feature is taken to be the nonbonding sp^2 hybridized electrons on oxygen which point towards the methyl group. In propylene, it is taken to be *cis* C—H bonding electrons. Since there are many complicating features in such molecules compared to the ethanelike cases, the models have been simplified to include only those few features which are thought to be most important; there can be no doubt that arguments and predictions for these molecules are on much weaker ground. Nevertheless, such models lead to a possible explanation for the difference of the effects-upon-the-barrier of halosubstitutions in the secondary position in propylene and in the analogous position in acetaldehyde. The inductive effect in propylene should be similar to that in ethane since the sigma electrons are all constrained in C—C and C—H bonds in both systems. In acetaldehyde, the inductive effect should be much more marked since the nonbonded electrons on oxygen are relatively more polarizable. This should reduce the effect of these electrons in attracting the methyl group into the proper conformation and also shield the halogen more effectively so that it has less tendency to repel the methyl group into the stable conformation. This behavior, coupled with reasonable resonance effects, leads to model-computed barriers in good agreement with experiment for both of these groups of molecules.

Considerations such as these have led to predictions for barriers for a number of molecules. These are shown in Table I.

TABLE I

Predicted Values for Barriers to Internal Rotation Based
on Electrostatic Models (78,79)

Molecule	Calculated V_3, kcal/mole	Observed V_3, kcal/mole
CH_3SeH	0.99	1.01
SiH_3CH_2F	1.99	
SiH_3CH_2Cl	2.44	2.55
SiH_3CH_2Br	2.45	
SiH_3CH_2I	2.32	
SiH_3SiH_3	1.08	
SiH_3SiHF_2	0.85	
SiH_3SiF_3	0.81	
$CH_3 \cdot CBr:CH_2$ [a]	2.38–2.54	2.7 ± 0.2
$CH_3 \cdot CF:C:O$	1.56	
$CH_3 \cdot CCl:C:O$	1.89	
$CH_3 \cdot CF:C:CH_2$	1.95	
$CH_3 \cdot CCl:C:CH_2$	2.28	
$SiH_3 \cdot CH:O$	0.85	
$SiH_3 \cdot CH:C:O$	0.96	
$SiH_3 \cdot CH:CH_2$	1.57	1.50
$CH_3 \cdot N:C:O$	<0.14	~0.05

[a] Two approximate structures have been used.

The major attraction of the semiempirical method based upon the integral Hellmann-Feynman formula is that it enables one to discuss barriers in terms of familiar concepts and to make calculations using simple mathematics. However, the method is based upon a number of unproved assumptions and the models for larger molecules are very far from being a complete representation of reasonable transition densities. With the present high level of activity in this field, it will presumably not be long before some more definitive information will be available about these points (468,469,479).

Fink and Allen (46) have made a series of very accurate SCF calculations on ethane, methylamine, methanol, hydrogen peroxide, hydrazine, and hydroxylamine. Their results indicate that the computed barrier values for the methyl tops are fairly constant over a range of improvements in the wavefunctions, but that hydrogen peroxide (and perhaps hydrazine) is more sensitive. This is equivalent to saying that, for the methyl tops, the *difference* between the negative change in two-electron energy, $-\Delta V_{ee}$, and the changes in one-electron energies and internuclear repulsion, $\Delta T + \Delta V_{ne} + \Delta V_{nn}$, is fairly constant from one extremum to another in the computed energy curves for a given molecule. This means that, over the range of improvements studied by Fink and Allen, improvements in energy effect both energy groups about equally and hence have little effect upon the barrier.

In attempting to describe the physical processes producing the barrier, Fink and Allen (46) make use of concepts associated with localized orbital energies. Their resulting physical picture describes the barrier as resulting from changes in nuclear-electronic and inter-electronic interactions between opposite ends of the molecule. However, since the actual transformation to localized orbitals has not been carried out by these authors, this physical picture must be regarded as speculative. Indeed, the localized orbital analysis of ethane by Pitzer (44) suggests that the situation is much more complicated.

Goodisman (83) has pointed out that the barrier in ethane may be characterized by knowing one number—the derivative of the barrier at some point between the staggered and eclipsed forms. This offers the attractive possibility of obtaining the barrier from the wavefunction of one state rather than two. The derivative of the barrier is a measure of the torque on the rotor. Use of the conventional Hellmann-Feynman theorem (not the same as the integral Hellmann-Feynman theorem) will extract this force from a wavefunction. By combining the Pitzer-Lipscomb wavefunctions for staggered and eclipsed ethane, Goodisman constructed an approximate wavefunction for ethane in an intermediate torsional state. Calculation of the torque led to a barrier height of about 60% of the experimental value. Attempts to improve the wavefunction by introducing some polarization effects and also by allowing the hydrogen $1s$ orbital exponents to vary have led to barrier values which are consistently too low (437). This is in accord with the observation that Hellmann-Feynman force calculations are extremely sensitive to wavefunction accuracy. Zülicke and Spangenberg (84) have tried a similar approach on hydrogen peroxide. Since this barrier is more complicated, they have made calculations at eleven different torsional angles using a simple, angle-independent wavefunction which describes the unshared electrons as being in sp^3 hybrids. They found a minimum-energy angle of about 120° and reasonable heights for the *cis* and *trans* barriers. However, in light of Goodisman's results, this good agreement with experiment would probably disappear with use of more realistic wavefunctions (474). Better indication of the usefulness of the method could be had by using the Hellmann-Feynman theorem on some of the more accurate wavefunctions for intermediate conformations in hydrogen peroxide which are now available.

Chemistry is a field which has thrived on the application of fairly simple, inexact models, and it is the search for such models which has motivated much of the activity in the internal rotation field. However, it seems likely that an *exact* description of the energetics consistent with accurate wavefunctions will be an extremely complex one involving different "simple" processes occurring simultaneously in all regions of the molecule. Furthermore, such a description will not be unique. One could use energies for localized orbitals or delocalized orbitals, or use the integral Hellmann-Feynman theorem and describe energies with respect to a transition density, or use the Hellmann-Feynman theorem and describe the situation in terms of forces. These equally valid formulations would appear quite different, and probably none would correspond to a process describable in terms of a single simple physical model. The situation is reminiscent of studies of the nature of the chemical bond (85). The problem in the theory of hindered internal rotation, then, seems to be to find usable models which are reasonably consistent with the more intimate knowledge now becoming available from accurate calculations. The fairly high degree of success which is enjoyed by some semiempirical methods suggests that it may be possible to focus on one or more *major* processes which can be parametrized to take other contributions into account implicitly. Knowledge gained from *ab initio* calculations will serve to put such simple methods on a surer footing and ensure that they are not extended into situations where they do not apply.

V. PREFACE TO TABLE II

1. The barrier for a given internal rotation is given in Table II on the basis of barrier symmetry and the atoms forming the bond about which torsion occurs. Thus the barrier for rotation of the methyl group in methyl nitrite (CH_3—ONO) is found in the section on threefold barriers under the subsection on C—VI bonds (since oxygen is in group VI of the periodic table). For rotation about the O—N bond in methyl nitrite, the barrier is tabulated in the "Other Barriers" section (since it is not expected to be purely three-, six-, or twelvefold) in the subsection "V—VI bonds."

2. The single bond about which rotation occurs is always explicitly shown in the formula given in the first column of the table.

3. Values of barrier heights given in the third column are sometimes given with error limits. Error limits without parentheses are quoted directly from experimental reports and may be misleading in some cases where only *some* error sources have been considered. An attempt has been made, however, to omit such misleading error limits. Error limits in parentheses have been made up by this author on the basis of discussions in the experimental reports or else on the basis of other information either on the molecule in question or on the method applied. Most energies are quoted without error limits. This does *not* imply that such energies are exactly known. The user of this table is strongly advised to refer to original sources before accepting any numbers quoted here. Where no obvious preference could be given to one of several results, all results are tabulated. Conformational energy differences are uniformly labeled ΔE even though some methods actually measure ΔH. However, the differences are usually slight compared to the experimental errors involved. Barrier heights are labeled V when no particular symmetry is implied. These heights are the usual activation energy, and a different symbol is used if free energy of activation ($\Delta F^\ddagger$) is meant. When symmetry exists or is assumed to exist, the symbol V_n is used and it is defined by the expression $V = \sum_n (V_n/2)(1 - \cos n\alpha)$, where α is the torsional angle.

4. Methods used to obtain barrier data are tabulated. The first methods listed are those used to obtain the most reliable data. Other methods which are not so accurate but which essentially agree with the more accurate results are listed next. Methods which are less reliable and which disagree are listed in parentheses. No effort has been made to compile a complete bibliography of all work done on a given barrier.

Hence, much early work does not appear. The code adopted for the experimental methods is:

Symbol	Method
T	Thermodynamic
DM	Dipole moment
$MW\begin{cases} i \\ s \end{cases}$	Microwave $\begin{cases} \text{intensities} \\ \text{splittings} \end{cases}$
IR	Infrared
Ram	Raman
NM(Q)R	Nuclear magnetic (quadrupole) resonance
ESR	Electron spin resonance
ED	Electron diffraction
NS	Neutron scattering
SA	Sound absorption

5. The state refers to that for which the barrier value applies, not necessarily that for which the measurement was made. For example, thermodynamic measurements on solids may yield barrier values appropriate for gases.

TABLE II

Experimental Barrier Measurements

A. Threefold Barriers

Formula	Name	Barrier height (V_3), kcal/mole	Method	State	Refs.
		1. C—C bonds			
CH_3—CH_3	Ethane	2.875 ± 0.125 [a]	T (IR)	Gas	96–99
		<5.3	NMR	Solid	120
CH_3—CH_2F	Ethyl fluoride	3.33 ± 0.05 [a]	IR, MW_s, MW_t	Gas	9,100–103
		($V_6 = -15$ cal/mole)			
CH_3—CHF_2	1,1-Difluoroethane	3.18	MW_s, IR, MW_t	Gas	101,103,104,443
CH_3—CF_3	1,1,1-Trifluoroethane	3.25 ± 0.2	T, IR, Ram (MW_t)	Gas	105–111
CF_3—CH_2F	1,1,2-Tetrafluoroethane	4.2 ± 0.15	IR	Gas	112
		4.58	Ram	Liq.	113
CF_3—CHF_2	Pentafluoroethane	3.51	MW_t, IR	Gas	457
CF_3—CF_3	Hexafluoroethane	3.92 [a]	T (ED)	Gas	114–116
CH_3—CH_2Cl	Ethyl chloride	3.68 [a]	MW_s, IR	Gas	19,102,104,117–119, 452
CH_3—$CHCl_2$	1,1-Dichloroethane	3.49 ± 0.2	T (IR)	Gas	105,118,121
CH_3—CCl_3	1,1,1-Trichloroethane	$2.91(\pm 0.2)$	T, IR (Ram)	Gas	109,123–125
		5.7	NMR	Solid	414
		5.8	NS	Solid, liq.	458
CCl_3—CCl_3	Hexachloroethane	10.8 [a]	ED	Gas	126,127
			IR, Ram	Solid	128
CH_3—CH_2Br	Ethyl bromide	3.68	MW_s, IR (MW_t)	Gas	102,130,131, 452
CH_3—CH_2I	Ethyl iodide	3.22 ± 0.5	MW_s	Gas	129

$CH_3—CF_2Cl$	1-Chloro-1,1-difluoroethane	3.70	IR, Ram	Gas–liq.	109
$CH_3—CFCl_2$	1,1-Dichloro-1-fluoroethane	3.47	IR, Ram	Gas–liq.	109
$CF_3—CH_2Cl$	1,1,1-Trifluoro-2-chloroethane	5.7–5.8	IR, Ram	Gas–liq.	113,122
$CF_3—CF_2Cl$	Chloropentafluoroethane	5.67	IR, Ram	Gas	415
$CF_3—CHCl_2$	1,1-Dichloro-2,2,2-trifluoroethane	2.23(??)ᶜ	IR, Ram	Gas–liq.	122
$CF_3—CH_2Br$	1-Bromo-2,2,2-trifluoroethane	6.21	Ram (IR)	Liq.	113,132
$CF_3—CF_2Br$	Bromopentafluoroethane	6.40	IR, Ram	Gas	415
$CF_3—CH_2I$	1-Iodo-2,2,2-trifluoroethane	6.13	Ram	Liq.	113
$CF_3—CF_2I$	Iodopentafluoroethane	7.09	IR, Ram	Gas	415
$CF_3—CHClBr$	1-Bromo-1-chloro-2,2,2-trifluoroethane	7.27	IR, Ram	Gas–liq.	133
$CH_3—C{\equiv}C—CH_3$	Dimethylacetylene	≤ 0.03	IR (T)	Gas	134–137
$CH_3—C{\equiv}C—CF_3$	1,1,1-Trifluoro-2-butyne	$0 < V_3 < 1.2$	MW	Gas	138
$CH_3—C{\equiv}C—CH_2Cl$	1-Chloro-2-butyne	<0.1	MW	Gas	139
$CH_3—C{\equiv}C—SiH_3$	Methylsilylacetylene	≤ 0.003	MW	Gas	15
$(CH_3)_3CC{\equiv}CH$	1,1-Dimethyl-3-butyne	≈ 4.0	MW, IR	Gas	426,452
$CH_3—CH_2CN$	Ethyl cyanide	3.05	MW_s (MW_i)	Gas	140,141
$(CH_3)_3CCN$	t-Butyl cyanide	≈ 4.0	MW_i, IR	Gas	426,452
$CH_3—CH_2—CH_3$	Propane	3.4 (methyls stagger methylene)	T (MW_s)	Gas	96b,143–145,159
$(CF_3—)_2CF_2$	Octafluoropropane	≈ 3	T	Solid, liq.	476

(continued)

TABLE II (continued)

Formula	Name	Barrier height (V_3), kcal/mole	Method	State	Refs.
$(CH_3—)_2CH—CH_3$	2-Methylpropane	3.9(±0.75) (methyls stagger methylene)	MW_i	Gas	146
		3.8–3.9	T	Gas	159,160
		2.6 (probably low)	NMR	Solid	414
$(CH_3—)_3C—CH_3$	2,2-Dimethylpropane	4.7	T	Gas	96b,159
		5.2	NS	Solid, liq.	458
(structure)	Methylcyclopropane	2.86(±0.15)	MW_s	Gas	480
$CH_3—CH—CF_2$ (cyclopropane)	2,2-Difluoro-1-methyl cyclopropane	2.26 ± 0.15	MW_s	Gas	481
$CH_3—CH_2CH_2—CH_3$	Butane	3.4	T	Gas	159
$CH_3—CH_2OH$	Ethanol	0.77 ± 0.1	MW_s (T)	Gas	148,149,428
$CH_3—CH_2SH$	Ethanethiol	3.31	T	Gas	311
$(CH_3—)_2CHOH$	2-Propanol	≈3.4	T	Gas	150
$(CH_3—)_2CHSH$	2-Propanethiol	3.95	T	Gas	151
		3.73	IR	Gas–liq.	152
$(CH_3—)_3CSH$	2-Methyl-2-propanethiol	5.1	T	Gas	312
$CH_3—CH_2SCH_3$	Methylthioethane	3.82	T	Gas	153
$CH_3—CH_2SCH_2—CH_3$	Ethyl sulfide	3.55	T	Gas	154
$(CH_3—)_2CHSCH_3$	Isopropylthiomethane	3.95	T	Gas	155
$(CH_3—)_3CSCH_3$	t-Butylthiomethane	4.9	T	Gas	156
$CH_3—CH_2COOH$	Propionic acid	2.36 ± 0.05 (barrier for planar cis form)	MW_s	Gas	157

$CH_3—CH_2CHO$	Propionaldehyde	2.28 ± 0.11 (barrier for planar *cis* form)	MW_s	Gas	158
$CH_3—CHO$	Acetaldehyde	1.16 ± 0.03 [b]	MW_s, IR, MW_i	Gas	9,104,142,161
$CH_3—CFO$	Acetyl fluoride	1.04 [b]	MW_s	Gas	162,163
$CH_3—CClO$	Acetyl chloride	1.30 ± 0.03 [b]	MW_s	Gas	164
$CH_3—CBrO$	Acetyl bromide	1.30 ± 0.03	MW_s	Gas	165
$CH_3—CIO$	Acetyl iodide	$1.30(\pm 0.07)$	MW_s	Gas	419
$CH_3—CCNO$	Acetyl cyanide	1.21 ± 0.03 [b]	MW_s	Gas	166,310
$(CH_3—)_2CO$	Acetone	0.78	MW_s (IR, T)	Gas	11,167–169
$CH_3—COCHCH_2$	Methyl vinyl ketone	$1.25(\pm 0.06)$ (double bonds are *trans*)	MW_s	Gas	260
$CH_3—COCCH$	Acetyl acetylene	1.07 ± 0.06	MW_s	Gas	170
$CH_3—COOH$	Acetic acid	0.48 ± 0.025	MW_s	Gas	171
$CF_3—CHO$	Trifluoroacetaldehyde	0.885 ± 0.075	MW_s (IR)	Gas	172,471
$CF_3—CFO$	Trifluoroacetyl fluoride	1.39 ± 0.21	IR	Gas	173
		1.55	IR, Ram	Gas	421
$CF_3—CClO$	Trifluoroacetyl chloride	1.6 ± 0.36	IR	Gas	427
$(CF_3—)_2CO$	Hexafluoroacetone	≈ 2.8	IR	Gas	427
		2.95	IR	Gas	452
		1.5	T	Solid, liq.	475
$CCl_3—COOH$	Trichloroacetic acid	≤ 5	NQR	Solid	174
$CH_3—CHCH_2$	Propylene	1.995 [b] $(V_6 = -37 \pm 6$ cal/mole)	MW_s (T, IR)	Gas	19,175–178,452
$CH_3—CFCH_2$	2-Fluoropropylene	2.34 $(V_6 = +11$ cal/mole)	IR (MW_3)	Gas	19,179
trans-$CH_3—CHCHF$	*trans*-1-Fluoropropylene	2.2 ± 0.1	MW_s	Gas	180
cis-$CH_3—CHCHF$	*cis*-1-Fluoropropylene	1.06 ± 0.05	MW_s	Gas	181
$CH_3—CHCF_2$	1,1-Difluoropropylene	1.25 ± 0.02	MW_s	Gas	182

(continued)

TABLE II (continued)

Formula	Name	Barrier height (V_3), kcal/mole	Method	State	Refs.
$CH_3—CCICH_2$	2-Chloropropylene	2.5–2.7 [b]	MW$_s$, IR, Ram	Gas	183–185
trans-$CH_3—CHCHCl$	trans-1-Chloropropylene	2.17 ± 0.1	MW$_s$	Gas	186
cis-$CH_3—CHCHCl$	cis-1-Chloropropylene	0.62(±0.03)	MW$_s$	Gas	187
$CH_3—CBrCH_2$	2-Bromopropylene	2.7(±0.2)	MW$_s$	Gas	433
trans-$CH_3—CHCHBr$	trans-1-Bromopropylene	2.12 ± 0.1	MW$_s$	Gas	188
cis-$CH_3—CHCHBr$	cis-1-Bromopropylene	0.23	MW$_s$	Gas	189
trans-$CH_3—CHCHCN$	trans-Crotonitrile	≥2.1	MW$_s$	Gas	190
cis-$CH_3—CHCHCN$	cis-Crotonitrile	1.4 (±0.07)	MW$_s$	Gas	191
$(CH_3—)_2CCH_2$	2-Methylpropylene	2.12 (stable form C_{2v} symmetry)	MW$_s$, IR, T	Gas	168,178,192
$CH_3—C{\overset{\displaystyle CH_2}{\diagup\hskip-0.5em\diagdown}}CH$	1-Methylcyclopropylene	1.39 ± 0.05	MW$_s$	Gas	469
trans-$CH_3—CHCH—CH_3$	trans-1-Methylpropylene	0.95	T	Gas	178,193
cis-$CH_3—CHCH—CH_3$	cis-1-Methylpropylene	0.65–0.7	MW$_s$ (T)	Gas	178,444
$CH_3—C(CH_2)CHCH_2$	2-Methyl-1,3-butadiene	≥2.1 (one isomer observed—prob. trans)	MW$_s$	Gas	194
$CH_3—CHCO$	Methyl ketene	1.18 ± 0.02 [b]	MW$_s$	Gas	195,196
$(CH_3—)_2CCO$	Dimethyl ketene	2.066(±0.1)	MW$_s$	Gas	483
$CH_3—CHCCH_2$	Methyl allene	1.59 (±0.08)	MW$_s$	Gas	197
$(CH_3—)_3CCHO$	Pivalaldehyde	(Methyl tops) 2.6, 3.5, 3.8	MW$_i$	Gas	198
$CH_3—CH_2CH_2F$	1-Fluoropropane	(Butyl top) 1.17–1.18 2.69 (trans) 2.87 (gauche)	MW$_s$ MW$_s$	Gas Gas	198 199

$CH_3-CH_2CH_2Cl$	1-Chloropropane	2.78 (*trans*) 2.96 (*gauche*)	IR	Gas	418
$CH_3-CH_2CH_2Br$	1-Bromopropane	2.36 (*trans*) 2.65 (*gauche*)	IR	Gas	418
$CH_3-CH_2CH_2I$	1-Iodopropane	2.47 (*trans*) 2.77 (*gauche*)	IR	Gas	418
$(CH_3-)_2CHCl$	2-Chloropropane	3.52 ≥ 3.45 (methyls stagger C—Cl bond)	T MW$_s$	Gas Gas	309 200
$(CH_3-)_2CHBr$	2-Bromopropane	4.26	IR	Gas	452
$(CH_3-)_2CHI$	2-Iodopropane	4.13	IR	Gas	452
$(CH_3-)_2CCl_2$	2,2-Dichloropropane	5.4	NMR	Solid	414
$(CH_3-)_3CF$	2-Methyl-2-fluoropropane	4.3	MW$_i$	Gas	146
$(CH_3-)_3CCl$	2-Chloro-2-methylpropane	3.7 (probably low)	NMR	Solid	414
$CH_3-CH_2CH_2NH_2$	Propylamine	≈ 3.4	Ram	Liq.	201
$CH_3-CH\overset{O}{\overset{\diagdown\diagup}{-}}CH_2$	Propylene oxide	2.57 ($V_6 = -26$ cal/mole)	IR, MW$_s$, MW$_i$	Gas	19,202,203,443
$CH_3-CH\overset{O}{\overset{\diagdown\diagup}{-}}CH-CH_3$	*cis*-2,3-Epoxybutane	1.61 ± 0.15	MW$_s$	Gas	204
$CH_3-CH\overset{O}{\overset{\diagdown\diagup}{-}}CH-CH_3$	*trans*-2,3-Epoxybutane	2.44 ± 0.15	MW$_s$	Gas	477
$CH_3-CH\overset{S}{\overset{\diagdown\diagup}{-}}CH_2$	Propylene sulfide	3.24 ± 0.16	MW$_s$, IR	Gas	19,205
$(CH_3-C_2H_4)_3SnCl$	Tripropyltin chloride	4.1 (for *trans* form of propyl)	IR	Liq.	206
$CH_3-CH_2CHCH_2$	1-Butene	4.04 (*cis* planar form) 3.21 (*gauche* form)	MW$_s$	Gas	338
$CH_3-CH_2C_6H_5$	Ethylbenzene	2.7	T	Gas	207
$CH_3-\dot{C}HCOOH$	Propionic acid radical	4 (±2)	ESR	Solid	208

(continued)

TABLE II (continued)

Formula	Name	Barrier height (V_3), kcal/mole	Method	State	Refs.
$CH_3—\dot{C}(COOH)_2$	Methyl malonic acid radical	<0.1 [d]	ESR	Solid	209
$CH_3—\dot{C}HCOO^-$	Propionic acid anion radical	3.6 ± 0.2 [d]	ESR	Solid	210
	2. Other C—IV bonds				
$CH_3—SiH_3$	Methyl silane	1.70 ± 0.1 [a]	MW$_s$, IR	Gas	16,211,293
$CH_3—SiH_2F$	Methylmonofluorosilane	1.56 ± 0.03 [a]	MW$_s$	Gas	212
$CH_3—SiHF_2$	Methyldifluorosilane	1.26 [a]	MW$_s$	Gas	213,214
$CH_3—SiF_3$	Methyltrifluorosilane	1.4 ± 0.2	MW, IR (MW$_i$)	Gas	17,111,215,216
$(CH_3—)_2SiH_2$	Dimethylsilane	1.65 ± 0.12	MW$_s$	Gas	218
$(CH_3—)_3SiH$	Trimethylsilane	1.83 ± 0.4 (methyls stagger Si—H)	MW$_i$	Gas	219
$(CH_3—)_4Si$	Tetramethylsilane	1.4 ± 0.1	T	Gas	220,221
$CH_2Cl—SiH_3$	Chloromethylsilane	2.55 ± 0.05	MW$_s$	Gas	222
$CCl_3—SiCl_3$	Trichloromethyl-trichlorosilane	4.3	ED	Gas	126
$CH_3—CHCH_2$	Vinyl silane	1.50 ± 0.03 [b]	MW$_s$	Gas	232
$CH_3—GeH_3$	Methylgermane	1.24 ± 0.025 [a]	MW$_s$, IR	Gas	17,211,224
$(CH_3—)_2GeH_2$	Dimethyl germane	1.18 ± 0.03	MW$_s$	Gas	484
$CH_3—GeCl_3$	Methyltrichlorogermane	1.3	IR, Ram	Liq.	225
$(CH_3—)_4Ge$	Tetramethylgermane	≈0.65	NMR	Solid	226
$CH_3—SnH_3$	Methylstannane	≈0.65	MW$_s$	Gas	17,227
$(CH_3—)_4Sn$	Tetramethylstannane	≈0.46	NMR	Solid	226
$(CH_3—)_4Pb$	Tetramethyllead	≈0.18	NMR	Solid	226

3. Other IV—IV bonds

		Order of magnitude, kcal[a]	IR, Ram		
SiH_3—SiH_3	Disilane	<5.3	IR, Ram	Gas	228,229
			NMR	Solid	120
SiH_3—SiH_2F	Disilanyl fluoride	1.05 ± 0.04	MW_s	Gas	231
$SiCl_3$—$SiCl_3$	Hexachlorodisilane	≈1.0[a]	ED (IR, Ram)	Gas	116,126,234,236
GeH_3—GeH_3	Digermane	1.49 ± 0.2[a]	IR, Ram	Gas	237,238

4. C—V bonds

		Order of magnitude, kcal[a]			
CH_3—NH_2	Methylamine	1.98 ± 0.01	MW_s (T)	Gas	239–244
		<5.4	NMR	Solid	120
$(CH_3—)_2NH$	Dimethylamine	3.62 ± 0.05	IR, MW_s	Gas	168,233,252c,452
$(CH_3—)_3N$	Trimethylamine	4.41 ± 0.03	IR, MW_t (T)	Gas	168,230,235,452
CH_3—N_3	Methylazide	0.71(±0.04)	MW_s	Gas	223
H_3C—NO	Nitrosomethane	$\begin{cases} 2.68 ± 0.57 \text{ (lower)} \\ 0.83 ± 0.14 \text{ (upper)} \end{cases}$ (two electronic states observed but not identified)	Absorption in visible	Gas	217
CH_3—NCH_2	N-Methylmethyleneimine	1.97(±0.1)[b]	MW_s	Gas	308,314
CH_3—$N\langle^{CH_2}_{\ \ }_{CH_2}\rangle$	N-Methyl ethyleneimine	3.46(±0.175)	MW_s	Gas	488
CH_3—NCO	Methyl isocyanate	0.049 ± 0.003	MW_s	Gas	315
CH_3—NCS	Methyl isothiocyanate	≈0.05	MW_s	Gas	278
CH_3—NSO	Methyl thionylamine	0.34 ± 0.02	MW_s	Gas	361
CCl_3—NO_2	Trichloronitromethane	2.4–3.4	ED	Gas	370,430
CH_3—PH_2	Methylphosphine	1.96(±0.1)	MW_s	Gas	365
CF_3—PH_2	Trifluoromethyl phosphine	2.36(±0.1)	MW_s	Gas	485

(continued)

TABLE II (*continued*)

Formula	Name	Barrier height (V_3), kcal/mole	Method	State	Refs.
(CH₃—)₃P	Trimethylphosphine	2.6(±0.5) (methyls stagger P—C bonds)	IR, Ram, ED	Gas	146,381
CH₃—AsF₂	Methyldifluoroarsine	1.32(±0.08)	MW$_s$	Gas	398
(CH₃—)₃As	Trimethylarsine	1.5–2.5	MW$_s$	Gas	405
		5. C—VI bonds			
CH₃—OH	Methyl alcohol	1.07(±0.02)	MW$_s$	Gas	244–250
		2–2.25	NMR	Solid	120
CH₃—OCl	Methyl hypochlorite	3.06 ± 0.15 [a]	MW$_s$	Gas	251
CH₃—O—CH₃	Dimethyl ether	2.7(±0.15) (methyls stagger opposite C—O bond)	MW$_s$, IR (T)	Gas	135,168,252a, 253–255,452
CH₃—ONO	Methyl nitrite	≈3	?	?	409
CH₃—ONO₂	Methyl nitrate	2.23(±0.11) (methyls stagger nearest NO₂ oxygen)	MW$_s$	Gas	257,258
CH₃—OCHO	Methyl formate	1.19 ± 0.04 (methyl group *cis* to and staggers C=O)	MW$_s$, IR	Gas	10b,104,259
CH₃—OCHCH₂	Methyl vinyl ether	3.83(±0.2) (stable form *cis*)	MW$_s$	Gas	261
CH₃—OCCH	Methoxyethyne	1.4	MW$_s$	Gas	464
CF₃—OO—CF₃	Bis(trifluoromethyl) peroxide				
CH₃—SH	Methanethiol	5.7	IR, Ram	Gas, liq.	486
		1.27 ± 0.03	MW$_s$ (MW$_s$, T)	Gas	17,262–264
(CH₃)₃C—SH	2-Methyl-2-propanethiol	1.36	T	Gas	312

$(CH_3-)_2S$	Dimethylsulfide	2.53 ± 0.1 [methyl–methyl interactions treated (methyls stagger C—S)]	MW_s	Gas	13a
		2.11 ± 0.1 (methyls assumed independent)	MW_s, IR (T)	Gas	12,13a, 168,254b,264, 267–271
$(CH_3-)_2SO$	Dimethyl sulfoxide	2.94	MW_s	Gas	429
$CH_3-SCH_2CH_3$	Methylthioethane	1.97	T	Gas	153
$CH_3-SCH(CH_3)_2$	1-Methyl-1-methylthio-ethane	2.0	T	Gas	155
$CH_3-SC(CH_3)_3$	1,1-Dimethyl-1-methyl-thioethane	2.0	T	Gas	156
CH_3-SCN	Methylthiocyanate	1.59(±0.08)	MW_s	Gas	272,273
CH_3-SCCH	Methylthioethyne	1.75	MW_s	Gas	465
$CH_3-SCHCH_2$	Methyl vinyl sulfide	3.23(±0.16) (prob. cis form)	MW_s	Gas	487
$CH_3-SS-CH_3$	Dimethyl disulfide	1.5	T	Gas	152,274
CF_3-SH	Trifluoromethanethiol	1.21	IR (T)	Gas	275,276
CH_3-SeH	Methyl selenol	1.01(±0.05)	MW_s, IR	Gas	89,277
$(CH_3-)_2Se$	Dimethylselinide	1.50 ± 0.02	MW_s	Gas	440

6. Other threefold barriers

PF_3-BH_3	Phosphorus trifluoride-borane	3.24 ± 0.15	MW_s	Gas	446
$(CH_3-)_3B$	Trimethyl borane	≈0	Ram, IR, ED	Gas	279,280
$[(CH_3-)_2BH]_2$	Tetramethyl diborane	≈1	ED	Gas	459
$(CH_3-)_2Zn$	Dimethyl zinc	≈0	IR	Gas	281
$(CH_3-)_2Hg$	Dimethyl mercury	≈0	IR	Gas	281
$(NH_3-)_2PdCl_2$	Dichlorodiamine palladium	≈1.2	IR	Solid	434
$(NH_3-)_2PdI_2$	Diiododiaminepalladium	≈0.6	IR	Solid	282
$(CH_3-)_4Au_2I_2$	Diiodotetramethylgold(III)	0.77 ± 0.2	IR	Solid	435

(continued)

TABLE II (*continued*)
B. Sixfold Barriers

Formula	Name	Barrier height (V_6), cal/mole	Method	State	Refs.
C_6H_5—CH_3	Methylbenzene	$V_6 = 13.94$	MW (T)	Gas	105,358,470
FC_6H_4—CH_3	1-Fluoro-4-methylbenzene	$V_6 \approx 13.82$	MW_s (T)	Gas	363,445
CH_3—BF_2	Methyldifluoroborane	$V_6 = 13.77 \pm 0.03$	MW_s	Gas	256
CH_3—NO_2	Nitromethane	$\left.\begin{array}{l} V_6 = 6.03 \pm 0.03 \\ V_{12} < 0.05 \end{array}\right\}$	MW_s	Gas	265
CF_3—NO_2	Trifluoronitromethane	$V_6 = 74.4 \pm 5$	MW_s	Gas	305
		$V \neq V_6$ (C—NO_2 non-planar)	ED	Gas	324

C. Twelvefold Barriers

Formula	Name	Barrier height (V_{12}), cal/mole	Method	State	Refs.
CF_3—SF_5	Trifluoromethyl sulfur pentafluoride	$V_{12}^\circ = 630 \pm 315$	MW_i	Gas	266
		V_{12} much lower	IR, Ram	Gas	292,298
1-CH_3—B_5H_8	1-Methylpentaborane	$V_{12} < 6$	MW_s	Gas	326

D. Other Barriers

Formula	Name	Data and notes, kcal/mole	Method	State	Refs.
		1. C—C bonds			
CH_2F—CH_2F	1,2-Difluoroethane	$\Delta E^\circ \leq \pm 0.2$	IR	Gas	283
		≈ -0.9	IR, Ram	Liq.	283
CHF_2—CHF_2	1,1,2,2-Tetrafluoroethane	$\Delta E^\circ = 1.16 \pm 0.1$	IR, Ram	Gas	284
CH_2Cl—CH_2Cl	1,2-Dichloroethane	$\Delta E^\circ = 1.2(\pm 0.1)$	DM, T, IR, Ram, ED	Gas	7,285–288
		$= 0.0(\pm 0.1)$	IR, Ram	Liq.	7,286,289–291
		$= 1.1 - 1.7^t$	NQR, Ram	Sol.	174
		$= 0.4$	IR	Soln. in hexane	289
		$V_3^g(gauche \to trans) = 3.2 \pm 0.5$	SA	Solns. in ether and acetone	6
		$V_3^g = 4.7$	IR	Liq.	294
		$V = 3.0(\pm 0.5)$	NQR	Solid	174,295
CH_2Cl—$CHCl_2$	1,1,2-Trichloroethane	$\Delta E^\circ = -2.6(\pm 0.3)$	IR, DM, T, ED	Gas	296,297,300,301
		$= -0.22(\pm 0.03)$	IR	Liq.	289,296,299
		$= -0.4$	IR	(1:2) Soln. in CCl_4	289
		$= 0.0$	IR	(1:2) Soln. in CH_3CN	289

(continued)

TABLE II (continued)

Formula	Name	Data and notes, kcal/mole	Method	State	Refs.
CH_2Cl—$CHCl_2$ (continued)	1,1,2-Trichloroethane	$V_2^g = 2.82$	DM, T	Gas	302
		$\left.\begin{array}{l} V(gauche \to trans) \\ \qquad \approx 5.0 \\ V(gauche \to gauche) \\ \qquad = 1.75 \end{array}\right\}$	T, ED, DM	Gas	300
$CHCl_2$—$CHCl_2$	1,1,2,2-Tetrachloroethane	$\Delta E^e = 0 \pm 0.2$	DM, IR	Gas	297,303
		$= -1.1 \pm 0.1$	IR, NMR, Ram	Liq.	289,304,306
		$= -0.9$	IR	Soln. n-heptane	289
		$= -1.4$	IR	Soln. CH_3NO_2	289
CH_2Br—CH_2Br	1,2-Dibromoethane	$\Delta E^e = 1.77 \pm 0.15$	IR	Gas	7,286,307,308
		0.74 ± 0.12	Ram, IR	Liq.	7,285,286,289–291,307,345
		$= 0.9$	IR	Soln. n-hexane	289
		$= 0.4$	IR	Soln. CH_3NO_2	289
		$V(gauche \to trans)$ $= 4.2 \pm 0.6$	SA	Soln. acetone	6
$CHBr_2$—CH_2Br	1,1,2-Tribromoethane	$\Delta E^e = -0.04$	NMR	Liq.	299,304
$CHBr_2$—$CHBr_2$	1,1,2,2-Tetrabromoethane	$\Delta E^e = -0.75$ to 0.85	IR, Ram, SA	Liq.	306,322
CH_2Cl—CH_2F	1-Chloro-2-fluoroethane	$\left.\begin{array}{l} V_1^g = 1.3 \\ \Delta E^e = 0.5 \end{array}\right\}$	DM	Gas	301
CF_2Cl—CH_2Cl	1,2-Dichloro-1,1-difluoro-ethane	$\Delta E^h = \pm 0.43 \pm 0.1$	IR, Ram	Gas, liq.	147
$CHCl_2$—CHF_2	1,1-Dichloro-2,2-difluoro-ethane	$\Delta E^e = -0.5$	NMR	Liq.	304
		$= -0.35$	IR	Liq.	304
		$V_1^g = 1.9$	DM	Gas	301
		$\Delta E^e = 0.7$	DM	Gas	301

CF_2Cl—CF_2Cl	1,2-Dichlorotetrafluoro-ethane	$\Delta E^\circ = 0.5 \pm 0.2$	IR (ED)	Gas	316,317
$CHCl_2$—$CHFCl$	1-Fluoro-1,2,2-trichloro-ethane	$E(\text{HHFClClCl})^1 = 0$ $E(\text{HClClHClF})$ $\quad = 0.4 \text{ or } 0.1$ $E(\text{HFClClClH})$ $\quad = 1.0 \text{ or } 0.1$	NMR	Liq.	304
$CHCl_2$—CF_2Cl	1,1-Difluoro-1,2,2-trichloroethane	$V_2{}^g = 1.9$ $E(\text{HFClClClF})^1 = 1.8$ $E(\text{HClClFClF}) = 0.0$	DM NMR	Gas Liq.	301 318
CF_2Cl—$CFCl_2$	1,1,2-Trichlorotrifluoro-ethane	$\Delta E^h = 0.35 \pm 0.15$ $\Delta E^\circ = -2.7 \pm 0.1,$ $\quad -2.3 \pm 0.3$	IR, Ram NMR	Gas Liq.	319 304
$CFCl_2$—$CHCl_2$	1-Fluoro-1,1,2,2-tetra-chloroethane	$\Delta E^h = 0.8(\pm 0.1)$ $\Delta E^\circ = -0.4(\pm 0.1)$	IR IR, NMR	Gas Liq.	320 304,320,412
$CFCl_2$—$CFCl_2$	1,2-Difluorotetrachloro-ethane	$\Delta E^\circ = 0.12 \pm 0.01$ $V(\text{Cl—Cl}, 2\text{Cl—F})^j$ $\quad = 9.15 \pm 0.1^k$	NMR	1:4 Soln. in $CFCl_3$	313,321
CH_2Cl—CH_2Br	1-Bromo-2-chloroethane	$\Delta E^\circ = 0.0$–0.5 $\Delta E^\circ = 1.43 \pm 0.1$ $\Delta E^\circ = 0.49 \pm 0.1$	ED IR Ram	Gas Gas Liq.	323 303 349
CH_2Br—CF_2Br	1,2-Dibromo-1,1-difluoro-ethane	$\Delta E^h = 1.03 \pm 0.1$	IR, Ram	Gas	147,325
CF_2Br—$CFBr_2$	1,1,2-Tribromotrifluoro-ethane	$E(\text{FFFBrBrBr})^1 = 0.76$ $E(\text{FFBrBrFBr}) = 0.0$ $V(3\text{Br—F})^j = 7.2 \pm 0.2^k$ $V(\text{Br—Br, Br—F, F—F})$ $\quad = 9.4 \pm 0.2^k$	NMR	Liq.	321
CF_2Br—CF_2Br	1,2-Dibromotetrafluoro-ethane	$\Delta E^\circ = 0.94 \pm 0.05$ $\Delta E^h = 0.92 \pm 0.05$	IR, Ram IR, Ram	Gas Liq.	313 313

(continued)

TABLE II (continued)

Formula	Name	Data and notes, kcal/mole	Method	State	Refs.
CF_2Br—$CHBrCl$	1-Chloro-1,2-dibromo-2,2-difluoroethane	$E(BrFClBrHF)$[1] $= 0$ $E(BrFClFHBr) =$ 0.17 ± 0.01 $E(BrBrClFHBr) =$ 0.25 ± 0.02 $V(Br$—Cl, Br—F, F—$H)$[1] ≥ 7.3[k] $V(Br$—Br, Cl—F, F—$H)$ ≥ 7.3[k] $V(Br$—F, Br—H, Cl—$F)$ $= 6.4 \pm 0.2$[k]	NMR	1:4 Soln. in CF_2Cl_2	321
CF_2Br—CCl_2Br	1,2-Dibromo-1,1-dichloro-difluoroethane	$E(FClBrClFBr)$[1] $= 0.32 \pm 0.02$ Other staggered conformers $= 0.0$ $V(Br$—Cl, Br—F, Cl—$F)$[j] $= 10.3 \pm 0.1$[k]	NMR	1:4 Soln. in $CFCl_3$	321
		11.0	NMR	Liq.	454
		$V(Br$—$Br, 2Cl$—$F)$[1] $\geq$ 11.5 ± 0.5[k]	NMR	1:4 Soln. in $CFCl_3$	321
		12.0 ± 0.2	NMR	Liq.	454

| CFBrCl—CFClBr | 1,2-Dibromo-1,2-dichloro difluoroethane | $E(\text{FBrClFBrCl})$[1] $= 0.0$
 $E(\text{FClClBr·BrF}) = 0.44$
 $E(\text{FFClClBr·Br}) = 0.44$
 $E(\text{FClClFBrBr}) = 0.12$
 $E(\text{FFBrClClBr}) = 0.0$
 $E(\text{FFClBrBrCl}) = 0.45$
 $V(\text{Br—Cl, Br—F, Cl—F})$[j] $= 9.6^k \pm 0.2$
 $V(\text{Br—Br, Cl—Cl, F—F})$ $\geq 11.5^k \pm 0.1$
 $V(\text{2Br—F, Cl—Cl}) = 9.4^k \pm 0.2$
 $V(\text{Br—Br, 2Cl—F}) \geq 10.1^k \pm 0.2$
 $V(\text{2Br—Cl, F—F}) \geq 10.1^k \pm 0.2$ | NMR | 1:20 Soln. in CS_2 | 321 |
| CF$_2$Br—CFClBr | 1-Chloro-1,2-dibromotri-fluoroethane | $E(\text{FFFBrFClBr})$[1] $= 0.0$
 $E(\text{FBrBrFClF}) =$ 0.31 ± 0.01
 $E(\text{FFBr·BrClF}) =$ 0.75 ± 0.03
 $V(\text{2Br—F, Cl—F})$[j] $= 7.4^k \pm 0.1$
 $V(\text{Br—Br, Cl—F, F—F})$ $\geq 8.3^k$
 $V(\text{Br—Cl, Br—F, F—F})$ $\geq 8.3^1$ | NMR | 1:2 Soln. in $CFCl_3$ | 304,321 |

(continued)

TABLE II (continued)

Formula	Name	Data and notes, kcal/mole	Method	State	Refs.
CF₂Br—CF₂Cl	1-Bromo-2-chlorotetra-fluoroethane	$\Delta E° = 0.60 \pm 0.2$	IR	Gas	316
		$\Delta E° = 0.52 \pm 0.2$	IR	Liq.	316
CH₂CN—CH₂CN	Succinonitrile	$\Delta E° = -0.36 \pm 0.05$	IR	Liq.	328
		$V_1{}^{\mathrm{g}} = 1.2 \pm 0.5$	DM	Soln. in toluene	329
CH₂F—CH₂OH	2-Fluoroethanol	$\Delta E° = -2.07 \pm 0.53$	IR	Dil. soln. in CCl₄	330
CH₂Cl—CH₂OH	2-Chloroethanol	$\Delta E° = -1.20 \pm 0.09$	IR	Dil. soln. in CCl₄	330
CH₂Br—CH₂OH	2-Bromoethanol	$= -0.95 \pm 0.02$	IR, Ram	Gas	331
		$\Delta E° = -1.25 \pm 0.08$	IR	Dil. soln. in CCl₄	330
CH₂I—CH₂OH	2-Iodoethanol	$\Delta E° = -0.81 \pm 0.09$	IR	Dil. soln. in CCl₄	330
C₂H₅—CH(CH₃)(OH)	2-Butanol	$\Delta E^{\mathrm{h}} = 0.80 \pm 0.06$	Temp. dep. of optical rotatory power	Dil. soln. cyclohexane	332
C₂H₅—CHO	Propionaldehyde	$\Delta E(gauche\text{–}cis) \approx 1$ $= 0.9 \pm 0.1$	NMR	Liq.	333
CH₂Cl—CH₂C₆H₅	β-Phenethyl chloride	$\Delta E° = 0.06$	MWₛ	Gas	158
CH₂Br—CH₂C₆H₅	β-Phenethyl bromide	$\Delta E° = 0.09$	NMR	Liq.	304
CH₃CHBr—CHBr (COC₆H₅)	α,β-Dibromobutyro-phenone	$\Delta E° = 0.96$ (at 295°K)	NMR	Liq.	304
		$= 1.06$ (at 363°K)	NMR	Soln. in dimethyl-formamide	304

Formula	Compound	Data	Method	State	Ref.
C_6H_5CHBr—$CHBr$ (COC_6H_5)	α,β-Dibromopropio-phenone	$\Delta E^e = 1.17$ (at $295°K$) $= 1.35$ (at $363°K$)	NMR	Soln. in di-methyl-formamide	304
$(C_6H_5)_2CH$—CH_2 (COC_6H_5)	β,β-Diphenylpropio-phenone	$\Delta E^e = -0.08$	NMR	Soln. in $CHCl_3$	304
$(p\text{-}BrC_6H_5)CHCl$—$CH(C_6H_5)(COC_6H_5)$	β-(4-Bromophenyl)-β-Chloro-α-Phenyl-propionone	$\Delta E^e = 1.04$	NMR	Solns. in CS_2, in phenol	304
C_2H_5—C_2H_5	n-Butane	$\Delta F^e \approx 0.61$	ED	Gas	334
		$V_3^{\,g} = 3.8$ (trans→gauche)	T	Gas	159
		4.2 ± 0.4 (trans→gauche)	SA	Liq.	462
C_2H_5—C_3H_7	n-Pentane	$\Delta E^e = 0.76 \pm 0.1$	Ram	Liq.	335,336
		$\Delta E^h = 0.45 \pm 0.06$	Ram	Liq.	335
		$V(gauche \to trans)$ $= 4.2 \pm 0.4$	SA	Liq.	462
C_2H_5—C_4H_9	n-Hexane	$\Delta E^h = 0.5 \pm 0.1$	Ram	Liq.	335
		$V(gauche \to trans)$ $= 3.1 \pm 0.4$	SA	Liq.	462
$(CH_3)_2CH$—CH_2—C_2H_5	2-Methylpentane	$\Delta E^h \approx 0.94 \pm 0.1$ V (from lower) $=$ 4.84 ± 0.25	SA	Liq.	337
CH_3CH_2—$CH(CH_3)$—CH_2CH_3	3-Methylpentane	$\Delta E^h = 0.94 \pm 0.1$ V (from lower) $=$ 5.04 ± 0.25	SA	Liq.	337
$(CH_3)_2CH$—$CH(CH_2)_3$ 2,3-Dimethylbutane		$\Delta E^h = 0.95 \pm 0.1$ V (from lower) $=$ 3.75 ± 0.14	SA	Liq.	337

(continued)

TABLE II (continued)

Formula	Name	Data and notes, kcal/mole	Method	State	Refs.
$(CH_3CH_2CH_2)_2CH_2$	n-Heptane	$V(trans \rightarrow gauche)$ $= 3.8 \pm 0.4$	SA	Liq.	462
$CH_2F-C_2H_5$	1-Fluoropropane	$\Delta E^\circ = -0.05 \pm 0.03$ $V(CH_3-F)^j = 10.1 \pm 4.4$ $V(CH_3-H) \approx 4.2 \pm 1.5$ $V_1 = 3.22 \pm 2.02$ $V_2 = -3.05 \pm 1.72$ $V_3 = 6.48 \pm 2.15$ $V_4 = -1.25 \pm 1.17$	$\Big\}$ MW_i	Gas	199
$CH_2Cl-C_2H_5$	1-Chloropropane	$\Delta E^\circ = -0.5$ $\Delta E^\circ = 0.05 \pm 0.06$ -0.3 ± 0.15	MW, ED (IR) Ram, IR Ram, IR	Gas Liq. Liq.	286,339–341 341 286,345
$CH_2Cl-CHClCH_3$	1,2-Dichloropropane	$\Delta E^\circ = 1.0 \pm 0.05$ $V_1{}^g \geq 4$	DM ED (DM)	Gas Gas	342 302,343
$CH_2Br-C_2H_5$	1-Bromopropane	$\Delta E^\circ = -0.28 \pm 0.1$ -0.44 ± 0.1	IR, Ram	Gas Liq.	286,341 286,345
$CH_2Br-CH_2-CH_2Br$	1,3-Dibromopropane	$\Delta E^\circ = -0.3 \pm 0.3$	Ram	Liq.	345
$CH_2Cl-CCl(CH_3)_2$	1,2-Dichloro-2-methyl-propane	$\Delta E^\circ \approx 1.0$ ≈ 0.0	IR IR, Ram	Gas Liq.	346
$(CH_3CHCl)_2-CH_2$	2,4-Dichloropentane	$\Delta E^\circ = 1.5 \pm 0.3$	NMR	Liq.	347,348
$(CH_3CHBr)_2-CH_2$	2,4-Dibromopentane	$\Delta E^\circ = 1.4 \pm 0.4$	NMR	Liq.	347,348
$(CH_3CHCN)_2-CH_2$	1,3-Dimethylglutaronitrile	$\Delta E^\circ = 1.2 \pm 0.3$	NMR	Liq.	347,348
CH_2Br-CH_2- CH_2CH_2Br	1,4-Dibromobutane	$\Delta E^\circ = -0.5 \pm 0.1$	Ram	Liq.	345
$(CH_3)_2CH-CHO$	3-Methylpropionaldehyde	$V_3 \geq 2.0$ $V_1 \approx 1.0$ $\Big\}$	ED	Gas	344

Formula	Name	Data	Method	Phase	Ref.
CH_2F—CFO	Fluoroacetyl fluoride	$\Delta E(trans\text{–}cis) = 0.91 \pm 0.1$	MW_i	Gas	463
$CHCl_2$—$COCl$	Dichloroacetylchloride	$\left.\begin{array}{l}\Delta E(gauche\text{–}cis) \approx 0.2 \\ \Delta E \approx 0.0\end{array}\right\}$	IR, Ram	Gas Liq.	350
CH_2Br—$COCl$	Bromoacetyl chloride	$\Delta E^e = 1.0 \pm 0.1$ (with respect to halogen atoms)	IR, Ram	Gas	351
CH_2Br—$COBr$	Bromoacetyl bromide	$\Delta E^e = 1.9 \pm 0.3$ (with respect to halogen atoms)	IR, Ram	Gas	351
CH_2F—$CHCH_2$	3-Fluoropropylene	$\left.\begin{array}{l}\Delta E(gauche\text{–}cis) = \\ 0.31 \pm 0.07 \\ V(\text{F–H})^1 \approx 2.0\end{array}\right\}$	MW_i, MW_s, NMR	Gas	352,431
CH_2Cl—$CClCH_2$	2,3-Dichloropropylene	$V_l^g = 2.08$	DM	Gas	302
CH_2CN—$CHCH_2$	3-Butenitrile	$\Delta E(cis\text{–}gauche) \approx (\pm)0.5$	IR, Ram, NMR	?	353
		$V = 2.18$	Ram	Liq.	353
CH_2CH—$CHCH_2$	1,3-Butadiene	$\left.\begin{array}{l}\Delta E(cis\text{–}trans) = 2.3 \\ V(\text{from } cis) = 2.6\end{array}\right\}$	T	Gas	354
CH_2CH—CCH_2CH_3	2-Methyl-1,3-butadiene	$V^{*1} = 28.1^m$	IR	Gas	355
CH_2CH—CHO	Acrolein	$V^{*1} = 35.7^m$	IR	Gas	355
		$V^{*1} = 21.2^m$	IR	Gas	355
		$V^{*1} = 28.6^m$	IR	Liq.	355
		$V > 2.3$	MW_i	Gas	369
CH_2CH—$CO(CH_3)$	Methyl vinyl ketone	$\left.\begin{array}{l}V(\text{from } cis) = 4.96 \\ \Delta E(cis\text{–}trans) = 2.06\end{array}\right\}$	SA	Liq.	441
$CH_2C(CH_3)$—CHO	2-Methyl acrolein	$V^{*1} = 15.7^m$	IR	Gas	355
		$V^{*1} = 31.9^m$	IR	Gas	355
OHC—CHO	Glyoxal	$V^{*1} = 13.7^m$	IR	Gas	355

(continued)

TABLE II (continued)

Formula	Name	Data and notes, kcal/mole	Method	State	Refs.
$OHC-C(CH_3)O$	Pyruvaldehyde	$V^{*1} = 12.4^m$	IR	Gas	355
$CH_3CHCH-CHO$	Crotonaldehyde	V (from cis) = 5.51; ΔE (cis–trans) = 1.93	SA	Liq.	441
$C_6H_5-CHCH-CHO$	Cinnamaldehyde	V (from cis) = 5.62; ΔE (cis–trans) = 1.5	SA	Liq.	441
$CH_2C(CH_3)-CHO$	Methacrolein	V (from cis) = 5.31; ΔE (cis–trans) = 3.07	SA	Liq.	441
$C_4H_3OCHCH-CHO$	Furacrolein	V (from cis) = 5.10; ΔE (cis–trans) = 1.2	SA	Liq.	441
$RCO-CHN_2$	Diazoketones	cis(O,N)→trans [trans→cis]			
		E_a $\quad$ $\Delta F^{\ddagger}$			
R = CH_3		15.5[15.5] 15.4[13.9]	NMR	40% by wt. in CDCl$_3$	455
R = C_2H_5		16.1[16.2] 15.3[13.5]			
R = $C_6H_5CH_2$		18.2[18.2] 15.3[13.4]			
R = CH_3O		12.5[12.5] 12.8[12.7]			
R = C_2H_5O		9.0[9.0] $\quad$ 13.3[13.2]			
R = CH_3		15.9[15.9] 15.4[14.1]	NMR	20% by wt. in CDCl$_3$	455
$O(CH_3)C-C(CH_3)O$	Biacetyl	$V^{*1} = 10.1^m$	IR	Gas	355
$CH_2CH-CFCH_2$	2-Fluoro-1,3-butadiene	$V^{*1} = 32.4^m$	IR	Gas	355,356
$CH_2CH-CClCH_2$	2-Chloro-1,3-butadiene	$V^{*1} = 31.7^m$	IR	Gas	355
$ClOC-COCl$	Oxalyl chloride	ΔE (cis–trans) = 2.8	IR, Ram	Liq.	327
C_6H_5-CHO	Benzaldehyde	$V^{*1} = 18.6^m$; $V_2 = 4.66$	IR	Gas	355

Formula	Compound	Data	Method	Phase	Ref.
		$V^{*1} = 26.7 \pm 0.8^{m}$, $V_2 = 6.69 \pm 0.2$, $\Delta F^{\ddagger} = 7.9$	IR	Liq.	355,357
			NMR	Soln. in CH_2CHCl	439
$C_6H_4F—CHO$	2-Fluorobenzaldehyde	$V_1 = 0.28$, $V_2 = 4.18$ } from O-trans	IR	Gas	456
$C_6H_4F—CHO$	3-Fluorobenzaldehyde	$V_1 = 1.37$, $V_2 = 4.14$ } from O-cis	IR	Gas	456
$C_6H_4F—CHO$	4-Fluorobenzaldehyde	$V_2 = 3.58$	IR	Gas	456
$2\text{-}C_6H_4Cl—CHO$	2-Chlorobenzaldehyde	$V_2^{\varepsilon} = 7.65$ (cis to oxygen), 6.0 (trans to oxygen)	IR	Liq.	357
$3\text{-}C_6H_4Cl—CHO$	3-Chlorobenzaldehyde	$V_1 = 0.7$, $V_2 = 3.85$ } from O-trans	IR	Gas	456
		$V_2^{\varepsilon} = 6.67$ (cis to oxygen), 7.71 (trans to oxygen)	IR	Liq.	357
$4\text{-}C_6H_4Cl—CHO$	4-Chlorobenzaldehyde	$V_1 = 0.66$, $V_2 = 4.04$ } from O-cis	IR	Gas	456
		$V_2^{\varepsilon} = 4.87$	IR	Soln. in nujol	357
$C_6H_4Br—CHO$	2-Bromobenzaldehyde	2.81	IR	Gas	456
		$V^{*1} = 0.98$ from O-trans	IR	Gas	456
$C_6H_4Br—CHO$	3-Bromobenzaldehyde	$V_1 = 1.44$, $V_2 = 4.02$ } from O-cis	IR	Gas	456
$C_6H_4Br—CHO$	4-Bromobenzaldehyde	$V_2 = 2.37$	IR	Gas	456

(continued)

TABLE II (*continued*)

Formula	Name	Data and notes, kcal/mole	Method	State	Refs.
$2\text{-}C_6H_4(CH_3)\text{—}CHO$	2-Methylbenzaldehyde	$V_2{}^g = 7.96$ (*cis* to oxygen) 6.77 (*trans* to oxygen)	IR	Liq.	357
		$V^{*1} = 0.66$ from *O-trans*	IR	Gas	456
$3\text{-}C_6H_4(CH_3)\text{—}CHO$	3-Methylbenzaldehyde	$V_2{}^g = 7.96$ (*cis* to oxygen) 6.77 (*trans* to oxygen)	IR	Liq.	357
		$V_1 = 1.71$ from *O-cis* $V_2 = 4.09$	IR	Gas	456
$4\text{-}C_6H_4(CH_3)\text{—}CHO$	4-Methylbenzaldehyde	$V_2 = 6.39$	IR	Liq.	357
		3.47	IR	Gas	456
$(CH_3)_2NC_6H_4\text{—}CHO$	p-N,N-Dimethylamino-benzaldehyde	$\Delta F^{\ddagger} = 10.8$	NMR	Soln. in CH_2Cl_2	439
$CH_3OC_6H_4\text{—}CHO$	p-Methoxybenzaldehyde	$\Delta F^{\ddagger} = 9.2$	NMR	Soln. in CH_2CHCl	439
$(C_4H_3O)\text{—}CHO$	2-Furanaldehyde	V (from $O,O\ cis$) $= 10\text{-}11$ V (from $O,O\ trans$) $= 11\text{-}12$	NMR	10% soln. in $(CH_3)_2O$	442
$C_5NH_5\text{—}CH_3$	Pyridine-4-aldehyde	$V_2 = 3.83$	IR	Gas	456
$C_5NH_5\text{—}CH_3$	Pyridine-3-aldehyde	$V^{*1} = 1.15$	IR	Gas	456
$C_6H_5\text{—}C(CH_3)O$	Acetophenone	$V_2 = 3.1$	IR	Gas	456
$C_6H_4F\text{—}C(CH_3)O$	4-Fluoroacetophenone	$V_2 = 3.5$	IR	Gas	456
$C_6H_4F\text{—}C(CH_3)O$	2-Fluoroacetophenone	$V^{*1} = 0.73$	IR	Gas	456
$C_6H_4F\text{—}C(CH_3)O$	3-Fluoroacetophenone	$V^{*1} = 0.9$	IR	Gas	456

Formula	Name	Value	Method	State	Ref.
C_4H_3O—CHO	Furan-2-aldehyde	$V_1 = 2.03$ } from $V_2 = 6.78$ } O-O-$trans$	IR	Gas	456
C_5H_4N—CHO	Pyridine-2-aldehyde	$V^{*1} = 1.16$	IR	Gas	456
C_6H_5—C_2H_5	Ethylbenzene	$V_3^g = 1.3$	T	Gas	207
$1,2$-C_6H_4—$(CH_3)_2$	1,2-Dimethylbenzene	$V_3^g \approx 2.0$	T	Gas	358
		$V = V_6^g = 1.85 \pm 0.08$	NS	Solid, liq.	359
$1,3$-C_6H_4—$(CH_3)_2$	1,3-Dimethylbenzene	$V_3^g \geq 1.0$	T	Gas	105
		Max. of V_3, $V_6 \approx 0.3 \pm 0.8$	T	Gas	358
$1,4$-C_6H_4—$(CH_3)_2$	1,4-Dimethylbenzene	$V_3^g \geq 1.0$	T	Gas	105
		$V = V_6^g = 0.35 \pm 0.02$	T	Gas	358
$1,3,5$-C_6H_3—$(CH_3)_3$	1,3,5-Trimethylbenzene	$V = V_6^g = 0.19 \pm 0.02$	T	Gas	105
C_6—$(CH_3)_6$	Hexamethylbenzene	$V_3^g = 3$–8	T	Gas	360
		$V_3^g = 1.07 \pm 0.09$ (<110°K) $V_3 = 1.35 \pm 0.10$ (>110°K)	NS	Solid	27,362
C_6H_5—C_6H_5	Biphenyl	$V_2 = 0.44 \pm 0.05$	NMR	Solns. in methyl-cyclo-hexane	
		$V_2 = 0.7 \pm 0.1$		In CHCl₃—CCl₄	364
		$V_2 = 1.1 \pm 0.25$ $V_4 = (\pm)0.1$ in each case		In CS₂	
C_3H_5—CHO	Cyclopropyl carboxalde-hyde	$V \geq 2.5$ $\Delta E(H$—H cis-$trans) = 0.12 \pm 0.24$	ED	Gas	26,366

(continued)

TABLE II (continued)

Formula	Name	Data and notes, kcal/mole	Method	State	Refs.
$^{-}OO\dot{C}H-CH_2-$ $CH_2-CH_2-COO^{-}$	Adipate anion radical	$V \approx 10$	ESR	Solid [a]	367
$C_6H_5-\dot{C}CH-C_6H_5{}^{-}$	Stilbene anion radical	$\Delta F^\ddagger \geq 9.0$	ESR	Soln. in 1,2-dimethoxy-ethane	368
		2. IV—V bonds			
$OH\dot{C}-NH_2$	Formamide	$V = 18 \pm 3$	NMR	Soln. in acetone	375
$OH\dot{C}-NHCH_3$	N-Methylformamide	$V \approx 14$	IR, Ram	Liq.	371
$OH\dot{C}-N(CH_3)_2$	N,N'-Dimethylformamide	$\Delta F^\ddagger = 22$ 24 $\left.\begin{array}{l} V = 9.6 \pm 1.5 \\ 7.0 \pm 3.0 \end{array}\right\}$	 NMR	 Liq.	376 374 372 376
$(HO)H\dot{C}-N(CH_3)_2{}^{+}$	Protonated N,N'-dimethylformamide	$V = 12.7 \pm 1.5$	NMR	In 100% H_2SO_4	372
$OH\dot{C}-N(C_2H_5)_2$	N,N'-Diethylformamide	$\Delta F^{\circ\ddagger} = 19.6 \pm 1.5$	NMR	Liq.	373
$OCH_3\dot{C}-NHCH_3$	N-Methylacetamide	$V \approx 14$	IR, Ram	Liq.	371
$OCH_3\dot{C}-N(CH_3)_2$	N,N'-Dimethylacetamide	$V = 12 \pm 2,\ \Delta F^\ddagger = 19$	NMR	Liq.	376
$OCH_3\dot{C}-N(C_2H_5)_2$	N,N'-Diethylacetamide	$\Delta F^{\circ\ddagger} = 16.9 \pm 1.2$	NMR	Liq.	373
$CH_3COCH_2O\dot{C}-$ $N(C_2H_5)_2$	N,N'-Diethylaceto-acetamide	$\Delta F^{\circ\ddagger} = 16.8 \pm 1.1$	NMR	Liq.	373
$OCH_3\dot{C}-N(C_3H_7)_2$	N,N'-Dipropylacetamide	$\Delta F^{\circ\ddagger} = 17.0 \pm 1.4$	NMR	Liq.	373
$OCH_3\dot{C}-N(C_3H_7)_2$	N,N'-Di(1-methylethyl)acetamide	$\Delta F^{\circ\ddagger} = 15.7 \pm 1.2$	NMR	Liq.	373

$OCH_3—N(C_6H_5)_2$	N,N'-Diphenylacetamide	$V = 7.3 \pm 0.5$, $\Delta F^{\ddagger} = 16.5$,	NMR	Soln. in CH_2Br_2	377
$OC_3H_7C—N(CH_3)_2$	N,N'-Dimethylpropion-amide	$V = 6.4 \pm 0.6$,	NMR	In CCl_4	377
		$V = 6\text{-}10$, $\Delta F^{\ddagger} = 16.7$	NMR		377
$OClC—N(CH_3)_2$	N,N'-Dimethylcarbonyl chloride	$V = 6\text{-}9$, $\Delta F^{\ddagger} = 16.5$	NMR	Soln. in CCl_4	377
$OHC—NHC_5H_5$	N-Benzylformamide	ΔE (cis-trans) $= 0.62 \pm 0.06$	IR	Soln. in CCl_4	378
$OHC—NCH_3(C_6H_5)$	N-Methyl-N'-benzyl-formamide	$V = 11 \pm 1.5$	NMR	Liq.	379
$OHC—NCH_3$ $(C_6H_4CH_3)$	N-Methyl-N'-methyl-benzylformamide	$V = 12 \pm 1.5$	NMR	Liq.	379
$OHC—NCH(C_6H_5)_2$ (CH_3)	N-Methyl-N'-diphenyl-methylformamide	$V = 11 \pm 1.5$	NMR	Liq.	379
$SCH_3C—NHC_6H_5$	N-Benzylthioacetamide	ΔE (cis-trans) $= 0.22 \pm 0.02$	IR	Soln. in CCl_4	378
$SHC—NHC_6H_5$	N-Benzylthioformamide	ΔE (cis-trans) $= -1.4 \pm 0.3$	IR	Soln. in CCl_4	378
$(CH_3)_2NC_6H_4—NO$	p-Nitrosodimethylaniline	$V = 11.2 \pm 1.1$	NMR	Soln. in acetone	438
			IR	Soln. in C_6H_6	438
$(CH_3)DN—C_6H_2$ $(NO_2)_3$	N-Methyl-N'-deutero-2,4,6-trinitroaniline	$V = 14.5 \pm 0.3$	NMR	Soln. in CH_2Cl_2	380
$C_2H_5—NO_2^-$	Nitroethane anion radical	$V_2^{g} = 1.4$	ESR	Aqueous soln.	25
$(CH_3)_2CH—NO_2^-$	2-Nitropropane anion radical	$V_2^{g} = 1.4$	ESR	Aqueous soln.	25

(continued)

TABLE II (continued)

Formula	Name	Data and notes, kcal/mole	Method	State	Refs.
$CH_2CH—NO_2$	Nitroethylene	$V = 6.5$ $V(\alpha\text{-deuterio}) = 6.0$	IR	Gas	450
$C_6H_5—NO_2$	Nitrobenzene	$V = 3.9 \pm 1.0$	MW_i	Gas	436
$[(CH_3)Si—NSi(CH_3)_2]_2$	Tetramethyl-N,N-bis-(trimethylcyclodisilizane)	$V \approx 5.0$	NMR	Solid	423
		3. IV—VI bonds			
$C_2H_5—ONO$	Ethyl nitrite	$V = 9.0 \pm 2$	NMR	3:4 soln. in toluene	384
$C_3H_7—ONO$	Propyl nitrite	$V = 9.0 \pm 2$	NMR	3:4 soln. in toluene	384
		$\Delta F^{\ddagger} \approx 10$ $\Delta E\ (cis\text{-}trans) = -0.13$	NMR	Liq.	374
$(CH_3)_2CH—ONO$	2-Methylethyl nitrite	$V = 6.0 \pm 2$	NMR	Liq.	384
$R—ONO$	Alkyl nitrites	$\Delta E\ (cis\text{-}trans) = -0.13$ $\Delta F^{\ddagger} \approx 10$	NMR	Liq.	374
$CH_2XCH_2—ONO$	Substituted ethyl nitrites	$X = F, \quad V \approx 10$ $X = Cl, \quad V = 6.8 \pm 2$ $X = Br, \quad V = 12.5 \pm 2$ $X = I, \quad V = 10.7 \pm 2$ $X = CN, \quad V = 13.0 \pm 2$ $X = CH_3, \quad V = 9.0 \pm 2$ $X = H, \quad V = 9.0 \pm 2$ $\Delta E\ (cis\text{-}trans) = 0.5–1$ for all X	NMR	Liq.	385
$C_2H_5—OH$	Ethanol	$V_3^g = 0.8$	T	Gas	148,149

$(CH_3)_2CH-OH$	2-Methyl ethanol	$V_3{}^g = 5.0°$	T	Gas	150
CF_3CH_2-OH	2,2,2-Trifluoroethanol	$\Delta E^e = -3.3$	IR	Dil. soln. in CCl$_4$	386
C_6H_5-OH	Phenol	Eq. conf. is planar			
		$V_2 = 3.28$	MW$_s$	Gas	387
		3.1 ± 0.3	MW$_s$	Gas	388
		3.47	IR	Gas	420
$C_{10}\ddot{H}_4(-\overset{+}{O}H)_4$	Naphthazarin semiquinone cation	$V \approx 4$	ESR	In H$_2$SO$_4$	389
HOC—OH	Formic acid	$\Delta E (cis\text{-}trans) = 2.0 \pm 0.3$	IR	Gas	390
		$V_1 = 2.1$ $V_2 = 9.9$ $\left.\right\}$ $V = 10.9??$ $V_3 = -0.1$ (see ref. 396)	IR	Gas	390
		$V = 13.4$	IR	Gas	396
		$V \approx 17$	MW$_i$	Gas	391
		$\Delta E (cis\text{-}trans) > 4$	MW	Gas	392
ClOC—OH	Chloroformic acid	$V = 13.8 \pm 1.6$	Rate of isomerization	Gas	460
CH_3CH_2-OCHO	Ethyl formate	$\Delta E (gauche\text{-}trans) = 0.186 \pm 0.06$ $V (trans\text{-}gauche) = 1.10 \pm 0.25$ $\left.\right\}$ $\Delta E^h = 2.5$	MW$_i$	Gas	393
$(CH_3)_2CHO-CHO$	2-Methylethyl formate	$V (\text{from lower}) = 5.88$	SA	Liq.	394
		$\Delta E (cis\text{-}trans) = 3.7 \pm 0.5$ $V (\text{from } trans) = 5.8 \pm 0.4$ $\left.\right\}$	SA	Dil. soln. in xylene and heptane	395

(continued)

TABLE II (*continued*)

Formula	Name	Data and notes, kcal/mole	Method	State	Refs.
$HO-CO_2^-$	Bicarbonate ion	$V_2 = 15.8$	IR	In solid KX matrices	396
$CH_3O-CHCH_2$	Methyl vinyl ether	ΔE (*cis–trans*)			382,397
		$= -1.15 \pm 0.25$	IR	Gas	
		$= -0.66 \pm 0.2$	IR	Soln. in $CHCl_2-CH_2Cl$	
C_2H_5-SH	Ethanethiol	$V_3{}^g = 1.64$	T	Gas	311
$(CH_3)_2CH-SH$	2-Propanethiol	$V_3{}^g = 1.39$	T	Gas	151
		$= 1.42$	IR	Gas	152
$(C_2H_5)_2-S$	Ethyl sulfide	$V_3{}^g = 1.75$	T	Gas	154

4. V—V bonds

Formula	Name	Data and notes, kcal/mole	Method	State	Refs.
NH_2-NH_2	Hydrazine	$V_2{}^g = 3.15 \pm 0.15$	MW_s	Gas	399,400
NF_2-NF_2	Tetrafluorohydrazine	$V > 3$	MW_i	Gas	401
		$V = 4-7$	NMR	Soln. in perfluoro-2,3-di-methyl-hexane	402
CH_3NH-NH_2	Methylhydrazine	$V_3{}^g = 3$	T	Gas	403
$CH_3NH-NHCH_3$	1,2-Dimethylhydrazine	$V_3{}^g = 3$	T	Gas	404
$(C_6H_5)_2N-$ $NHC_6H_2(NO_2)_3$	2,2-Diphenyl-1-(2,4,6-trinitrophenyl)-hydrazine	$V = 12.5 \pm 0.2$	NMR	Soln. in CH_2Cl_2	380
$(CH_3)_2N-NO$	N,N'-Dimethylnitrosamine	$V = 23$	NMR	Liq.	406
		$\Delta F^{\ddagger} = 25$	NMR	Liq.	374

5. V—VI bonds

HO—NO	Nitrous acid	$\Delta E(cis\text{–}trans) = 0.39 \pm 1$	IR	Gas	407
		$V(\text{from } trans) =$			
		11.57 ± 0.2	IR	Gas	407
		$V(\text{from } cis) = 8.7 \pm 1$	IR	Gas	408
		$= 9.7 \pm 0.7$	IR	Solid	408
		$\Delta E(cis\text{–}trans) = 0.5$	IR	Gas	68
		$V(\text{from } trans) \approx 12$	IR	Gas	68
CH$_3$O—NO	Methyl nitrite	$\Delta E(cis\text{–}trans)$			
		$= 0.6 \pm 0.2$	IR	Gas	382
		$= 0.8$	NMR	Liq.	383
		$V = 10.5 \pm 2$	NMR	Liq.	383
		$= 9.0 \pm 2$	NMR	3:4 soln. in toluene	384

6. VI—VI bonds

HO—NO$_2$	Nitric acid	$V_2 = 7.8 \pm 0.1$	IR, Ram (T)	Gas	410,411
FO—NO$_2$	Fluorine nitrate	$V_2 = 10.23 \pm 1.0$	IR	Gas, solid	451
ClO—NO$_2$	Chlorine nitrate	$V_2 = 7.93 \pm 1.0$	IR	Gas, solid	451
HO—OH and DO—OD	Hydrogen peroxide	$W(\alpha) = W_0 + W_1 \cos \alpha + W_2 \cos 2\alpha + W_3 \cos 3\alpha$	IR	Gas	20,413,422
		$W_0 = 2.25,\ W_1 = 2.84,\ W_2 = 1.82,\ W_3 = 0.125$			
		$V(cis) = 7.03,$			
		$V(trans) = 1.10$			
		$\alpha_0 = 111.5°$ (energy min.)			
		Structural uncertainties, see ref. 21			

(continued)

TABLE II (*continued*)

Formula	Name	Data and notes, kcal/mole	Method	State	Refs.
HS—SH	Hydrogen persulfide	$V_2^g = 6.8$	IR	Gas	275,406
CH_3S—SCH_3	Dimethyl disulfide	$V = 6.8$	T	Gas	152,274
		7. Other barriers			
$CH_3(C_6H_5)N$—$B(CH_3)_2$	[Methylphenylamino]-dimethylborane	$V = 15 \pm 3$	NMR	Liq.	424
$C_6H_5(Cl)B$—$N(CH_3)_2$	[Dimethylamino]-chloro-phenylborane	$V = 18 \pm 2$	NMR	Liq.	425
C_6H_5B—$[N(CH_3)_2]_2$	Di[dimethylamino]-phenylborane	$V < 10$	NMR	Liq.	425

a The stable conformation has been determined to be staggered.

b The stable conformation has been determined to be that having a C—H (or Si—H) bond eclipsing the double bond.

c The barrier is probably incorrect on the basis of comparisons with barriers for related molecules.

d The species was produced by x-irradiation of a single crystal.

e ΔX is here defined as X (*gauche*)–X (*trans*).

f The *trans* form is the only form in the solid state.

g The assumed potential form is not dictated by molecular symmetry and is an approximation.

h The identity of the more stable form is not known.

i To reconstruct the conformation represented by the code, draw a Newman projection for a staggered ethane frame and place the symbols for the elements on the bonds in the sequence given and in the clockwise direction.

j The paired symbols represent atoms which eclipse each other at the peak of the barrier.

k This energy has been obtained by subtracting 0.5 kcal/mole from the free energy of activation. See reference 321.

l $V^* = V_1 + 4V_2 + 9V_3$.

m The stable conformation is *trans* planar.

References

1. J. D. Kemp and K. S. Pitzer, *J. Chem. Phys.*, *4*, 749 (1936); *J. Am. Chem. Soc.*, *59*, 53 (1937).
2. E. B. Wilson, Jr., *Advances in Chemical Physics*, Vol. II, I. Prigogine, Ed., Interscience, New York, 1959, p. 367.
3. K. S. Pitzer and W. D. Gwinn, *J. Chem. Phys.*, *10*, 428 (1942); K. S. Pitzer, *ibid.*, *14*, 239 (1946); J. E. Kilpatrick and K. S. Pitzer, *ibid.*, *17*, 1064 (1949); K. S. Pitzer, *ibid.*, *5*, 469 (1937).
4. J. S. Rowlinson, *Nature*, *162*, 820 (1948).
5. D. Telfair, *J. Chem. Phys.*, *10*, 167 (1942).
6. J. E. Piercy, *J. Chem. Phys.*, *43*, 4066 (1965).
7. S. Mizushima, *Structure of Molecules and Internal Rotation*, Academic Press, New York, 1954.
8. C. C. Lin and J. D. Swalen, *Rev. Mod. Phys.*, *31*, 841 (1959).
9. P. H. Verdier and E. B. Wilson, Jr., *J. Chem. Phys.*, *29*, 340 (1958).
10. D. G. Burkhard, *J. Chem. Phys.*, *21*, 1541 (1953); C. R. Quade and C. C. Lin, *ibid.*, *38*, 540 (1963).
11. J. D. Swalen and C. C. Costain, *J. Chem. Phys.*, *31*, 1562 (1959).
12. L. Pierce and M. Hayashi, *J. Chem. Phys.*, *35*, 479 (1961).
13. K. D. Moller and H. G. Andresen, *J. Chem. Phys.*, *37*, 1800 (1962); *39*, 17 (1963).
14. D. R. Herschbach and J. D. Swalen, *J. Chem. Phys.*, *29*, 761 (1958).
15. W. H. Kirchoff and D. R. Lide, Jr., *J. Chem. Phys.*, *43*, 2203 (1965).
16. D. Kivelson, *J. Chem. Phys.*, *22*, 1733 (1954); *23*, 2236 (1955).
17. B. Kirtman, *J. Chem. Phys.*, *37*, 2516 (1962).
18. D. R. Lide, Jr. and D. K. Coles, *Phys. Rev.*, *80*, 911 (1950).
19. W. G. Fately and F. A. Miller, *Spectrochim. Acta*, *19*, 611 (1963).
20. R. H. Hunt, R. A. Leacock, C. W. Peters, and K. T. Hecht, *J. Chem. Phys.*, *42*, 1931 (1965).
21. W. R. Busing and H. A. Levy, *J. Chem. Phys.*, *42*, 3054 (1965).
22. J. A. Pople, W. G. Schneider, and H. J. Bernstein, *High Resolution Nuclear Magnetic Resonance*, McGraw-Hill, New York, 1959.
23. A. Allerhand, H. S. Gutowsky, J. Jonas, and R. A. Meinzer, *J. Am. Chem. Soc.*, *88*, 3185 (1966).
24. C. Heller, *J. Chem. Phys.*, *36*, 175 (1962).
25. E. W. Stone and A. H. Maki, *J. Chem. Phys.*, *37*, 1326 (1962).
26. L. S. Bartell and J. P. Guillory, *J. Chem. Phys.*, *43*, 647 (1965).
27. J. J. Rush and T. I. Taylor, *J. Chem. Phys.*, *44*, 2749 (1966).
28. E. Gorin, J. Walter, and H. Eyring, *J. Am. Chem. Soc.*, *61*, 1876 (1939).
29. A. Eucken and K. Schäffer, *Naturwiss.*, *8*, 122 (1939); A. Langseth, H. J. Bernstein, and B. Bak, *J. Chem. Phys.*, *8*, 713 (1940).
30. K. S. Pitzer, *J. Am. Chem. Soc.*, *70*, 2140 (1948).
31. M. Cignitto and T. Allen, *J. Phys. Chem.*, *68*, 1292 (1964).
32. J. G. Aston, S. Iserow, G. J. Szasz, and R. M. Kennedy, *J. Chem. Phys.*, *12*, 336 (1944).
33. F. A. French and R. S. Rasmussen, *J. Chem. Phys.*, *14*, 389 (1946).
34. Y. Amako and P. A. Giguere, *Can. J. Chem.*, *40*, 765 (1962).

35. E. N. Lassettre and L. B. Dean, *J. Chem. Phys.*, *16*, 151, 553 (1948); *17*, 317 (1949).
36. G. B. Kistiakowsky, J. R. Lacher, and W. W. Ransom, *J. Chem. Phys.*, *6*, 900 (1938).
37. L. Oosterhoff, Thesis, University of Leiden, 1949; *Discussions Faraday Soc.*, *10*, 79 (1951).
38. A. Tang, *J. Chinese Chem. Soc.* (*Taiwan*), *18*, 1 (1951); *19*, 33 (1952); *Sci. Sinica* (*Peking*), *3*, 279 (1954).
39. N. W. Luft, *J. Chem. Phys.*, *21*, 179, 754 (1953); *22*, 1260, 1814 (1954); *Trans. Faraday Soc.*, *49*, 118 (1953).
40. W. L. Clinton, *J. Chem. Phys.*, *33*, 632 (1960).
41. L. Pauling, *Proc. Natl. Acad. Sci. U.S.*, *44*, 211 (1958); *The Nature of the Chemical Bond*, 3rd ed., Cornell University Press, Ithaca, New York, 1960, p. 130.
42. R. A. Scott and H. A. Scheraga, *J. Chem. Phys.*, *42*, 2209 (1965); *44*, 3054 (1966).
43. R. M. Pitzer and W. N. Lipscomb, *J. Chem. Phys.*, *39*, 1995 (1963).
44. R. M. Pitzer, *J. Chem. Phys.*, *41*, 2216 (1964).
45. E. Clementi and D. R. Davis, *J. Chem. Phys.*, *45*, 2593 (1966).
46. W. H. Fink and L. C. Allen, *J. Chem. Phys.*, *46*, 2261 (1967); *46*, 2285 (1967); *47*, 895 (1967).
47. J. B. Pedley, *Trans. Faraday Soc.*, *57*, 1492 (1961); *58*, 23 (1962).
48. H. E. Simmons and J. K. Williams, *J. Am. Chem. Soc.*, *86*, 3222 (1964).
49. R. S. Mulliken, *J. Phys. Chem.*, *56*, 295 (1952).
50. H. Eyring, *J. Am. Chem. Soc.*, *54*, 3191 (1932).
51. F. J. Adrian, *J. Chem. Phys.*, *28*, 608 (1958).
52. C. A. Coulson and D. Stocker, *Mol. Phys.*, *2*, 397 (1959).
53. E. A. Mason and M. M. Kreevoy, *J. Am. Chem. Soc.*, *77*, 5808 (1955).
54. K. E. Howlett, *J. Chem. Soc.*, *1957*, 4353; *1960*, 1055.
55. V. Magnasco, *Nuovo Cimento*, *24*, 425 (1962).
56. J. Barton, *J. Chem. Soc.*, *1948*, 340.
57. J. van Dranen, *J. Chem. Phys.*, *20*, 1982 (1952).
58. W. F. Libby, *J. Chem. Phys.*, *35*, 1527 (1961).
59. R. Hoffmann, *J. Chem. Phys.*, *39*, 1397 (1963).
60. H. Cambron-Brüderlein and C. Sandorfy, *Theoret. Chim. Acta*, *4*, 224 (1966).
61. W. G. Penney, *Proc. Roy. Soc.* (*London*) *Ser. A*, *144*, 166 (1934).
62. H. G. Hecht, D. M. Grant, and H. Eyring, *Mol. Phys.*, *3*, 577 (1960).
63. G. M. Harris and F. E. Harris, *J. Chem. Phys.*, *31*, 1450 (1959).
64. M. Karplus, *J. Chem. Phys.*, *30*, 11 (1959); *33*, 316 (1960).
65. H. Eyring, D. M. Grant, and H. Hecht, *J. Chem. Ed.*, *39*, 466 (1962).
66. H. G. Hecht, *Theoret. Chim. Acta*, *1*, 133 (1963).
67. U. Kaldor and I. Shavitt, *J. Chem. Phys.*, *44*, 1823 (1966).
68. L. Jones, R. Badger, and G. Moore, *J. Chem. Phys.*, *19*, 1599 (1951).
69. B. Bak, *J. Chem. Phys.*, *24*, 918 (1956).
70. O. J. Sovers and M. Karplus, *J. Chem. Phys.*, *44*, 3033 (1966).
71. K. S. Pitzer and E. Catalano, *J. Am. Chem. Soc.*, *78*, 4844 (1956).
72. R. L. McCullough and P. E. McMahon, *Trans. Faraday Soc.*, *60*, 2089 (1964).

73. E. A. Magnusson and H. Shull, *Proc. Intern. Symp. Mol. Struct. Spectry.*, *Tokyo, 1962*, p. C405.
74. R. E. Wyatt and R. G. Parr, *J. Chem. Phys.*, *41*, 3262 (1964); *43*, S217 (1965); *44*, 1529 (1966).
75. R. G. Parr, *J. Chem. Phys.*, *40*, 3726 (1964).
76. H. J. Kim and R. G. Parr, *J. Chem. Phys.*, *41*, 2892 (1964).
77. K. Ruedenberg, *J. Chem. Phys.*, *41*, 588 (1964).
78. J. P. Lowe and R. G. Parr, *J. Chem. Phys.*, *43*, 2565 (1965); *44*, 3001 (1966).
79. J. P. Lowe, *J. Chem. Phys.*, *45*, 3059 (1966).
80. M. Karplus and R. G. Parr, *J. Chem. Phys.*, *38*, 1547 (1963).
81. Personal letter from J. P. Lowe to E. B. Wilson, Jr., February 10, 1966.
82. H. Eyring, G. H. Stewart, and R. P. Smith, *Proc. Natl. Acad. Sci. U.S.*, *44*, 259 (1958).
83. J. Goodisman, *J. Chem. Phys.*, *44*, 2085 (1966).
84. L. Zülicke and H. J. Spangenberg, *Theoret. Chim. Acta*, *5*, 139 (1966).
85. K. Ruedenberg, *Rev. Mod. Phys.*, *34*, 326 (1962).
86. B. L. Trus and L. C. Cusachs, paper presented at Symposium on Molecular Structure and Spectroscopy, Columbus, Ohio, September 1966, paper B-10.
87. H. J. Bernstein, *J. Chem. Phys.*, *17*, 262 (1949).
88. A. Lowenstein and T. M. Connor, *Z. Electrochem.*, *67*, 280 (1963).
89. C. N. Thomas, *Bull. Am. Phys. Soc.*, *11*, 235 (1966). (Value incorrectly reported in abstract. Revised value of $V_3 = 1010$ cal/mole given at meeting.)
90. J. A. Pople and D. P. Santry, *Mol. Phys.*, *7*, 269 (1963–1964).
91. J. A. Pople and G. A. Segal, *J. Chem. Phys.*, *43*, S136 (1965).
92. I. Miyagawa, *J. Chem. Soc. Japan*, *75*, 1162, 1169, 1173, 1177 (1954).
93. W. E. Palke and W. N. Lipscomb, *J. Am. Chem. Soc.*, *88*, 2384 (1966).
94. A. Veillard, *Theoret. Chim. Acta*, *5*, 413 (1966).
95. G. B. Penney and G. B. B. M. Sutherland, *Trans. Faraday Soc.*, *30*, 898 (1954); *J. Chem. Phys.*, *2*, 492 (1934).
96. K. S. Pitzer, *Discussions Faraday Soc.*, *10*, 79 (1951); *Chem. Rev.*, *27*, 39 (1940).
97. G. B. Kistiakowsky, J. R. Lacher, and F. Stitt, *J. Chem. Phys.*, *7*, 289 (1939).
98. D. R. Lide, Jr., *J. Chem. Phys.*, *29*, 1426 (1958).
99. L. G. Smith, *J. Chem. Phys.*, *17*, 139 (1949).
100. G. Sage and W. Klemperer, *J. Chem. Phys.*, *39*, 371 (1963).
101. D. R. Herschbach, *J. Chem. Phys.*, *25*, 358 (1956).
102. D. R. Lide, Jr., *J. Chem. Phys.*, *30*, 37 (1959).
103. D. C. Smith, R. A. Saunders, J. R. Nielsen, and E. E. Ferguson, *J. Chem. Phys.*, *20*, 847 (1952).
104. W. G. Fately and F. A. Miller, *Spectrochim. Acta*, *17*, 857 (1961).
105. C. A. Wulff, *J. Chem. Phys.*, *39*, 1227 (1963).
106. H. S. Gutowsky and H. B. Levine, *J. Chem. Phys.*, *18*, 1297 (1950).
107. H. W. Thompson and R. B. Temple, *J. Chem. Soc.*, *1948*, 1428.
108. H. Russell, D. R. V. Golding, and D. M. Yost, *J. Am. Chem. Soc.*, *66*, 16 (1944).
109. D. C. Smith, G. M. Brown, J. R. Nielsen, R. M. Smith, and C. Y. Liang, *J. Chem. Phys.*, *20*, 473 (1952).

110. J. R. Nielsen, H. H. Claasen, and D. C. Smith, *J. Chem. Phys.*, *18*, 1471 (1950).

111. H. T. Minden and B. P. Dailey, *Phys. Rev.*, *82*, 338 (1951).

112. A. Danti and J. L. Wood, *J. Chem. Phys.*, *30*, 582 (1959).

113. C. R. Ward and C. H. Ward, *J. Mol. Spectry.*, *12*, 289 (1964).

114. D. E. Mann and E. K. Plyler, *J. Chem. Phys.*, *21*, 1116 (1953).

115. E. L. Pace, *J. Chem. Phys.*, *16*, 74 (1948).

116. D. A. Swick and I. L. Karle, *J. Chem. Phys.*, *23*, 1499 (1955).

117. R. H. Schwendeman and G. D. Jacobs, *J. Chem. Phys.*, *36*, 1245 (1962).

118. L. W. Daasch, C. Y. Liang, and J. R. Nielsen, *J. Chem. Phys.*, *22*, 1293 (1954).

119. R. S. Wagner and B. P. Dailey, *J. Chem. Phys.*, *23*, 1355 (1955).

120. T. P. Das, *J. Chem. Phys.*, *27*, 763 (1957).

121. J. C. M. Li and K. S. Pitzer, *J. Am. Chem. Soc.*, *78*, 1077 (1956).

122. J. R. Nielsen, C. Y. Liang, and D. C. Smith, *J. Chem. Phys.*, *21*, 1060 (1953).

123. K. S. Pitzer and J. L. Hollenberg, *J. Am. Chem. Soc.*, *75*, 2219 (1953).

124. M. Z. El-Sabban, A. G. Meister, and F. F. Cleveland, *J. Chem. Phys.*, *19*, 855 (1951).

125. T. R. Rubin, B. H. Levedahl, and D. M. Yost, *J. Am. Chem. Soc.*, *66*, 279 (1944).

126. Y. Morino and E. Hirota, *J. Chem. Phys.*, *28*, 185 (1958).

127. Y. Morino and M. Iwasaki, *J. Chem. Phys.*, *17*, 216 (1949).

128. S. Mizushima, Y. Morino, T. Simanouti, and K. Kuratani, *J. Chem. Phys.*, *17*, 838 (1949).

129. T. Kasuya, *J. Phys. Soc. Japan*, *15*, 1273 (1960).

130. C. Flanagan and L. Pierce, *J. Chem. Phys.*, *38*, 2963 (1963).

131. R. S. Wagner, B. P. Dailey, and N. Solimene, *J. Chem. Phys.*, *26*, 1593 (1957).

132. J. R. Nielsen and R. Theimer, *J. Chem. Phys.*, *27*, 891 (1957).

133. R. Theimer and J. R. Nielsen, *J. Chem. Phys.*, *27*, 887 (1957).

134. P. R. Bunker and H. C. Longuet-Higgins, *Proc. Roy. Soc.* (*London*), *Ser. A*, *280*, 340 (1964).

135. G. B. Kistiakowsky and W. W. Rice, *J. Chem. Phys.*, *8*, 618 (1940).

136. D. W. Osborne, C. S. Garner, and D. M. Yost, *J. Chem. Phys.*, *8*, 131 (1940).

137. B. L. Crawford, Jr. and W. W. Rice, *J. Chem. Phys.*, *7*, 437 (1939).

138. B. Bak, L. Hansen, and J. Rastrup-Andersen, *J. Chem. Phys.*, *21*, 1612 (1953).

139. V. W. Laurie and D. R. Lide, Jr., *J. Chem. Phys.*, *31*, 939 (1959).

140. V. W. Laurie, *J. Chem. Phys.*, *31*, 1500 (1959).

141. R. G. Lerner and B. P. Dailey, *J. Chem. Phys.*, *26*, 678 (1957).

142. R. W. Kilb, C. C. Lin, and E. B. Wilson, Jr., *J. Chem. Phys.*, *26*, 1695 (1957).

143. K. S. Pitzer, *J. Chem. Phys.*, *12*, 310 (1944).

144. J. D. Kemp and C. J. Egan, *J. Am. Chem. Soc.*, *60*, 1521 (1938).

145. D. R. Lide, Jr., *J. Chem. Phys.*, *33*, 1514 (1960).

146. D. R. Lide, Jr., and D. E. Mann, *J. Chem. Phys.*, *29*, 914 (1958).

147. H. P. Bucker and J. R. Nielsen, *J. Mol. Spectry.*, *11*, 47 (1963).

148. G. M. Barrow, *J. Chem. Phys.*, *20*, 1739 (1952).

149. S. C. Schumann and J. G. Aston, *J. Chem. Phys.*, *6*, 480 (1938).

150. S. C. Schumann and J. G. Aston, *J. Chem. Phys.*, 6, 485 (1938).
151. J. P. McCullough, H. L. Finke, D. W. Scott, M. E. Gross, J. F. Messerly, R. E. Pennington, and G. Waddington, *J. Am. Chem. Soc.*, 76, 4796 (1954).
152. G. A. Crowder and D. W. Scott, *J. Mol. Spectry.*, 16, 122 (1965).
153. D. W. Scott, H. L. Finke, J. P. McCullough, M. E. Gross, K. D. Williamson, G. Waddington, and H. M. Huffman, *J. Am. Chem. Soc.*, 73, 261 (1951).
154. D. W. Scott, H. L. Finke, W. N. Hubbard, J. P. McCullough, G. D. Oliver, M. E. Gross, C. Katz, K. D. Williamson, G. Waddington, and H. M. Huffman, *J. Am. Chem. Soc.*, 74, 4656 (1952).
155. J. P. McCullough, H. L. Finke, J. F. Messerly, R. E. Pennington, I. A. Hossenlopp, and G. Waddington, *J. Am. Chem. Soc.*, 77, 6119 (1955).
156. D. W. Scott, W. D. Good, S. S. Todd, J. F. Messerly, W. T. Berg, I. A. Hossenlopp, J. L. Lacina, A. Osborne, and J. P. McCullough, *J. Chem. Phys.*, 36, 406 (1962).
157. O. L. Stiefvater, paper presented at Symposium on Molecular Structure and Spectroscopy, Columbus, Ohio, September 1966, paper V10.
158. S. S. Butcher and E. B. Wilson, Jr., *J. Chem. Phys.*, 40, 1671 (1964).
159. K. S. Pitzer, *J. Chem. Phys.*, 5, 473 (1937).
160. J. G. Aston, R. M. Kennedy, and S. C. Schumann, *J. Am. Chem. Soc.*, 62, 2059 (1940).
161. R. W. Kilb and C. C. Lin, *Bull. Am. Phys. Soc.*, 1, 198 (1956).
162. L. Pierce and L. C. Krisher, *J. Chem. Phys.*, 31, 875 (1959).
163. L. Pierce, *Bull. Am. Phys. Soc.*, 1, 198 (1956).
164. K. M. Sinnott, *J. Chem. Phys.*, 34, 851 (1961); *Bull. Am. Phys. Soc.*, 1, 198 (1956).
165. L. C. Krisher, *J. Chem. Phys.*, 33, 1237 (1960).
166. L. C. Krisher and E. B. Wilson, Jr., *J. Chem. Phys.*, 31, 882 (1959).
167. R. Nelson and L. Pierce, *J. Mol. Spectry.*, 18, 344 (1965).
168. W. G. Fately and F. A. Miller, *Spectrochim. Acta*, 18, 977 (1962).
169. S. C. Schumann and J. G. Aston, *J. Chem. Phys.*, 6, 485 (1938).
170. O. L. Stiefvater and J. Sheridan, *Proc. Chem. Soc.*, 1963, 368.
171. W. J. Tabor, *J. Chem. Phys.*, 27, 974 (1957).
172. R. E. Dodd, H. L. Roberts, and L. A. Woodward, *J. Chem. Soc.*, 1957, 2783.
173. K. R. Loos and R. C. Lord, *Spectrochim. Acta*, 21, 119 (1965).
174. M. Buyle-Bodin, *Ann. Phys. (Paris)*, 10, 533 (1955).
175. E. Hirota, *J. Chem. Phys.*, 45, 1984 (1966).
176. D. R. Herschbach and L. C. Krisher, *J. Chem. Phys.*, 28, 728 (1958).
177. D. R. Lide, Jr., and D. E. Mann, *J. Chem. Phys.*, 27, 868 (1957).
178. J. E. Kilpatrick and K. S. Pitzer, *J. Res. Natl. Bur. Std.*, 37, 163 (1946).
179. L. Pierce and J. M. O'Reilly, *J. Mol. Spectry.*, 3, 536 (1959).
180. S. Segal, *J. Chem. Phys.*, 27, 989 (1957).
181. R. A. Beaudet and E. B. Wilson, Jr., *J. Chem. Phys.*, 37, 1133 (1962).
182. V. W. Weiss, P. Beak, and W. H. Flygare, *J. Chem. Phys.* 46, 981 (1967).
183. M. L. Unland, V. Weiss, and W. H. Flygare, *J. Chem. Phys.*, 42, 2138 (1965).
184. R. Meyer, paper presented at Symposium on Molecular Structure and Spectroscopy, Columbus, Ohio, September 1966, Paper J4.
185. H. Hunziker and Hs. H. Günthard, *Spectrochim. Acta*, 21, 51 (1965).
186. R. A. Beaudet, *J. Chem. Phys.*, 37, 2398 (1962).

187. R. A. Beaudet, *J. Chem. Phys.*, *40*, 2705 (1964).
188. R. A. Beaudet, paper presented at Symposium on Molecular Structure and Spectroscopy, Columbus, Ohio, September 1966, paper V12.
189. R. A. Beaudet, personal communication (indirect).
190. V. W. Laurie, *J. Chem. Phys.*, *32*, 1588 (1960).
191. R. A. Beaudet, *J. Chem. Phys.*, *38*, 2548 (1963).
192. V. W. Laurie, *J. Chem. Phys.*, *34*, 1516 (1961).
193. L. Guttman and K. S. Pitzer, *J. Am. Chem. Soc.*, *67*, 324 (1945).
194. D. R. Lide, Jr., and M. Chen., *Chem. Phys.*, *40*, 252 (1964)
195. B. Bak, J. J. Christiansen, K. Kunstmann, L. Nygaard, and J. Rastrup-Andersen, *J. Chem. Phys.*, *45*, 883 (1966).
196. B. Bak, D. Christensen, J. Christiansen, L. Hansen-Nygaard, and J. Rastrup-Andersen, *Spectrochim. Acta*, *18*, 1421 (1962).
197. D. R. Lide, Jr., and D. E. Mann, *J. Chem. Phys.*, *27*, 874 (1957).
198. A. M. Ronn and R. C. Woods III, *J. Chem. Phys.*, *45*, 3831 (1966).
199. E. Hirota, *J. Chem. Phys.*, *37*, 283 (1962).
200. F. L. Tobiason and R. H. Schwendeman, *J. Chem. Phys.*, *40*, 1014 (1964).
201. Landolt-Börstein, *Zahlenwerte und Funktionen*, Vol. I, 2, I, Springer, Berlin, 1951.
202. D. R. Herschbach and J. D. Swalen, *J. Chem. Phys.*, *29*, 761 (1958).
203. J. D. Swalen and E. R. Herschbach, *J. Chem. Phys.*, *27*, 100 (1957).
204. M. Sage, *J. Chem. Phys.*, *35*, 142 (1961).
205. S. S. Butcher, *J. Chem. Phys.*, *38*, 2310 (1963).
206. P. Taimsalu and J. L. Wood, *Spectrochim. Acta*, *20*, 1043 (1964).
207. F. G. Brickwedde, M. Moskow, and R. B. Scott, *J. Chem. Phys.*, *12*, 547 (1946),
208. A. Horsfield, J. R. Morton, and D. H. Whiffen, *Mol. Phys.*, *5*, 115 (1962); *4*, 425 (1961).
209. C. Heller, *J. Chem. Phys.*, *36*, 175 (1962).
210. I. Miyagawa and K. Itoh, *J. Chem. Phys.*, *36*, 2157 (1962).
211. J. E. Griffiths, *J. Chem. Phys.*, *38*, 2879 (1963).
212. L. Pierce, *J. Chem. Phys.*, *29*, 383 (1958).
213. J. D. Swalen and B. P. Stoicheff, *J. Chem. Phys.*, *28*, 671 (1958).
214. L. C. Krisher and L. Pierce, *J. Chem. Phys.*, *32*, 1619 (1960).
215. R. L. Collins and J. R. Nielsen, *J. Chem. Phys.*, *23*, 351 (1955).
216. J. Sheridan and W. Gordy, *J. Chem. Phys.*, *19*, 965 (1951).
217. R. N. Dixon and H. W. Kroto, *Proc. Roy. Soc. (London)*, Ser. A, *283*, 423 (1965).
218. L. Pierce, *J. Chem. Phys.*, *34*, 498 (1961).
219. L. Pierce and D. H. Petersen, *J. Chem. Phys.*, *33*, 907 (1960).
220. K. Shimizu and H. Murata, *J. Mol. Spectry.*, *5*, 44 (1960).
221. J. G. Aston, R. M. Kennedy, and G. H. Messerly, *J. Am. Chem. Soc.*, *63*, 2343 (1941).
222. R. H. Schwendeman and G. D. Jacobs, *J. Chem. Phys.*, *36*, 1251 (1962).
223. W. S. Salathiel and R. F. Curl, Jr., *J. Chem. Phys.*, *44*, 1288 (1966).
224. V. W. Laurie, *J. Chem. Phys.*, *30*, 1210 (1959).
225. J. A. Aronson and J. R. Durig, *Spectrochim. Acta*, *20*, 219 (1964).
226. G. W. Smith, *J. Chem. Phys.*, *42*, 4229 (1965).

227. P. Cahill and S. S. Butcher, *J. Chem. Phys.*, *35*, 2255 (1962).
228. H. S. Gutowsky and E. O. Stejskal, *J. Chem. Phys.*, *22*, 939 (1954).
229. G. W. Bethke and M. K. Wilson, *J. Chem. Phys.*, *26*, 1107 (1957).
230. D. R. Lide, Jr., and D. E. Mann, *J. Chem. Phys.*, *28*, 572 (1958).
231. A. P. Cox and R. Varma, *J. Chem. Phys.*, *44*, 2619 (1966).
232. J. M. O'Reilly and L. Pierce, *J. Chem. Phys.*, *34*, 1176 (1961).
233. V. W. Laurie and J. E. Wallrab, paper presented at Symposium on Molecular Structure and Spectroscopy, Columbus, Ohio, September 1966, paper AA6.
234. Y. Morino, *J. Chem. Phys.*, *24*, 164 (1956).
235. J. G. Aston, M. L. Sagenkahn, G. J. Szasz, G. W. Moessen, and H. F. Zuhr, *J. Am. Chem. Soc.*, *66*, 1171 (1944).
236. M. Katayama, T. Simanouti, Y. Morino, and S. Mizushima, *J. Chem. Phys.*, *18*, 506 (1950).
237. J. E. Griffiths and G. E. Walrafen, *J. Chem. Phys.*, *40*, 321 (1964).
238. D. A. Dows and R. M. Hexter, *J. Chem. Phys.*, *24*, 1029,1117 (1956).
239. T. Nishikawa, *J. Phys. Soc. Japan*, *12*, 668 (1957).
240. D. R. Lide, Jr., *J. Chem. Phys.*, *27*, 343 (1957).
241. D. Kivelson and D. R. Lide, Jr., *J. Chem. Phys.*, *27*, 353 (1957).
242. T. Itoh, *J. Phys. Soc. Japan*, *11*, 264 (1956).
243. T. Nishikawa, T. Itoh, and K. Shimoda, *J. Chem. Phys.*, *23*, 1735 (1955).
244. J. G. Aston and F. L. Gittler, *J. Am. Chem. Soc.*, *77*, 3175 (1955).
245. D. G. Burkhard and D. M. Dennison, *Phys. Rev.*, *84*, 408 (1951).
246. K. T. Hecht and D. M. Dennison, *J. Chem. Phys.*, *26*, 48 (1957).
247. T. Nishikawa, *J. Phys. Soc. Japan*, *11*, 781 (1956).
248. J. D. Swalen, *J. Chem. Phys.*, *23*, 1739 (1955).
249. E. V. Ivash and D. M. Dennison, *J. Chem. Phys.*, *21*, 1804 (1953).
250. P. Ventkateswarlu and W. Gordy, *J. Chem. Phys.*, *23*, 1200 (1955).
251. J. S. Rigden and S. S. Butcher, *J. Chem. Phys.*, *40*, 2109 (1964).
252. J. P. Perchard, M. T. Forel, and M. L. Josien, *J. Chim. Phys.*, *61*, 632 (1964); *61*, 645 (1964); *61*, 652 (1964).
253. P. H. Kasai and R. J. Myers, *J. Chem. Phys.*, *30*, 1096 (1959).
254. U. Blukis, P. H. Kasai, and R. J. Myers, *J. Chem. Phys.*, *38*, 2753 (1963).
255. R. M. Kennedy, M. Sagenkahn, and J. G. Aston, *J. Am. Chem. Soc.*, *63*, 2267 (1941).
256. R. E. Naylor, Jr., and E. B. Wilson, Jr., *J. Chem. Phys.*, *26*, 1057 (1957).
257. W. B. Dixon and E. B. Wilson, Jr., *J. Chem. Phys.*, *35*, 191 (1961).
258. J. D. Ray, *J. Chem. Phys.*, *40*, 3440 (1964).
259. R. F. Curl, Jr., *J. Chem. Phys.*, *30*, 1529 (1959).
260. P. D. Foster, V. M. Rao, and R. F. Curl, Jr., *J. Chem. Phys.*, *43*, 1064 (1965).
261. P. Cahill, L. P. Gold, and N. Owen, *J. Chem. Phys.*, in press.
262. T. Kojima and T. Nishikawa, *J. Phys. Soc. Japan*, *12*, 680 (1957).
263. N. Solimene and B. P. Dailey, *J. Chem. Phys.*, *23*, 124 (1955).
264. J. L. Binder, *J. Chem. Phys.*, *17*, 499 (1949).
265. E. Tannenbaum, R. J. Myers, and W. D. Gwinn, *J. Chem. Phys.*, *25*, 42 (1956).
266. P. Kisluik and G. A. Silvey, *J. Chem. Phys.*, *20*, 517 (1952).
267. V. H. D. Rudolph, H. Dreizler, and W. Maier, *Z. Naturforsch.*, *15a*, 742 (1960).

268. J. P. McCullough, W. N. Hubbard, F. R. Frow, I. A. Hossenlopp, and G. Waddington, *J. Am. Chem. Soc.*, *79*, 561 (1957).
269. D. W. Osborne, R. N. Doescher, and D. M. Yost, *J. Am. Chem. Soc.*, *64*, 169 (1942).
270. D. W. Osborne, R. N. Doescher, and D. M. Yost, *J. Chem. Phys.*, *8*, 506 (1940).
271. H. Dreizler and H. D. Rudolph, *Z. Naturforsch.*, *17a*, 712 (1962).
272. S. Nakagawa, S. Takahashi, T. Kojima, and C. C. Lin, *J. Chem. Phys.*, *43*, 3583 (1965).
273. S. Nakagawa, T. Kojima, S. Takahashi, and C. C. Lin, *J. Mol. Spectry.*, *14*, 201 (1964).
274. W. N. Hubbard, D. R. Douslin, J. P. McCullough, D. W. Scott, S. S. Todd, J. F. Messerly, I. A. Hossenlop, A. George, and G. Waddington, *J. Am. Chem. Soc.*, *80*, 3547 (1958).
275. R. L. Redington, *J. Mol. Spectry.*, *9*, 469 (1962).
276. R. E. Dininny and E. L. Pace, *J. Chem. Phys.*, *32*, 805 (1960).
277. A. B. Harvey and M. K. Wilson, *J. Chem. Phys.*, *45*, 678 (1966).
278. S. Siegel, Thesis, Harvard University, 1958.
279. L. S. Bartell and B. L. Carroll, *J. Chem. Phys.*, *42*, 3076 (1965).
280. L. A. Woodward, J. R. Hall, R. N. Dixon, and N. Sheppard, *Spectrochim. Acta*, *15*, 249 (1959).
281. H. S. Gutowsky, *J. Chem. Phys.*, *17*, 128 (1949).
282. R. C. Leech, D. M. Powell, and N. Sheppard, *Spectrochim. Acta*, *21*, 559 (1965).
283. P. Klaboe and J. R. Nielsen, *J. Chem. Phys.*, *33*, 1764 (1960).
284. P. Klaboe and J. R. Nielsen, *J. Chem. Phys.*, *32*, 899 (1960).
285. S. Mizushima, Y. Morino, I. Wanatabe, T. Simanouti, and S. Yamaguchi, *J. Chem. Phys.*, *17*, 591 (1949).
286. Y. A. Pentin and V. M. Tatetevskii, *Dokl. Akad. Nauk SSSR*, *108*, 290 (1956).
287. H. J. Bernstein, *J. Chem. Phys.*, *17*, 258 (1949).
288. J. Ainsworth and J. Karle, *J. Chem. Phys.*, *20*, 425 (1952).
289. N. Sheppard and J. J. Turner, *Proc. Roy. Soc. (London)*, *Ser. A*, *252*, 506 (1959).
290. D. H. Rank, R. E. Kagarise, and D. W. E. Axford, *J. Chem. Phys.*, *17*, 1354 (1949).
291. H. Gerding and P. G. Merman, *Rec. Trav. Chim.*, *61*, 523 (1942).
292. D. F. Eggers, Jr., H. E. Wright, and D. W. Robinson, *J. Chem. Phys.*, *35*, 1045 (1961).
293. R. W. Kilb and L. Pierce, *J. Chem. Phys.*, *27*, 108 (1957).
294. I. Ichishima, H. Kamiyama, T. Shimanouchi, and S. Mizushima, *J. Chem. Phys.*, *29*, 1190 (1958).
295. T. Tokuhiro, *J. Chem. Phys.*, *41*, 438 (1964).
296. K. Kuratani and S. Mizushima, *J. Chem. Phys.*, *22*, 1403 (1954).
297. J. R. Thomas and W. D. Gwinn, *J. Am. Chem. Soc.*, *71*, 2785 (1949).
298. J. E. Griffiths, paper presented at Symposium on Molecular Structure and Spectroscopy, Columbus, Ohio, September 1966, paper M7.
299. F. E. Mahlerbe and H. J. Bernstein, *J. Am. Chem. Soc.*, *74*, 1859 (1952).

300. R. H. Harrison and K. A. Kobe, *J. Chem. Phys.*, *26*, 1411 (1957).
301. A. D. Giacomo and C. P. Smyth, *J. Am. Chem. Soc.*, *77*, 1361 (1955).
302. R. A. Oriani and C. P. Smyth, *J. Chem. Phys.*, *16*, 930 (1948).
303. J. Powling and H. J. Bernstein, *J. Am. Chem. Soc.*, *73*, 1815 (1951).
304. H. S. Gutowsky, G. G. Belford, and P. E. McMahon, *J. Chem. Phys.*, *36*, 3353 (1962).
305. W. M. Tolles, E. T. Handeman, and W. D. Gwinn, *J. Chem. Phys.*, *43*, 3019 (1965).
306. R. E. Kagarise and D. H. Rank, *Trans. Faraday Soc.*, *48*, 394 (1952).
307. H. J. Bernstein, *J. Chem. Phys.*, *18*, 897 (1950).
308. K. V. L. N. Sastry and R. F. Curl, Jr., *J. Chem. Phys.*, *41*, 77 (1964).
309. G. Y. Kabo and D. N. Andrewskii, *Neftekhimiya*, *3*, 764 (1963); *Chem. Abstr.*, *60*, 2388e (1964).
310. L. C. Krisher, *J. Chem. Phys.*, *33*, 304 (1960).
311. J. P. McCullough, D. W. Scott, H. L. Finke, M. E. Gross, K. D. Williamson, R. E. Pennington, G. Waddington, and H. M. Hoffman, *J. Am. Chem. Soc.*, *74*, 2801 (1952).
312. J. P. McCullough, D. W. Scott, H. L. Finke, W. N. Hubbard, M. E. Gross, C. Katz, R. E. Pennington, J. F. Messerly, and G. Waddington, *J. Am. Chem. Soc.*, *75*, 1818 (1953).
313. R. E. Kagarise and L. W. Daasch, *J. Chem. Phys.*, *23*, 113 (1955).
314. J. T. Yardley, J. Hinze, and R. F. Curl, Jr., *J. Chem. Phys.*, *41*, 2562 (1964).
315. R. F. Curl, Jr., V. M. Rao, K. V. L. N. Sastry, and J. A. Hodgeson, *J. Chem. Phys.*, *39*, 3335 (1963).
316. R. E. Kagarise, *J. Chem. Phys.*, *26*, 380 (1957).
317. M. Iwasaki, S. Nagase, and R. Kojima, *J. Chem. Phys.*, *22*, 959 (1954).
318. R. N. Fessenden and J. S. Waugh, *J. Chem. Phys.*, *37*, 1466 (1962).
319. P. Klaboe and J. R. Nielsen, *J. Mol. Spectry.*, *6*, 379 (1961).
320. R. E. Kagarise, *J. Chem. Phys.*, *29*, 680 (1958).
321. R. A. Newmark and C. H. Sederholm, *J. Chem. Phys.*, *43*, 602 (1965).
322. R. E. Kagarise, cited by K. Krebs and J. Lamb, *Proc. Roy. Soc. (London)*, *Ser. A*, *244*, 558 (1958).
323. M. Iwasaki, S. Nagase, and R. Kojima, *Bull. Chem. Soc. Japan*, *30*, 230 (1957).
324. I. L. Karle and J. Karle, *J. Chem. Phys.*, *36*, 1969 (1962).
325. R. E. Kagarise, *J. Chem. Phys.*, *24*, 1264 (1956).
326. Cohen, E. A. and R. A. Beaudet, paper presented at Symposium on Molecular Structure and Spectroscopy, Columbus, Ohio, September 1966, paper V4.
327. B. D. Saksena and R. E. Kagarise, *J. Chem. Phys.*, *19*, 987 (1951).
328. W. E. Fitzgerald and G. J. Janz, *J. Mol. Spectry*, *1*, 49 (1957).
329. G. L. Lewis and C. P. Symth, *J. Chem. Phys.*, *7*, 1085 (1939).
330. P. J. Krueger and H. D. Mettee, *Can. J. Chem.*, *42*, 326 (1964).
331. S. Mizushima, T. Shimanouchi, T. Miyazawa, K. Abe, and M. Yasume, *J. Chem. Phys.*, *19*, 1477 (1951).
332. H. J. Bernstein and E. E. Pedersen, *J. Chem. Phys.*, *17*, 885 (1949).
333. R. J. Abraham and J. A. Pople, *Mol. Phys.*, *3*, 609 (1960).
334. L. S. Bartell and D. A. Kohl, *J. Chem. Phys.*, *39*, 3097 (1963).
335. N. Sheppard and G. J. Szasz, *J. Chem. Phys.*, *17*, 86 (1949).
336. G. J. Szasz, N. Sheppard, and D. H. Rank, *J. Chem. Phys.*, *16*, 704 (1948).

337. J. A. Chen and A. A. Petrauskas, *J. Chem. Phys.*, *30*, 304 (1959).
338. E. Hirota, personal communication (indirect).
339. T. N. Sarachman, *J. Chem. Phys.*, *39*, 469 (1963).
340. Y. Morino and K. Kuchitsu, *J. Chem. Phys.*, *28*, 175 (1958).
341. C. Komaki, I. Ichishima, K. Kuratani, T. Miyazawa, T. Shimanouchi, and S. Mizushima, *Bull. Chem. Soc. Japan*, *28*, 330 (1955).
342. Y. Morino, I. Miyagawa, and T. Haga, *J. Chem. Phys.*, *19*, 791 (1951).
343. W. W. Wood and V. Schomaker, *J. Chem. Phys.*, *20*, 555 (1952).
344. J. P. Guillory and L. S. Bartell, *J. Chem. Phys.*, *43*, 654 (1965).
345. J. Goubeau and H. Pajenkamp, *Acta Phys. Austriaca*, *3*, 283 (1949).
346. M. Hayashi, I. Ichishima, T. Shimanouchi, and S. Mizushima, *Spectrochim. Acta*, *10*, 1 (1957).
347. P. E. McMahon and W. C. Tincher, *J. Mol. Spectry.*, *15*, 180 (1965).
348. P. E. McMahon, *J. Mol. Spectry.*, *16*, 221 (1965).
349. J. K. Wilmshurst and H. J. Bernstein, *Can. J. Chem.*, *35*, 734 (1957).
350. A. Miyake, I. Nakagawa, T. Miyazawa, I. Ichishima, T. Shimanouchi, and S. Mizushima, *Spectrochim. Acta*, *13*, 161 (1958).
351. I. Nakagawa, I. Ichishima, K. Kuratani, T. Miyazawa, T. Shimanouchi, and S. Mizushima, *J. Chem. Phys.*, *20*, 1720 (1952).
352. E. Hirota, *J. Chem. Phys.*, *42*, 2071 (1965).
353. G. H. Griffith and L. A. Harrah, paper presented at Symposium on Molecular Structure and Spectroscopy, Columbus, Ohio, September 1966, paper D4.
354. J. G. Aston, G. Szasz, H. W. Woolley, and F. G. Brickwedde, *J. Chem. Phys.*, *14*, 67 (1946).
355. W. G. Fately, R. K. Harris, F. A. Miller, and R. E. Witkowski, *Spectrochim. Acta*, *21*, 231 (1965).
356. D. R. Lide, Jr., *J. Chem. Phys.*, *37*, 2074 (1962).
357. H. G. Silver and J. L. Wood, *Trans. Faraday Soc.*, *60*, 5 (1964).
358. K. S. Pitzer and D. W. Scott, *J. Am. Chem. Soc.*, *65*, 803 (1943).
359. J. J. Rush, *Bull. Am. Phys. Soc.*, *10*, 492 (1965).
360. M. Frankosky and J. G. Aston, *J. Phys. Chem.*, *69*, 3126 (1965).
361. V. M. Rao, J. T. Yardley, and R. F. Curl, Jr., *J. Chem. Phys.*, *42*, 284 (1965).
362. R. C. Leech, D. B. Powell, and N. Sheppard, *Spectrochim. Acta*, *22*, 1 (1966).
363. D. W. Scott, J. F. Messerly, S. S. Todd, I. A. Hossenlopp, D. R. Douslin, and J. P. McCullough, *J. Chem. Phys.*, *37*, 867 (1962).
364. R. J. Kurland and W. B. Wise, *J. Am. Chem. Soc.*, *86*, 1877 (1964).
365. T. Kojima, E. L. Breig, and C. C. Lin, *J. Chem. Phys.*, *35*, 2139 (1961).
366. R. H. Schwendeman and H. N. Volltrauer, paper presented at Symposium on Molecular Structure and Spectroscopy, Columbus, Ohio, September 1966, paper AA5.
367. M. Kashiwagi and Y. Kurita, *J. Chem. Phys.*, *39*, 3165 (1963).
368. R. Chang and C. S. Johnson, Jr., *J. Chem. Phys.*, *41*, 3272 (1964).
369. R. Wagner, J. Fine, J. W. Simmons, and J. H. Goldstein, *J. Chem. Phys.*, *26*, 634 (1957).
370. R. E. Knudsen, C. F. George, and J. Karle, *J. Chem. Phys.*, *44*, 2334 (1966).
371. T. Miyazawa, *Bull. Chem. Soc. Japan*, *34*, 691 (1961).
372. G. Fraenkel and C. Franconi, *J. Am. Chem. Soc.*, *82*, 4478 (1960).
373. R. M. Hammaker and B. A. Gugler, *J. Mol. Spectry.*, *17*, 356 (1965).

374. W. D. Phillips, C. E. Looney, and C. P. Spaeth, *J. Mol. Spectry.*, *1*, 35 (1957).
375. B. Sunners, L. H. Piette, and W. G. Schneider, *Can. J. Chem.*, *38*, 681 (1961).
376. H. S. Gutowsky and C. H. Holm, *J. Chem. Phys.*, *25*, 1228 (1956).
377. J. C. Woodbrey and M. T. Rogers, *J. Am. Chem. Soc.*, *84*, 13 (1962).
378. I. Suzuki, M. Tsuboi, T. Shimanouchi, and S. Mizushima, *Spectrochim. Acta*, *16*, 471 (1960).
379. C. Franconi, *Z. Electrochem.*, *65*, 645 (1961).
380. J. Heidberg, J. A. Weil, G. A. Janusonis, and J. K. Anderson, *J. Chem. Phys.*, *41*, 1033 (1964).
381. L. S. Bartell and L. O. Brockway, *J. Chem. Phys.*, *32*, 512 (1960).
382. N. L. Owen and N. Sheppard, *Proc. Chem. Soc.* (*London*), *1963*, 264.
383. P. Gray and L. W. Reeves, *J. Chem. Phys.*, *32*, 1878 (1960).
384. L. H. Piette and W. A. Anderson, *J. Chem. Phys.*, *30*, 899 (1959).
385. B. Bak, P. O. Jensen, and K. Schaumburg, *J. Mol. Spectry.*, *20*, 1 (1966).
386. P. J. Krueger and H. D. Mettee, *Can. J. Chem.*, *42*, 340 (1964).
387. H. Forest and B. P. Dailey, *J. Chem. Phys.*, *45*, 1736 (1966).
388. T. Kojima, *J. Phys. Soc. Japan*, *15*, 284 (1960).
389. J. R. Bolton, A. Carrington, and P. F. Todd, *Mol. Phys.*, *6*, 169 (1963).
390. T. Miyazawa and K. S. Pitzer, *J. Chem. Phys.*, *30*, 1076 (1959).
391. R. G. Lerner, B. P. Dailey, and J. P. Friend, *J. Chem. Phys.*, *26*, 680 (1957).
392. D. R. Lide, Jr., in *Ann. Rev. of Phys. Chem.*, *15*, 225 (1964).
393. J. M. Riveros and E. B. Wilson, Jr., *J. Chem. Phys.*, *46*, 4605 (1967).
394. D. Tabuchi, *J. Chem. Phys.*, *28*, 1014 (1958).
395. J. E. Piercy and S. V. Subrahmanyam, *J. Chem. Phys.*, *42*, 1475 (1965).
396. D. L. Bernitt, K. O. Hartman, and I. C. Hisatsune, *J. Chem. Phys.*, *42*, 3553 (1965).
397. N. L. Owen and N. Sheppard, *Trans. Faraday Soc.*, *60*, 634 (1964).
398. L. J. Nugent and C. D. Cornwall, *J. Chem. Phys.*, *37*, 523 (1962).
399. T. Kasuya, *Sci. Papers Inst. Phys. Chem. Res.* (*Tokyo*), *56*, 1 (1962).
400. T. Kasuya and T. Kojima, *J. Phys. Soc. Japan*, *18*, 364 (1963).
401. D. R. Lide, Jr., and D. E. Mann, *J. Chem. Phys.*, *31*, 1129 (1959).
402. C. C. Colburn, F. A. Johnson, and C. Haney, *J. Chem. Phys.*, *43*, 4526 (1965).
403. J. G. Aston, H. L. Fink, G. J. Janz, and K. E. Russell, *J. Am. Chem. Soc.*, *73*, 1939 (1951).
404. J. G. Aston, G. J. Janz, and K. E. Russell, *J. Am. Chem. Soc.*, *73*, 1943 (1951).
405. D. R. Lide, Jr., *Spectrochim. Acta*, *15*, 473 (1959).
406. G. Winnewisser, M. Winnewisser, and W. Gordy, *Bull. Am. Phys. Soc.*, *2*, 312 (1966).
407. G. E. McGraw, D. L. Bernitt, and I. C. Hisatsune, *J. Chem. Phys.*, *45*, 1392 (1966).
408. R. T. Hall and G. C. Pimentel, *J. Chem. Phys.*, *38*, 1889 (1963).
409. W. D. Gwinn, cited by Ray, J. D., *J. Chem. Phys.*, *40*, 3440 (1964).
410. G. E. McGraw, D. L. Bernitt, and I. C. Hisatsune, *J. Chem. Phys.*, *42*, 237 (1965).
411. H. Cohn, C. K. Ingold, and H. G. Poole, *J. Chem. Soc.* (*London*), *1952*, 4272.

412. R. J. Abragam and H. J. Bernstein, *Can. J. Chem.*, *39*, 39 (1961).
413. R. L. Redington, W. B. Olson, and P. C. Cross, *J. Chem. Phys.*, *36*, 1311 (1962).
414. E. O. Stejskal, D. E. Woessner, T. C. Farrar, and H. S. Gutowsky, *J. Chem. Phys.*, *31*, 55 (1959).
415. O. Risgin and R. C. Taylor, *Spectrochim. Acta*, *15*, 1036 (1959).
416. H. Murata, *J. Chem. Phys.*, *21*, 181 (1953).
417. J. Dale, *Tetrahedron*, *22*, 3373 (1966).
418. K. Radcliffe and J. L. Wood, *Trans. Faraday Soc.*, *62*, 1678 (1966).
419. M. J. Moloney and L. C. Krisher, *J. Chem. Phys.*, *45*, 3277 (1966).
420. H. D. Bist and D. R. Williams, *Bull. Am. Phys. Soc.*, *11*, 826 (1966).
421. P. M. Wilt and E. A. Jones, *Bull. Am. Phys. Soc.*, *11*, 826 (1966).
422. R. H. Hunt and R. A. Leacock, *J. Chem. Phys.*, *45*, 3141 (1966).
423. H. Levy and W. E. Grizzle, *J. Chem. Phys.*, *45*, 1954 (1966).
424. G. E. Ryschkewitsch, W. S. Brey, Jr., and A. Saji, *J. Am. Chem. Soc.*, *83*, 1010 (1963).
425. P. A. Barfield, M. F. Lappert, and J. Lee, *Proc. Chem. Soc.* (*London*), *1961*, 421.
426. L. J. Nugent, D. F. Mann, and D. R. Lide, Jr., *J. Chem. Phys.*, *36*, 965 (1962).
427. C. V. Berney, *Spectrochim. Acta*, *21*, 1809 (1965).
428. J. Michielsen-Effinger, *Bull. Classe Sci., Acad. Roy. Belg.*, *50*, 645 (1964).
429. H. Dreizler and G. Dendl, *Z. Naturforsch.*, *20a*, 1431 (1965).
430. J. Karle, *J. Chem. Phys.*, *45*, 4149 (1966).
431. R. E. Rondeau and L. H. Harrah, *J. Mol. Spectry.*, *21*, 332 (1966).
432. R. J. Abraham, L. Cavalli, and K. G. R. Pachler, *Mol. Phys.*, *11*, 471 (1966).
433. H. P. Benz, A. Bauder, and Hs. H. Günthard, *J. Mol. Spectry.*, *21*, 165 (1966).
434. R. C. Leech, D. B. Powell, and N. Sheppard, *Spectrochim. Acta*, *22*, 1 (1966).
435. R. C. Leech, D. B. Powell, and N. Sheppard, *Spectrochim. Acta*, *22*, 1931 (1966).
436. Y. Hanyu, C. O. Britt, and J. E. Boggs, *J. Chem. Phys.*, *45*, 4725 (1966).
437. J. Goodisman, *J. Chem. Phys.*, *45*, 4689 (1966).
438. D. D. MacNichol, R. Wallace, and J. C. D. Brand, *Trans. Faraday Soc.*, *61*, 1 (1965).
439. F. A. L. Anet and M. Ahmad, *J. Am. Chem. Soc.*, *86*, 119 (1964).
440. J. F. Beecher, *J. Mol. Spectry.*, *21*, 414 (1966).
441. M. S. de Groot and J. Lamb, *Proc. Roy. Soc.* (*London*), *Ser A*, *242*, 36 (1957).
442. K. I. Dahlqvist and S. Forsén, *J. Phys. Chem.*, *69*, 4062 (1965).
443. A. S. Esbitt and E. B. Wilson, Jr., *Rev. Sci. Instr.*, *34*, 901 (1963).
444. T. N. Sarachman and D. R. Lide, Jr., private communication.
445. H. D. Rudolph and H. Seiler, *Z. Naturforsch.*, *20a*, 1682 (1965).
446. R. L. Kuczkowski and D. R. Lide, Jr., *J. Chem. Phys.*, *46*, 357 (1967).
447. M. M. Zaalberg, *Theoret. Chim. Acta*, *5*, 79 (1966); *6*, 362 (1966).
448. R. C. Woods, *J. Mol. Spectry.*, *21*, 4 (1966); *22*, 49 (1966).
449. D. Coffee, Jr. and J. E. Boggs, paper presented at Symposium on Molecular Structure and Spectroscopy, Columbus, Ohio, September 1966, paper V1.
450. K. R. Loos and H. H. Günthard, *J. Chem. Phys.*, *46*, 1201 (1967).

451. R. H. Miller, D. L. Bernitt, and I. C. Hisatsune, *Spectrochim. Acta, 23A*, 223 (1967).
452. K. D. Möller, A. R. DeMeo, D. R. Smith, and L. H. London, *J. Chem. Phys.*, *47*, 2609 (1967).
453. L. Pedersen and K. Morokuma, *J. Chem. Phys.*, *46*, 3941 (1967).
454. R. L. Vold and H. S. Gutowsky, *J. Chem. Phys.*, *47*, 2495 (1967).
455. F. Kaplan and G. K. Meloy, *J. Am. Chem. Soc.*, *88*, 950 (1966).
456. F. A. Miller, W. G. Fateley, and R. E. Witkowski, *Spectrochim. Acta, 23A*, 891 (1967).
457. A. B. Tipton, C. O. Britt, and J. E. Boggs, *J. Chem. Phys.*, *46*, 1606 (1967).
458. J. J. Rush, *J. Chem. Phys.*, *46*, 2285, (1967).
459. L. S. Bartell, B. L. Carroll, and J. P. Guillory, *Tetrahedron Letters, 13*, 705 (1964).
460. R. J. Jensen and G. C. Pimentel, *J. Phys. Chem.*, *71*, 1803 (1967).
461. W. E. Palke and R. M. Pitzer, *J. Chem. Phys.*, *46*, 3948 (1967).
462. J. E. Piercy and M. G. S. Rao, *J. Chem. Phys.*, *46*, 3951 (1967).
463. E. Saegebarth and E. B. Wilson, Jr., *J. Chem. Phys.*, *46*, 3088 (1967).
464. D. Den Engelsen, H. A. Dijkerman, and J. Kerssen, *Rec. Trav. Chim.*, *84*, 1357 (1965).
465. D. Den Engelsen, *J. Mol. Spectry.*, *22*, 426 (1967).
466. B. P. Van Eijck, *Rec. Trav. Chim.*, *85*, 1129 (1966).
467. J. R. Hoyland, *Chem. Phys. Letters, 1*, 247 (1967).
468. W. H. Fink and L. C. Allen, *J. Chem. Phys.*, *46*, 3270 (1967).
469. M. K. Kemp and W. H. Flygare, *J. Am. Chem. Soc.*, *89*, 3925 (1967).
470. H. D. Rudolph, H. Dreizler, A. Jaeschke, and P. Windling, *Z. Naturforsch.*, *22a*, 940 (1967).
471. R. C. Woods, *J. Chem. Phys.*, *46*, 4789 (1967).
472. C. R. Quade, *J. Chem. Phys.*, *47*, 1073 (1967).
473. R. M. Pitzer, *J. Chem. Phys.*, *47*, 965 (1967).
474. J. Goodisman, *J. Chem. Phys.*, *47*, 334 (1967).
475. A. C. Plaush and E. L. Pace, *J. Chem. Phys.*, *47*, 44 (1967).
476. E. L. Pace and A. C. Plaush, *J. Chem. Phys.*, *47*, 38 (1967).
477. M. R. Emptage, *J. Chem. Phys.*, *47*, 1293 (1967).
478. R. M. Lees and J. G. Baker, paper presented at Symposium on Molecular Structure and Spectroscopy, Columbus, Ohio, September, 1967, paper G3.
479. M. P. Melrose and R. G. Parr, *Theoret. Chim. Acta, 8*, 150 (1967).
480. R. G. Ford and R. A. Beaudet, paper presented at Symposium on Molecular Structure and Spectroscopy, Columbus, Ohio, September, 1967, paper T7.
481. R. G. Ford and R. A. Beaudet, paper presented at Symposium on Molecular Structure and Spectroscopy, Columbus, Ohio, September, 1967, paper T8.
482. G. Dellepiane and G. Zerbi, *J. Mol. Spectry.*, *24*, 62 (1967).
483. H. Dreizler, H. D. Rudolph, and D. Sutter, paper presented at Symposium on Molecular Structure and Spectroscopy, Columbus, Ohio, September, 1967, paper Ω6.
484. E. C. Thomas and V. W. Laurie, paper presented at Symposium on Molecular Structure and Spectroscopy, Columbus, Ohio, September, 1967, paper G1.

485. I. Y. M. Wang and J. E. Boggs, paper presented at Symposium on Molecular Structure and Spectroscopy, Columbus, Ohio, September, 1967, paper T6.
486. D. D. Wertz and J. R. Durig, paper presented at Symposium on Molecular Structure and Spectroscopy, Columbus, Ohio, September, 1967, paper X9.
487. R. E. Penn and R. F. Curl, Jr., paper presented at Symposium on Molecular Structure and Spectroscopy, Columbus, Ohio, September, 1967, paper T4.
488. M. D. Harmony and M. Sancho, *J. Chem. Phys.*, *47*, 1911 (1967).

Nucleophilic Substitution at Sulfur

E. Ciuffarin and A. Fava

Institute of General Chemistry, University of Pisa, Italy

CONTENTS

I. INTRODUCTION

Sulfur atoms bound to atoms or groups that can exist as anions are readily attacked by nucleophiles. As a consequence nucleophilic substitution is by far the most common type of reaction among sulfur compounds.

A number of recent reviews (1–3) have extensively surveyed the literature on sulfur reactions, some dealing also with their mechanistic aspect, notably in connection with scission of the sulfur–sulfur bond (2,4,5). The present review will consider nucleophilic substitutions at di-, tri-, and tetracoordinated sulfur specifically from a mechanistic point of view. Thus the literature will not be extensively covered, since the papers examined are mostly recent papers pertinent to the mechanism. Ours will probably be a biased view of the subject since, in many instances, we had the chance to present unpublished results from our laboratories, thus offering our own view besides the various authors' opinions and data.

II. DICOORDINATED SULFUR

Elemental sulfur reacts with cyanide to form the thiocyanate ion (6,7). The rate law for this reaction has been found to be

$$\text{Rate} = k[S_8][CN^-]$$

and the formation of the thiocyanate ion has been interpreted as a nucleophilic attack of cyanide ion on the sulfur ring. The open-chain product reacts very rapidly with excess cyanide until, step by step, only thiocyanate remains as a final product [eqs. (1) and (2)].

$$^-S-(S)_n-SCN + CN^- \longrightarrow {}^-S-(S)_{n-1}-SCN + SCN^- \tag{2}$$

Sequences like the above are very common. Destruction by the action of nucleophiles is, in fact, the most characteristic feature of polythiocompounds. Foss has interpreted the degradation of polythionates by sulfide or cyanide in the same terms (8) [eqs. (3) and (4)]:

$$^-O_3S-(S)_n-S_2O_3{}^- + nCN^- + H_2O \longrightarrow$$
$$nSCN^- + S_2O_3{}^{2-} + SO_4{}^{2-} + 2H^+ \tag{3}$$

$$^-O_3S-(S)_n-S_2O_3{}^- + nSO_3{}^{2-} + H_2O \longrightarrow$$
$$(n+1)S_2O_3{}^{2-} + SO_4{}^{2-} + 2H^+ \tag{4}$$

For example [eqs. (5)–(8)]:

$$\overset{1}{^-O_3S}-\overset{2}{S}-\overset{3}{S}-\overset{3}{S}-\overset{2}{S}-\overset{1}{S}O_3{}^- + CN^- \longrightarrow {}^-O_3\overset{1}{S}-\overset{2}{S}-\overset{3}{S}-\overset{3}{S}-CN + {}^-\overset{2}{S}-\overset{1}{S}O_3{}^-$$
$$\tag{5}$$

$$\overset{1}{^-O_3S}-\overset{2}{S}-\overset{3}{S}-\overset{3}{S}-CN + CN^- \longrightarrow {}^-O_3\overset{1}{S}-\overset{2}{S}-\overset{3}{S}-CN + \overset{3}{S}CN^- \tag{6}$$

$$\overset{1}{^-O_3S}-\overset{2}{S}-\overset{3}{S}-CN + CN^- \longrightarrow {}^-O_3\overset{1}{S}-\overset{2}{S}-CN + \overset{3}{S}CN^- \tag{7}$$

$$\overset{1}{^-O_3S}-\overset{2}{S}-CN \xrightarrow{H_2O} \overset{1}{S}O_4{}^{2-} + \overset{2}{S}CN^- \tag{8}$$

Foss's brilliant studies date back to 1945, and, in effect, he deserves the credit for recognizing the importance of nucleophilic substitution reactions in sulfur chemistry. He also correctly predicted the site of attack by nucleophiles on polythionates [eq. (5)] (see below). The site of reaction of polysulfides and polythionates has been recently examined by Davis (9). An empirical correlation has been given between the activation energy of a displacement reaction and the length of the sulfur–sulfur bond being broken in the reaction. Two types of bonds, —S—S— and —S—SO₃⁻, were considered in a number of reactions to establish the correlation which predicts that the longer bond will be more easily broken. The prediction has been verified by data presented in the same paper (9). However successful, the correlation is likely to be fortuitous. In fact, in the case of tetrathionate which contains both —S—S— and —S—SO$_3^-$ bonds, it is the —S—S— bond, *the shorter*, which is more easily broken. If a difference between the two types of bonds is causing the correlation to fail, there is no sound basis to use both types of bonds in establishing the correlation itself.

It seems likely that the leaving group is also important in determining which bond is broken. Thiosulfate, being a weaker base than sulfite, will be more easily displaced by a nucleophile. An analysis of the thiosulfate-catalyzed decomposition of tetra-, penta-, and hexathionate (10a) shows that it is the weaker base which is displaced faster even if that means breaking the shorter bond.

The mechanism proposed is S_N2 or a bimolecular displacement on sulfur. A detailed analysis has been carried out, among others, on the reactions of trithionate (10) and tetrathionate (11) with anions [eqs. (3) and (4)] and on the sulfite–thiosulfate (12) and sulfide–thiosulfate (13) exchange reactions [eqs. (9) and (10)].

$$SO_3{}^{2-} + {}^-S{-}\overset{*}{S}O_3{}^- \longrightarrow {}^-\overset{*}{S}O_3{-}S^- + SO_3{}^{2-} \qquad (9)$$

$$\overset{*}{H}S^- + {}^-S{-}SO_3{}^- \rightleftharpoons \overset{*}{H}SS^- + SO_3{}^{2-} \qquad (10)$$

The latter reaction is complicated by two other simultaneous equilibria [eqs. (11) and (12)].

$$HSS^- + H\overset{*}{S}{}^- \rightleftharpoons HS^- + H\overset{*}{S}S^- \qquad (11)$$

$$H\overset{*}{S}S^- + {}^-S{-}SO_3{}^- \rightleftharpoons HS\overset{*}{S}S^- + SO_3{}^{2-} \qquad (12)$$

All these inorganic species are the simplest homologs of organic thiosulfates, disulfides, and thiols. Reactions similar to those shown above can be studied with organic compounds in aqueous or nonaqueous media in order to investigate more thoroughly the mechanism and geometry of the transition state of a nucleophilic substitution on divalent sulfur. The following reactions have been studied (14–18):

$$RS{-}SO_3{}^- + \overset{*}{S}O_3{}^{2-} \longrightarrow RS{-}\overset{*}{S}O_3{}^- + SO_3{}^{2-} \qquad (13)$$

$$RS{-}\overset{*}{S}R + RS^- \longrightarrow RS{-}SR + R\overset{*}{S}{}^- \qquad (14)$$

$$RS{-}SCN + R'{-}NH_2 \longrightarrow RS{-}NH{-}R' + H^+SCN^- \qquad (15)$$
$$\Big\downarrow R'{-}NH_2$$
$$R'NH_3{}^+SCN^-$$

[Equation (15) is similar to the reaction of $S(SCN)_2$ (19) and $S_2(SCN)_2$ (20) with amines.] The results are given in Table I.

TABLE I

Comparison of the Rates of S_N2 Reactions at Sulfur and Carbon

	Relative rates			
R	RS—SO$_3$$^-$ + SO$_3$$^{2-}$ [a]	RS—SR + RS$^-$ [b]	RS—SCN + Bu—NH$_2$ [c]	RCH$_2$—X + Y$^-$ [d]
Me	100 [e]			100 [d]
Et	50			40
Bu		40 [f]		
i-Pr	0.7		1.0 [g]	3.0
t-Bu	0.0006	0.00015	0.000125	0.0011

   [a] Water, 25°C, pH 7.9.
   [b] Methanol, 25°C.
   [c] Toluene, 20°C.
   [d] Various leaving groups and nucleophiles; average values from reference 21.
   [e] $k = 2.2 \times 10^{-1}$ liters mole^{-1} sec^{-1}.
   [f] Relative to the values in the first column.
   [g] $k = 14.0$ liters mole^{-1} sec^{-1}.

The rate profile for the reactions shown in Table I as R is varied is typical of a steric effect; moreover, a comparison of these results with those obtained for similar displacement on carbon (see Table I) shows that steric requirements are extremely similar. It is well known that S_N2 substitutions at carbon involve backside attack and a linear transition state. Therefore it could be inferred that for S_N2 reactions at

divalent sulfur the transition state is also linear. However, if a linear transition state for nucleophilic displacement on carbon can be justified on a quantum-mechanical basis, since carbon has all its available orbitals filled and an entering and leaving group must therefore occupy the same p orbital, sulfur has vacant d orbitals of not much higher energy which could be used in the transition state. A transition state with d-orbital participation need not be linear.

It is worth noting at this point that the rate profile is similar for two types of reactions with completely different charge distributions in the transition state [eqs. (13) and (15)]. In the transition state for the exchange reaction between a thiosulfate monoester and sulfite, the leaving and entering groups bear negative charges. Thus, they repel

$$
\begin{array}{c}
\text{R} \\
| \\
\text{SO}_3\text{------S------SO}_3 \\
-(1+\delta) \quad (2\delta-1) \quad -(1+\delta) \\
\textbf{(1)}
\end{array}
$$

each other, and, simply on electrostatic grounds, tend to stay as far apart as possible. On the other hand, neutral reagents give rise to charged products in the reaction between sulfenyl thiocyanate and amines. Consequently, entering and leaving groups are likely to have opposite charges in the transition state, and would thus tend to stay as close together as possible. If d orbitals were easily available to form sp^3d hybrids, the reaction angle might be expected to be smaller than

$$
\begin{array}{c}
\text{R} \\
| \\
\text{Bu—NH}_2\text{------S------SCN} \\
(\delta+) \qquad\quad (\delta-)
\end{array}
$$

180°. However, if such were the case, the steric effect should be much smaller than in the s-alkylthiosulfate–sulfite reaction. As we have seen, the rate profile is so similar in the two reactions that we could draw the conclusion that either there is no d-orbital participation, or it is of such character as to be without effect on the geometry of the transition state.

The preceding results are quite interesting if one compares them with those obtained for another second row element, silicon, in nucleophilic substitutions. For example, 1-silabicyclo-[2,2,2]-octyl chloride has been found to react with hydride ion at the same rate as the open-chain compound (22). Since backside attack is impossible in the bridge-head compound, this finding requires the reaction angle to be < 180°, which has been accounted for in terms of a sp^3d pentacovalent transition state. On the other hand, nucleophilic displacement at carbon can occur only from the backside and the bridgehead compound is highly unreactive (23,81).

Another study has revealed that the electronic requirements of the reactive site for substitution at sulfur are very small. For illustration we may consider the rates of exchange of two series of compounds in the following reaction (14):

$$
\text{RS—SO}_3^- + \overset{*}{\text{S}}\text{O}_3{}^{2-} \longrightarrow \text{RS—}\overset{*}{\text{S}}\text{O}_3^- + \text{SO}_3{}^{2-} \tag{16}
$$
$$
\text{R} = p\text{-X—C}_6\text{H}_4\text{—;} \quad p\text{-X—C}_6\text{H}_4\text{—CH}_2\text{—}
$$

A summary of the relative rates is given in Table II. The Hammett's ρ is $+0.58$ for the benzyl and $+0.85$ for the phenyl series. Thus the electronic effects are small and the sign of ρ suggests that the reaction center in the transition state is only slightly more negative than the initial state. The effect of *para* substitution has been recently examined

TABLE II

Relative Rates of Exchange (in water, 20°C, pH = 7.87, μ = 0.16)

$$RS\!-\!SO_3^- + \overset{*}{S}O_3{}^{2-} \longrightarrow RS\!-\!\overset{*}{S}O_3^- + SO_3{}^{2-}$$

X	R = p-X—C$_6$H$_4$—CH$_2$—	R = p-X—C$_6$H$_4$—
CH$_3$O	0.92	0.68
CH$_3$	0.94	0.76
H	1.00	1.00
F		1.55
Cl	1.67	1.70
CN	2.81	
NO$_2$	3.40	5.22

(24) for the chloride ion-catalyzed hydrolysis of phenyl sulfenyl chlorides. A value of $\rho = +1.04$ has been found. However the mechanism is not a simple nucleophilic displacement since the hydrolysis has been found to be a third-order reaction [eq. (17)].

$$\text{Rate} = k[\text{RSCl}][\text{H}_2\text{O}][\text{Cl}^-] \qquad (17)$$

$$\text{RSCl} + \text{H}_2\text{O} \xrightarrow{\text{Cl}^-} \text{RSOH} + \text{HCl}$$

$$\text{RSOH} + \text{RSCl} \xrightarrow{\text{fast}} \text{RSOSR} + \text{HCl}$$

In a transition state formed with full d-orbital participation (**2**)

$$^{(-)}SO_3\!-\!\!\overset{(-)}{\underset{\underset{\text{R}}{\mid}}{S}}\!\!-\!SO_3{}^{(-)}$$

(2)

the central sulfur should be considerably more negative than in the initial state. In a linear transition state formed without d-orbital participation and with synchronous bond forming and breaking (**1**), the increase in charge on the central sulfur in going from the initial state to the activated complex will be small, its sign being determined by the extent of bond forming and breaking in the transition state. Thus the electronic effects are at least consistent with the view that there is little, if any, d-orbital participation. From another point of view, one could argue that the small value of ρ is a consequence of the fact that the negative charge of the reaction center does not need to be dispersed by electron-withdrawing groups in the organic moiety, since it can be very easily accommodated by a low-lying d orbital. It is interesting to note, however, in this connection, that for nucleophilic substitution at saturated carbon the values of ρ for the corresponding phenylethyl and benzyl series are $+0.59$ and $+0.785$, respectively* (25–27). On the other hand, where the seat of substitution is silicon and d-orbital participation in nucleophilic substitution is clearly evident, the values range from $\rho = +2.70$ (alkaline methanolysis of $R_3SiOMen\dagger$) to $\rho = +4.30$ (alkaline ethanolysis of $R_3SiH\dagger$) (28).

The analogy based on the similarity in behavior of sulfur and saturated carbon in the substitution reaction fails in one respect. Nucleophilic substitutions occurring at carbon via an S_N1 mechanism are very common, while for sulfur there is as yet *no unambiguous example of a reaction proceeding through an electron-deficient intermediate* with the following mechanism:

$$RS\text{—}X \rightleftarrows RS^+ + X\text{:}^- \tag{18}$$

$$RS^+ + Y\text{:}^- \longrightarrow RS\text{—}Y \tag{19}$$

Kharash considers the formation of sulfenium ions well established (3) not only for Friedel-Crafts reactions, i.e., reactions in the presence of a very powerful electrophilic reagent like $AlCl_3$, but also for simple nucleophilic substitutions like those we are considering, even though he himself suggests that detailed studies are required to clarify the precise mechanism. In actual fact all nucleophilic substitutions at divalent sulfur for which kinetic studies exist can be accurately described by the following expression,

$$\text{Rate} = [RS\text{—}X][Y\text{:}]$$

where Y: is a common nucleophile. Moreover, every kinetic feature such as salt and solvent effects, agrees with a mechanism whose slow step is a direct nucleophilic displacement. Thus the bimolecular mechanism prevails exclusively in the simple substitution reaction occurring in the absence of powerful electrophilic agents.

---

* However the coincidence is likely to be fortuitous since the solvent was water for substitutions at sulfur and acetone for substitutions at carbon.

† R_3 = α-naphthylphenylmethyl.

This situation is not unique for sulfur, but is common to other second or higher row elements such as silicon,* germanium, and phosphorus. A much faster bimolecular mechanism or a much slower monomolecular mechanism or both can explain the inherent difficulty of detecting an S_N1 mechanism for second row elements. It is well known that S_N2 reactions occur with much greater ease for second than for first row elements due to their greater radius and thus greater ability to disperse a negative charge and their polarizability. While d-orbital participation would also lower the energy of the transition state for second row elements thus facilitating the reaction, for sulfur it does not seem necessary to invoke this explanation. As stated earlier d-orbital participation is well established for silicon, but this is not the case for sulfur where all the available evidence can be accounted for without resorting to d-orbital participation. The reluctance of second row elements to react by the S_N1 mechanism could be due to a less favorable formation of electron-deficient species. Conjugation between the reaction center and the alkyl or aryl moiety may be less likely to occur due to the difference of atomic dimensions between first and second row elements with consequent smaller orbital overlap. For sulfur this explanation has been brought forth in support of many experimental facts. However, the matter is far from being settled. For example it is known that the experimental resonance energy obtained from heats of combustion is much greater for thiophene (3) than for furane (4) (30), suggesting that $2p$–$3p$ is better than $2p$–$2p$ overlap. This view is supported by recent MO calculations relative to the same compounds. The overlap integrals are as follows (31):

Bond	Distance, Å	Overlap	Overlap integral
C—O	1.40	$2p$–$2p$	0.150
C—S	1.74	$2p$–$3p$	0.174

(3)   (4)

It is interesting to remark that the geometry of the transition state for substitution at divalent sulfur corresponds to a trigonal bipyramid

* Note, however, that Sommer has recently found evidence for the formation of an ionic intermediate during the racemization of optically active α-naphthyl-phenylmethyl-chlorosilane (29).

with axial entering and leaving groups, while the equatorial positions are taken by the organic moiety and the lone electron pairs (**5**).

$$X—\overset{\displaystyle ..}{\underset{\displaystyle R}{S}}—Y$$

(5)

There are many examples of stable molecules isoelectronic with **5** which conform to this geometry, for example, ClF_3, SF_4, PCl_5, PF_5, and PCl_2F_3. The closest analog is chlorine trifluoride which has, in fact, a flat, T-shaped structure, with the Cl—F axial bonds longer and,

$$F——Cl——F$$
$$\underset{\displaystyle F}{|}$$

(6)

therefore, weaker than the equatorial one.* The bonding in these molecules can be described in terms of three sp^2 (equatorial) and two pd (axial) bonds. Alternatively, an "orbital-deficient" model can be used, omitting the d orbital and assuming that four orbitals of the central atom suffice, three of them forming sp^2 equatorial hybrids while the fourth orbital extends over both axial groups in a three-center bond (32). In either picture the equatorial orbitals are the more electronegative (having greater s character) and will prefer the more electropositive substituents. Thus it is not surprising that the transition state for substitution at sulfur has the geometry shown in **5**, since this fills the requirements of putting the more electropositive ligands in equatorial positions. The two lone pairs are obviously ligands of extreme electropositive character and the organic moiety is probably also more electropositive than either leaving or entering group.

Since steric factors, even though quite large (Table I), are unable to change the geometry of the transition state, it seems likely that in the case of the divalent sulfur, d-orbital participation in the equatorial bonds is minor.† Further studies, however, are desirable to more firmly assess this point.

It seems useful to include in this section on divalent sulfur a discussion of a few reactions which have been described in the literature

---

* In every known example the axial bonds are found to be longer than equivalent equatorial ones.

† For a recent discussion on the bonding in sulfur compounds see reference 33.

as electrophilic substitutions at sulfur. An electrophilic attack has been proposed for the acid hydrolysis of disulfides (34) and for the acid-catalyzed interaction of disulfides with olefins (35). In 1958 an article by Benesch and Benesch appeared proposing a chain mechanism for the reaction of cystine (RS—SR) and bis-(2,4-dinitrophenyl)-cystine (R'S—SR') (36).

Initiation:

$$RS—SR + H^+ \rightleftharpoons RS^+ + RSH \tag{20}$$

Propagation:

$$RS^+ + R'S—SR' \rightleftharpoons RS—SR' + R'S^+ \tag{21}$$

With this evidence on hand, Parker and Kharash in 1959 suggested that many reactions of disulfide could be interpreted as electrophilic substitutions (4). However more recent data (37), although incomplete, suggest the mechanism is not that proposed by Benesch and Benesch, or at least that electrophilic substitutions are not as general as proposed by Parker and Kharash.

The isotopic exchange between thiophenol and diphenyl disulfide

$$RS—SR + R\overset{*}{S}H \longrightarrow RS—\overset{*}{S}R + RSH \tag{22}$$

is not catalyzed by any acid but specifically by HCl, HBr, and HI, and the relative effects are 1, 100, and 10,000, respectively. Perchloric acid does not have any effect, its greater acid strength notwithstanding. Moreover, the reaction is second order in acid. Since the solvent used was acetic acid, the hydrogen halides are not dissociated, being mostly in the form of ion pairs. Therefore, the second order in acid means that both proton and anion are present in the transition state. Parallelism between the catalytic effect of the hydrogen halides and the nucleophilicity of the corresponding anions is further evidence that the transition state contains the nucleophile. Thus, these so called "electrophilic" reactions are probably nucleophilic displacements on the conjugate acid of disulfides [eq. (23)].

$$RS\overset{+}{—}SR + X^- \xrightarrow{\text{slow}} RSX + SR \tag{23}$$
$$\underset{H}{|} \qquad\qquad\qquad \underset{H}{|}$$

This mechanism, of course, cannot be kinetically distinguished from a four-center mechanism, as represented by structure 7.

$$\begin{array}{cc} RS\text{———}SR \\ \vdots \quad\quad \vdots \\ H\text{———}X \end{array}$$

(7)

Clearly these data exclude the mechanism proposed by Benesch and Benesch, but they do not provide a conclusive basis for formulating the correct mechanism. For instance in the proposed nucleophilic displacement the order in thiophenol should be zero. Instead the reaction is actually slowed down by addition of thiol, the order in thiol being between -0.2 and -0.4, depending on the disulfide and the thiol used.

The viewpoint that an alleged electrophilic substitution may actually be a nucleophilic substitution assisted by electrophiles is corroborated by recent reports (38,39). Kice and Guaraldi have discussed the scission of various types of S—S bonds in terms of what they-called a "concomitant electrophilic and nucleophilic catalysis" [eq. (24)]. The electrophilic agent, E, can be $ArSO^+$, ArS^+, a carbonium ion, H^+, or any other electrophilic species able to activate the S—S bond in the sense that it transforms a poor leaving group, RS—, into a

$$RS\text{—}SR + E^+ \rightleftharpoons R\overset{+}{\underset{|}{S}}\text{—}SR \xrightarrow{N} products \qquad (24)$$
$$\text{E}$$

good one, $R\overset{+}{S}(E)$—, capable of being expelled in a subsequent nucleophilic attack. They found that this type of catalysis is occurring not only for substitutions at divalent sulfur (39), but also for reactions at the sulfinyl sulfur (SO—S) bonds and suggest the phenomenon is quite general (38).

III. TRICOORDINATED SULFUR

The more interesting compounds in this group are the sulfoxides, R_1R_2SO; sulfites, $R_1O\text{—}SO\text{—}OR_2$; sulfinates, $R_1\text{—}SO\text{—}OR_2$; and sulfonium salts, $R_1R_2R_3S^+X^-$. Their structure may be generally represented as shallow pyramids with sulfur as the apex. Angles between the bonds can vary between 93 and 114° (40). Inversion of the pyramid occurs only at high temperatures. Indeed only very recently an example

$$\overset{a}{\underset{b}{\diagdown}}\overset{}{\underset{c}{S}}\text{—}: \rightleftharpoons :\text{—}\overset{a}{\underset{c}{S}}\overset{}{\underset{b}{\diagup}} \qquad (24)$$

has been reported of a facile inversion of the pyramid: the racemization of *tert*-butylethylmethylsulfonium perchlorate at 50°C (41).

Studies of the substitution reactions of tricoordinated sulfur compounds have been made much easier because of their susceptibility to the classical methods of stereochemistry. When the sulfur atom is bonded to three different groups the molecule is asymmetric and can be resolved in its enantiomeric forms which at room temperature are usually stable.

The first stereochemical study of a nucleophilic substitution at tricoordinated sulfur dates back to 1925 (42). Phillips showed that reaction of $(-)$ethyl p-toluensulfinate with n-butyl alcohol yields $(+)n$-butyl p-toluensulfinate of opposite configuration. Thus, trans-esterification of sulfinic esters occurs with a Walden-like inversion of configuration, suggesting a direct nucleophilic displacement (S_N2).

$$(-)p\text{-}CH_3C_6H_4\text{—}\underset{\underset{O}{\|}}{S}\text{—O—Et} + n\text{-BuOH} \longrightarrow (+)p\text{-}CH_3C_6H_4\text{—}\underset{\underset{O}{\|}}{S}\text{—O—Bu-}n \quad (25)$$

It was only in 1958, however, that the problem of nucleophilic substitution at tricoordinated sulfur has been revived by a study of Bunton, Tillet, and de la Mare on the hydrolysis of organic sulfites (43). From the kinetic data and experiments performed in water enriched in ^{18}O it was concluded that the reaction [eq. (26)] is a nucleophilic dis-

$$
\begin{array}{c}
\underset{\underset{CH_2\text{—O}}{|}}{CH_2\text{—O}}\!\!\diagdown\!\!S{=}O + {}^{18}OH^- \xrightarrow{\text{slow}} \underset{\underset{CH_2\text{—O}}{|}}{CH_2\text{—OH}}\!\!\diagdown\!\!SO_2^- \\[6pt]
\xrightarrow[\text{fast}]{{}^{18}OH^-} \underset{\underset{CH_2\text{—OH}}{|}}{CH_2\text{—OH}} + S^{18}O_3^{2-} \quad (26)
\end{array}
$$

placement by hydroxyl ions on sulfur. The authors preferred a direct nucleophilic displacement to a mechanism involving the formation of intermediates similar to those formed in the hydrolysis of carboxylic

$$
\begin{array}{ccc}
\underset{CH_2O}{\overset{CH_2O}{|}}\!\!\diagdown\!\!S{=}O + {}^{18}OH^- & & \underset{CH_2O}{\overset{CH_2O}{|}}\!\!\diagdown\!\!S{=}^{18}O + OH^- \\
\diagdown\diagdown & & \diagup\diagup \\
\underset{CH_2O}{\overset{CH_2O}{|}}\!\!\diagdown\!\!\underset{{}^{18}OH}{\overset{O^-}{S}} & \rightleftharpoons & \underset{CH_2O}{\overset{CH_2O}{|}}\!\!\diagdown\!\!\underset{{}^{18}O^-}{\overset{OH}{S}} \qquad (27) \\
\diagdown & & \diagup \\
& \underset{CH_2O}{\overset{CH_2OH}{|}}\!\!\diagdown\!\!SO_2^- \longrightarrow \text{products} &
\end{array}
$$

esters. They came to this conclusion despite the small enrichment (0.016%) found in the unreacted ester recovered after 50% hydrolysis from the reaction in $H_2{}^{18}O$, which they considered to be of no significance. However, Davis (44) considers the enrichment as indicative of a multistage nucleophilic reaction as shown in eq. (27).

The stereochemical course of nucleophilic substitutions at tricoordinated sulfur has been studied in detail very recently (45) in the case of the solvolysis of 5-acetoxy-9-oxa-1-thionabicyclo[3,3,1]nonane perchlorate [eq. (28)]. When the solvent was enriched in ^{18}O, the product, 1-thia-cyclooctane-5-one-1-oxide (9), contained two types of oxygen. The oxygen in position 1 had the same degree of enrichment as the solvent, while the oxygen in position 5 was identical in isotopic composition with the starting material. These data suggest for sulfonium salts an inversion mechanism at the tricoordinated sulfur [eq. (28)].

(8)   (9)

This mechanism has been further proven by Johnson (46) by means of the following sequence of stereochemical changes. Alkylation of cis-4-(4-chlorophenyl)-thian-1-oxide (10) with triethyloxonium fluoborate yielded the cis sulfonium salt (11). Alkaline hydrolysis of 11 yielded the trans isomer of the starting material. Alkylation of 13 and subsequent hydrolysis gave the starting material. This sequence established without any doubt that nucleophilic attack by hydroxyl ions on the sulfur atom of alkoxysulfonium salts occurs with complete inversion.

Stereospecific inversion at sulfur has been found by Mislow et al. also for other types of tricoordinated sulfur compounds, namely, sulfinic esters and sulfoxides. Pure levorotatory menthyl-*p*-toluensulfinate reacts with ethyl-magnesium bromide to give pure dextrorotatory ethyl-*p*-toluensulfoxide (47).

This type of reaction has been found to be 100% stereospecific (47) and occurs with inversion of configuration as demonstrated by optical rotatory dispersion studies and reaction sequences (47,48).

$$
\underset{p\text{-}CH_3C_6H_4}{\overset{O}{\underset{}{\cdots S}}} OC_{10}H_{19} \xrightarrow{\text{Et-Mg-Br}} \underset{C_6H_4CH_3\text{-}p}{\overset{O}{\underset{}{C_2H_5 - S \cdots}}} \qquad (30)
$$

By using this type of reaction Mislow and co-workers have prepared a series of optically pure *p*-tolyl alkylsulfoxides. They then studied absolute configurations (47) and the circumstances that produce racemization (49) in the various members of the series. It was found that, while the uncatalyzed racemization reaction requires temperatures of about 180° both for neat reagents and in solution (50–52), fast racemization obtains at room temperature in the presence of hydro-

$$
(+)Ar\text{---}SO\text{---}R \xrightarrow{HCl_{(aq)}} (\pm)Ar\text{---}SO\text{---}R \qquad (31)
$$

chloric acid (49). The specificity of hydrochloric acid is remarkable. In analogous conditions in the presence of hydrofluoric acid, after 11 days no racemization could be detected. Apparently only an acid whose conjugate base is a good nucleophile is able to induce racemization of sulfoxides.

The inertness of hydrofluoric acid seems to be inconsistent with recent findings by Kice and Guaraldi (53) indicating the nucleophilicity of fluoride to be comparable to that of chloride ion ($Cl^-/F^- = 3$) versus sulfynil sulfur. However, the two observations may not be directly comparable since the reaction investigated by Kice, the nucleophile-catalyzed hydrolysis of sulfinyl sulfones [eq. (32)], involves

$$
ArSO\text{---}SO_2Ar + Nu^- \longrightarrow ArSO\text{---}Nu + ArSO_2^-
$$
$$
\downarrow H_2O \qquad\qquad (32)
$$
$$
ArSO_2H
$$

attack on the sulfinyl function rather than on its conjugate acid, as the racemization [eq. (34)] likely does (see below).

As for hydrobromic and hydroiodic acid, while they could be expected to be more effective in promoting racemization, their catalytic power cannot be tested since their interaction with sulfoxides leads essentially to sulfide as the reaction product (see below).

Mislow suggested that the mechanism of the HCl-catalyzed racemization involves the formation of a dichloride intermediate (R_2SCl_2) through the equilibrium given by eq. (33), since such com-

$$R_2SO + 2HCl \overset{\longrightarrow}{\longleftarrow} R_2SCl_2 + H_2O \tag{33}$$

pounds and equilibria had been previously described and postulated (54–56). In its support he found that the rate of exchange of ^{18}O with the solvent (2:1 v/v mixture dioxane and $12M$ HCl in $H_2{}^{18}O$) equals the rate of racemization. The intermediate could be formulated either as **14**, a species with a plane of symmetry, or as the ion pair **15**. In the latter case, rapid exchange of the chlorines via a transition state similar to **14** would produce racemization.

(14)   (15)

It is interesting to note that the steric effects on the racemization reaction exactly parallel those found for nucleophilic substitution at dicoordinated sulfur (Table III) suggesting that the slow step for

TABLE III

Relative Rates of Nucleophilic Substitutions on Divalent Sulfur

$$R\text{—}S\text{—}SO_3{}^- + \overset{*}{S}O_3{}^{2-} \longrightarrow R\text{—}S\text{—}\overset{*}{S}O_3 + SO_3{}^{2-}$$

and of the Racemization of Alkyl-p-Tolylsulfoxides Catalyzed by HCl

$$d\text{-Ar}\underset{\underset{O}{\|}}{\text{—S—R}} \longrightarrow dl\text{-Ar}\underset{\underset{O}{\|}}{\text{—S—R}}$$

R	Me	Et	i-Pr	t-Bu
Substitution	100[a]	50	0.7	0.0006
Racemization	100[b]	26	0.7	0.0003

[a] $k = 2.2 \times 10^{-1}$ liter mole^{-1} sec^{-1} (25°C).
[b] $k = 3.7 \times 10^{-2}$ sec^{-1} (25°C).

racemization involves a nucleophilic displacement with a linear transition state. In view of the specificity of hydrochloric acid it seems a reasonable hypothesis to formulate the slow step as in eq. (34) where **16** could either be an intermediate or a transition state leading to the

$$
\underset{\underset{\text{Ar}}{\overset{\text{OH}}{\underset{|}{R}}}{\overset{|}{\text{S}^{+}}}:\ +\ \text{Cl}^{-} \xrightarrow{\text{slow}} \underset{\underset{\text{Ar}\quad\text{Cl}}{}}{\overset{R\ \ \text{OH}}{\text{S}—:}} \tag{34}
$$

(16)

chlorosulfonium ion (**15**).* Whatever the detailed mechanism, the racemization reaction can hardly be formulated without resorting to a dichloride species, either **14** or **15**. However, this seems to conflict with some recent findings on the chlorinolysis of sulfides (59) which demand the formation of a dichloride species and its ionization product, both of which undergo very fast carbon–sulfur cleavage:

$$
\text{Ar—S—R} + \text{Cl}_2 \rightleftarrows \underset{\underset{\text{R—S cleavage}}{}}{\text{Ar—S}\overset{R}{\underset{\text{Cl}}{}}\text{Cl}} \rightleftarrows \underset{\underset{\text{R—S cleavage}}{}}{[\text{Ar—S}\overset{R}{}\text{Cl}]^{+}\ \text{Cl}^{-}}
$$

products

Since the racemization of sulfoxides occurs without any cleavage, it is clear that the same intermediate cannot be invoked to account for both sets of experiments. It may be that structurally different dichloride

---

* The racemization of sulfoxides is not the only example of hydrochloric acid-catalyzed racemization for tricoordinate sulfur. In 1959 Herbandson and Dickerson found that menthyl arenesulfinates, ArSO_2Men, in nitrobenzene epimerize in the presence of HCl and chloride ions as cocatalysts (57). The mechanism proposed in this case was a displacement by Cl^- aided by HCl as shown in eq. (35).

$$
R—\!\!\left\langle\bigcirc\right\rangle\!\!—\underset{\underset{\text{Cl}^{-\frac{1}{2}}}{}}{\overset{\overset{\text{O}\ \ \text{H}\cdots\text{Cl}^{-\frac{1}{2}}}{\|}}{\text{S}}}\cdots\text{O—Men} \rightleftarrows R—\!\!\left\langle\bigcirc\right\rangle\!\!—\underset{\underset{\text{Cl}}{}}{\overset{\overset{\text{O}}{\|}}{\text{S}}}—:\ +\ \text{HO—Men} \tag{35}
$$

$\updownarrow$ fast exchange with Cl^- and epimerization

In sharp contrast, however, in the same medium the HCl-catalyzed racemization of sulfoxides [eq. (31)] is markedly depressed in the presence of ionic chloride (58). Evidently the two reactions occur by different mechanisms.

species are formed in the two cases, which are characterized by different reactivities. Obviously more data are necessary to obtain a clearer picture.

The formation of an intermediate similar to **16** can be invoked also in the reduction of sulfoxides by iodide ion (60,61). The reaction has been found to be first order in H^+ and first order in iodide thus suggesting it involves a nucleophilic attack by iodide on the conjugate acid of the sulfoxide followed by a fast formation of products [eq. (36)].*

$$I^- + Ar_2S^+\!\!-\!OH \rightleftarrows Ar_2S\!\!\begin{array}{c} \diagup I \\ \diagdown OH \end{array} \qquad (36)$$

$$(17)$$

$$17 + I^- \xrightarrow[\text{fast}]{H^+} Ar_2S + I_2 + H_2O$$

This scheme is in agreement with the finding that in the reaction of optically active p-tolyl benzylsulfoxide with HI, the rate of racemization equals the rate of reduction (62).

In spite of the fact that many questions appear to remain open, it is likely that nucleophilic substitution at tricoordinated sulfur occur via a bimolecular displacement and that entering and leaving group and reaction center are collinear.

The latter notion may not be general, however. The reaction of p-tolyl sulfoxide with N-sulfinylsulfonamide (TsNSO) to give a sulfilimine [eq. (37)] has been described by Day and Cram as a nucleophilic substitution at sulfinyl sulfur occurring via a cyclic intermediate or transition state (**18**) involving two molecules of TsNSO (63). If this

$$CH_3\!\!-\!\!\underset{\displaystyle}{\bigcirc}\!\!-\!\!\overset{\displaystyle O}{\underset{\displaystyle \|}{S}}\!\!-\!CH_3 \xrightarrow[-SO_2]{p\text{-}CH_3C_6H_4SO_2NSO} CH_3\!\!-\!\!\underset{\displaystyle}{\bigcirc}\!\!-\!\!\overset{\displaystyle NSO_2C_6H_4CH_3\text{-}p}{\underset{\displaystyle \|}{S}}\!\!-\!CH_3 \qquad (37)$$

were the case, entering and leaving groups could not occupy the axial positions of a trigonal bipyramid. To accommodate the exatomic ring, entering and leaving groups could both be in equatorial positions or one could occupy an axial and the other an equatorial position. Because of the similarity of entering and leaving groups it has been

---

* This reduction has been performed in acetic acid. However, reduction of dimethyl sulfoxide by iodide in water (62b) displays second order in H^+ suggesting

$$+OH_2$$

that the substrate is doubly protonated ($CH_3\!\!-\!\!\overset{|}{S}{}^+\!\!-\!CH_3$). A similar intermediate could perhaps be invoked also in the acid catalyzed racemization of sulfoxides (see above).

suggested (63) that they both occupy equatorial positions. However the evidence presented so far is not conclusive, and a more detailed investigation is desirable.

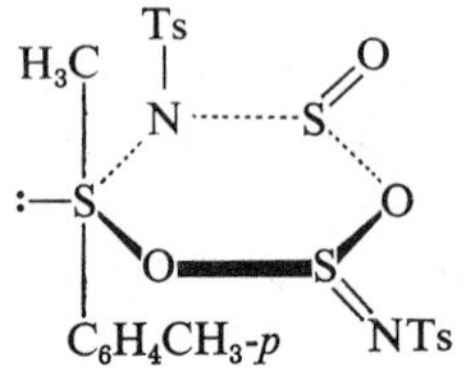

(18)

IV. TETRACOORDINATED SULFUR

The amount of data concerning nucleophilic substitution at tetra-coordinated sulfur is very limited. In this category hydrolysis of mono- and diesters of sulfuric acid is perhaps the most thoroughly studied reaction.

Hydrolysis of dialkyl esters occurs exclusively with C—O fission except for the reaction of ethylene sulfate and dimethyl sulfate with alkali (64); however, the extent of S—O fission is very limited, amount-ing to 14% and less than 1%, respectively. The fraction occurring with S—O fission is probably a bimolecular displacement on sulfur.

Hydrolysis of alkyl monoesters occurs at carbon in basic solution, but at sulfur in acid. Kaiser, Panar, and Westheimer proposed a uni-molecular mechanism for the acid-catalyzed hydrolysis (64). The operation of this mechanism has not yet been proved. However, very recently Batts (65) submitted strong evidence in its favor [eq. (38)]. The formation of a zwitterion **(19)** is supported by the large increase in

$$RO-SO_3^- \; \rightleftarrows \; R\overset{+}{\underset{\underset{H}{|}}{O}}-SO_3^- \xrightarrow{\text{slow}} ROH + SO_3 \qquad (38)$$

(19)

rate (a factor of 10^7) when the solvent is changed from pure water to aqueous dioxane (2% water). The large entropy of activation in moist dioxane suggests the formation of uncharged and slightly solvated species from a charged and highly solvated intermediate.

In recent papers Kice (66) and Benkovich (67) have arrived at the same mechanism for the acid-catalyzed hydrolysis of aryl sulfate mono-esters which occurs about 500 times as fast as for alkyl monosulfates (68). However, the preliminary formation of a zwitterion is not a necessary condition for the hydrolysis to occur. When the alkoxide moiety is a particularly good leaving group, the A-1 mechanism may operate in a "spontaneous" hydrolysis in neutral medium, as in the case of p-nitrophenyl sulfate (67).

$$NO_2-\langle\bigcirc\rangle-O-SO_3^- \longrightarrow NO_2-\langle\bigcirc\rangle-O^- + SO_3 \qquad (39)$$

A mechanism similar to eq. (39) with elimination of SO_2 has been suggested in an equation for the acid-catalyzed hydrolysis of monoesters of sulfurous acid (68). These are the only examples of nucleophilic substitutions at sulfur which are likely to take place via a unimolecular mechanism. However, even in these cases the reaction does not pass through an electron-deficient intermediate.*

The alkaline hydrolysis of sulfonates has also been studied in some detail. The reaction has been found to be a nucleophilic displacement on sulfur by OH^-. The mechanism has been formulated as a synchronous one-step process (20) by Christman and Oae with the aid of ^{18}O studies (70). The starting material recovered after 50% reaction did not show any enrichment suggesting that there is no intermediate as proposed by Bunton (71), or that it decomposes to give only products.

$$\begin{array}{c} Ar \\ | \\ HO\cdots\cdots S\cdots\cdots OR \\ O^{\diagup} \quad {}^{\diagdown}O \\ (20) \end{array}$$

The most interesting results pertaining to nucleophilic substitutions at tetracoordinated sulfur are probably those concerning the stereochemistry of these reactions. A nucleophile can attack a sulfonate ester both at carbon and sulfur, the amount of C—O and S—O scission depending on the ester and the nucleophile used. For example, 4-nitrophenyl 4-methylphenylsulfonate and 4-nitrophenyl 2,4,6-trimethylphenylsulfonate react with piperidine with 90 and 58% S—O cleavage,

$$Ar-SO_2-OAr' + Y^- \begin{array}{c} \nearrow Ar-SO_2-Y + Ar'O^- \\ \\ \searrow Ar-SO_3^- + Ar'Y \end{array} \qquad (40)$$

* Actually there is one example where an electron-deficient intermediate has been suggested in an A-1 reaction. This is the case of the acid-catalyzed hydrolysis of p-anysilsulfinyl sulfone, $CH_3C_6H_4SOSO_2C_6H_4OCH_3$, for which Kice and Guaraldi have postulated the formation of $CH_3OC_6H_4SO^+$ as an intermediate (69).

respectively [eq. (40)], the rate of attack on sulfur being about the same in the two cases (72). The obvious conclusion can be drawn that the steric hindrance offered by the two *ortho* methyl groups is negligible. The importance of this finding can be perceived if one compares the

$$X = H, CH_3 \tag{41}$$

behavior of analogous carbon compounds where two *ortho* methyl groups are known to afford enormous steric hindrance to nucleophilic attack (73).

TABLE IV
Effect of *ortho*-Methyl Groups on Nucleophilic Substitutions at Sulfur

(1) $Ar-S-Cl \xrightarrow{R_2NH^a} Ar-S-NR_2$

(2) $(+)Ar-SO-R \xrightarrow{HCl} (\pm)Ar-SO-R^g$

(3) $Ar-SO_2-OAr \xrightarrow{R_2NH^b} Ar-SO_2-NR_2$

(4) $Ar-SO-Cl \xrightarrow{R_2NH^a} Ar-SO-NR_2$

(5) $Ar-S-Cl \xrightarrow{H_2O} Ar-S-OH \xrightarrow[\text{fast}]{ArSCl} ArSO-S-Ar$

	Relative rates			
Reaction	Ar = $4\text{-}CH_3C_6H_4-$	$2\text{-}CH_3C_6H_4-$	$2,4,6\text{-}(CH_3)_3C_6H_2-$	Ref.
1	100[f,c]	—	50[d]	74
2	100[h,e]	57	0.52	49
3	100[j]	—	28	72
4	100[i]	—	75	74
5	100	65	14	24

[a] Diisopropylamine.
[b] Piperidine.
[c] $4\text{-}ClC_6H_4-$.
[d] $2,6\text{-}(CH_3)_2-C_6H_3-$.
[e] C_6H_5-.
[f] $k = 46$ liter mole^{-1} sec^{-1}.
[g] $R = p$-tolyl.
[h] $k = 8 \times 10^{-5}$ sec^{-1}.
[i] $k = 9.2 \times 10^{-2}$ liter mole^{-1} sec^{-1}.
[j] $k = 3.64$ liter mole^{-1} sec^{-1}.

An explanation can be offered on the basis of conformational analysis (72). A perspective view of the sulfonate ester along the S—C bond in its *most stable conformation* is **21**. (The horizontal bars represent the benzene ring.) Reaction of sulfonic esters with nucleophiles should not suffer appreciable steric retardation since the presence of the methyl groups does not interfere with the incoming nucleophile.

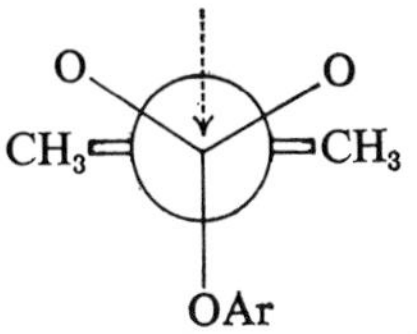

(21)

The study of the effect exerted by *ortho* methyl groups has been extended to di- and tricoordinated sulfur. A summary of the results is given in Table IV. The effect of two *ortho* methyl groups is negligible in the case of dicoordinate sulfur (74,24). The most stable conformation being **22**, there is no steric hindrance to the backside attack of a nucleophile. As for tricoordinated sulfur, two types of behavior are observed.

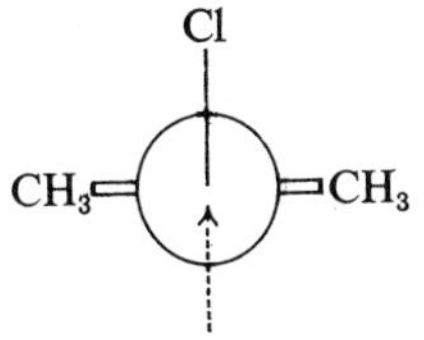

(22)

The presence of two methyl groups decreases the rate about 200-fold in the racemization of sulfoxides (49), whereas the rate is only slightly affected for the reaction of sulfinyl chloride with amines (74). A reasonable explanation of these results may be offered in terms of stability and population of conformers. For sulfinyl chloride the three possible conformations may be represented as **23–25**. Probably the most

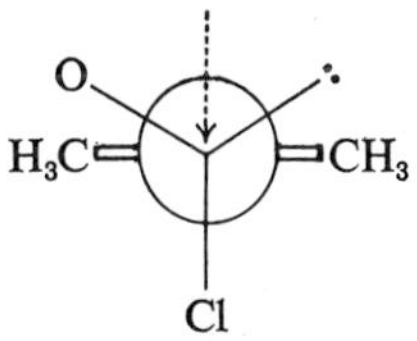

(23) 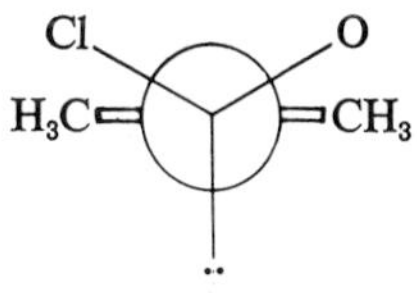   **(24)** 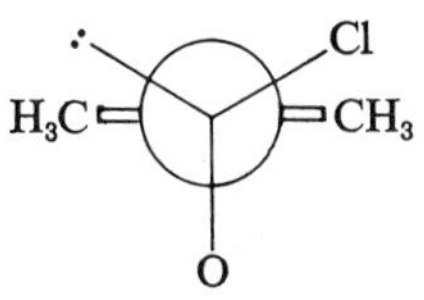   **(25)**

populated conformation is **23**, chlorine being the bulkiest group. In this conformation, attack by a nucleophile is not greatly influenced by the presence of the *ortho* groups.* In the case of the racemization of sulfoxides, *if* the reaction is a nucleophilic attack by chloride ion on the conjugate acid of the sulfoxide (see however the preceding section), the possible conformations are **26–28**. Since in the case at hand R is the largest group (R = *p*-tolyl), the most stable conformation is likely to be **26**. In this conformation the path for attack by the nucleophile is somewhat hindered.

(26)   (27)   (28)

Although no data exist as to the effect of two *ortho* methyl groups on the reaction of sulfonyl chloride with nucleophiles, it might be interesting to note that steric hindrance is also an important factor for this reaction as demonstrated by Scott et al. (75). Sulfonyl chlorides are, in fact, very stable to displacement by weak bases like water or alcohols, if one compares it to the great reactivity of sulfinyl chlorides under the same circumstances (75,76).

V. CYCLIC SYSTEMS

The common feature of sulfur-containing cyclic systems is that nucleophilic attack at sulfur, be it di-, tri-, or tetracoordinated, is always faster than for corresponding open-chain analogs. Moreover the increase in rate, as compared to acyclic systems, is much greater for five- than for six-membered rings. The reason for this behavior is not as yet fully understood since the more simple explanation, i.e., the relief of internal strain, is not always supported by thermochemical data. Unfortunately bond length and distances from x-ray measurements, which would help to clarify the picture, are known only in very few instances.

* A much greater effect has been found by Darwish and Noreyko on the base catalyzed solvolysis of *p*-methoxyneophyl benzenesulfinates (74b). Addition of two *ortho* methyl groups decreases the rate about 20 times. However, the neophyl group is so bulky that it probably interferes with the *ortho* methyl groups even in a conformation similar to **23**.

Reaction of *n*-butyl mercaptide with 1,2-dithiacyclopentane [eq. (42)] is about 10^4 times as fast as with its open chain analog [eq. (43)] (17). In this case the large increase in rate can be explained in terms of internal strain. The dihedral angle (C—S—S—C) for an open-chain

$$\text{(pentagon with S—S)} + n\text{-}BuS^- \longrightarrow n\text{-}BuS\text{—}S(CH_2)_3\text{—}S^- \tag{42}$$

$$(n\text{-}BuS)_2 + n\text{-}Bu\overset{*}{S}{}^- \longrightarrow n\text{-}Bu\overset{*}{S}\text{—}SBu\text{-}n + n\text{-}BuS^- \tag{43}$$

disulfide is 90° (77). In the pentatomic cyclic compound the angle is much smaller, probably close to 26°, this being the value found by x-ray (78) measurements for a compound of the same ring type,

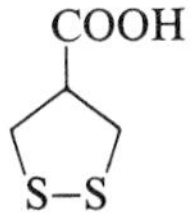

COOH

(pentagon structure with S—S)

(29)

1,2-dithiacyclopentan-4-carboxylic acid (**29**). When the dihedral angle is small there is considerable repulsion between the nonbonding electron pairs of the two sulfur atoms. The existence of internal strain is confirmed by thermochemical data. The enthalpy of oxidation of 1,3-propan-dithiol to give a pentatomic cyclic [eq. (44)] compound is about 4–5 kcal/mole greater than the enthalpy of oxidation of 1,4-butan-dithiol [eq. (45)] to give the hexatomic cyclic compound or of thiols to give linear disulfides [eq. (46)] (79).

$$HS\text{—}(CH_2)_3\text{—}SH \longrightarrow \text{(pentagon with S—S)} \tag{44}$$

$$HS\text{—}(CH_2)_4\text{—}SH \longrightarrow \text{(hexagon with S—S)} \tag{45}$$

$$2n\text{-}BuSH \longrightarrow n\text{-}BuS\text{-}SBu\text{-}n \tag{46}$$

Similar results have been obtained for tricoordinated sulfur. Ethylene sulfite (**30**) hydrolyzes 360 times as fast as the corresponding noncyclic compound, dimethyl sulfite (**31**), which in turn reacts at about

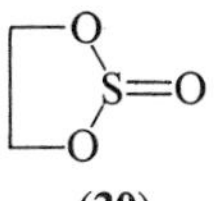

(structure with O and S=O) H_3C—O, H_3C—O with S=O   (ring with O and S=O)

(30) 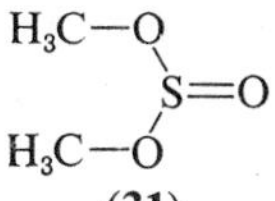   **(31)** 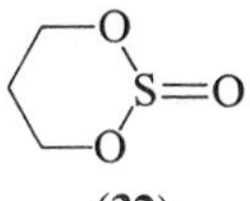   **(32)**

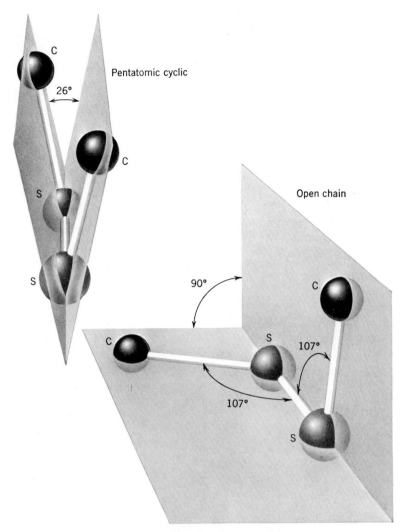

Fig. 1. **Dihedral angles of disulfides.**

the same rate as the cyclic hexatomic compound, trimethylene sulfite (32) (80,81). Also, *o*-phenylene sulfite (33) reacts 1.5×10^3 times as fast as diphenyl sulfite (34) (82).

The nonbonding interactions between the electron pairs of oxygen atoms in positions 1 and 3 have been considered responsible for the

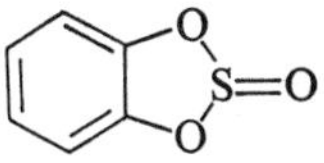

(33) 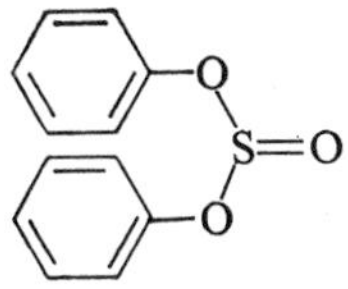   (34)

fast hydrolysis of ethylene sulfite (80). However, Davis, in a semi-quantitative calculation, found that 1,3 interelectronic repulsions are small between two oxygen atoms, and he concluded that thermodynamic strain will result only if there are at least four oxygen atoms about the sulfur atom (81).

The heats of hydrolysis of ethylene and dimethyl sulfite are almost the same, therefore suggesting that there is no internal strain in the cyclic system (81,83). If the interpretation of the thermochemical data has any validity in the present case, one reason for the increase in rate could be the geometry of the rigid cyclic systems conferring greater favor to attack by a nucleophilic reagent.

It is interesting to note that the rate difference between the cyclic and the open-chain compounds for tricoordinated sulfur is not as large as that found for either di- or tetracoordinated sulfur, the largest rate increase having been found for tetracoordinated sulfur. For example, o-phenylene sulfate (35) hydrolyzes 2×10^7 times as fast as diphenyl sulfate (36) (84). Thermochemical data indicate that in this case there is

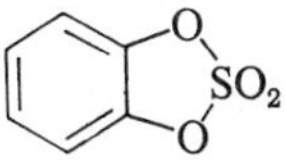

(35) 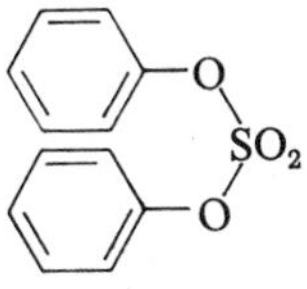   (36)

internal strain, the heat of hydrolysis being 6 kcal/mole greater for the cyclic compound (85). This large increase in rate for cyclic sulfates can only be compared to that found for cyclic phosphates, where ethylene phosphate hydrolyzes 10^7 times as fast as dimethyl phosphate (86). The enthalpy of hydrolysis of ethylene phosphate is 5.5 kcal/mole as great as for the cyclic analog (85).

The high reactivity of the five-membered cyclic sulfates persists when one of the oxygens is substituted by a methylene group; the sultone (37) reacts much faster, 6.8×10^5 times as fast as its open-chain analog 39 (87).

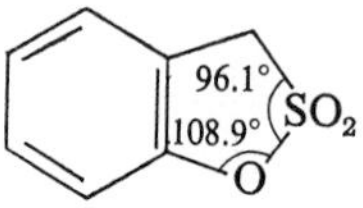

(37) 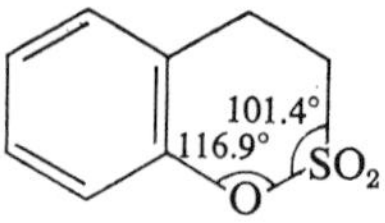   (38)

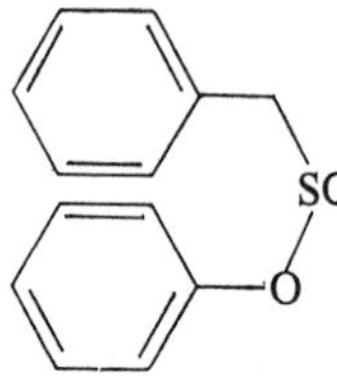

(39)

Hydrolysis of **37** in D_2O—OD^- solution shows that the methylene group undergoes extensive exchange suggesting that a possible mechanism for hydrolysis is via formation of sulfene (**40**) (88). However the formation of a sulfene is not necessary for the rapid hydrolysis of a sultone since the five-membered sultone (**41**), which cannot give rise to an intermediate sulfene, undergoes hydrolysis at about the same rate as **37** (88). On the other hand, the six-membered sultone (**38**) reacts only three times as fast as **39** (87). From these findings it appears that the large increase in rate relatively to the open-chain analogs is characteristic of the pentatomic cycle.

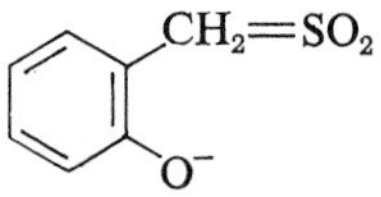

(40) 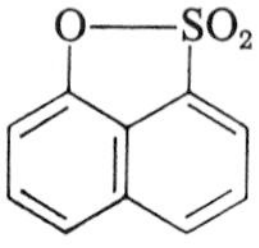   (41)

Very recent x-ray measurements (89) on **37** and **38** have shown that the C—S—O and S—O—C ring angles are 5.3 and 8.0° smaller in the five-than in the six-membered rings. Thus angle strain in the five-membered ring could be the origin of the difference in rate for the two sultones.

The large rate increase for five-membered esters of phosphorus compared to their open-chain analogs has been discussed by Westheimer (90–92). He presented evidence that the rate increase can be explained by considering the strain release in the formation of a trigonal bipyramidal intermediate (90). His results, although not conclusive, can be explained by assuming that the bipyramidal intermediate can undergo a "pseudo-rotation" (91). In certain instances the pseudo-rotation is prevented by the formation of considerable strain (the C—P—O angle expands towards 120° in the diequatorial position) or by the unfavorable energy of alkyl groups in the apical position. His highly attractive hypothesis could be extended to explain the results obtained for the sulfur cyclic compounds. It is currently being tested by Kaiser and co-workers (93).

References

1. N. Kharash, S. J. Potempa, and H. L. Wehrmeister, *Chem. Rev.*, *39*, 269 (1946).
2. O. Foss, *Organic Sulfur Compounds*, Vol. I, Pergamon Press, London, 1961, Chap. 9.
3. N. Kharash, *Organic Sulfur Compounds*, Vol. I, Pergamon Press, London, 1961, Chap. 32 and many chapters by various authors in the same volume.
4. A. J. Parker and N. Kharash, *Chem. Rev.*, *59*, 583 (1959).
5. R. E. Davis, *Survey of Progress in Chemistry*, Academic Press, New York, 1964, pp. 189–238.
6. P. D. Bartlett and R. E. Davis, *J. Am. Chem. Soc.*, *80*, 2513 (1958).
7. O. Foss, *Acta Chem. Scand.*, *4*, 404 (1950).
8. O. Foss, *Kgl. Norske Videnskab. Selskabs, Skrifter, 1945*, No. 2, p. 29.
9. R. E. Davis, J. B. Louis, and A. Cohen, *J. Am. Chem. Soc.*, *88*, 1 (1966).
10. A. Fava and G. Pajaro, *Ann. Chim. (Rome)*, *44*, 545 (1954); *J. Chim. Phys.*, *51*, 594 (1954).
10a. A. Fava, *Gazz. Chim. Ital.*, *83*, 87 (1953).
10b. A. Fava and S. Bresadola, *J. Am. Chem. Soc.*, *77*, 5792 (1955).
11. R. E. Davis, *J. Phys. Chem.*, *62*, 1599 (1958).
12. D. P. Ames and J. E. Willard, *J. Am. Chem. Soc.*, *73*, 164 (1951).
13. E. Ciuffarin and W. A. Pryor, *J. Am. Chem. Soc.*, *86*, 3621 (1964).
14. A. Fava and A. Ceccon, unpublished data.
15. A. Fava and A. Iliceto, *J. Am. Chem. Soc.*, *80*, 3478 (1958).
16. A. Fava and G. Pajaro, *J. Am. Chem. Soc.*, *78*, 5203 (1956).
17. A. Fava, A. Iliceto, and E. Camera, *J. Am. Chem. Soc.*, *79*, 833 (1957).
18. U. Tonellato, unpublished data.
19. H. Lecher and A. Goebel, *Ber.*, *55*, 1483 (1922).
20. A. Baroni, *Atti R. Acad. Lincei*, *23*, 139 (1936).
21. A. Streitwieser, Jr., *Chem. Rev.*, *56*, 572 (1956).
22. L. H. Sommer and O. F. Bennett, *J. Am. Chem. Soc.*, *79*, 1008 (1957); *81*, 251 (1959).
23. P. D. Bartlett and L. H. Knox, *J. Am. Chem. Soc.*, *61*, 3184 (1939).
24. L. Di Nunno, G. Modena, and G. Scorrano, *Ric. Sci.*, *36*, 825 (1966).
25. G. Baddeley and G. M. Bennett, *J. Chem. Soc.*, *1935*, 1819.
26. P. J. C. Fierens, H. Hannaert, J. Van Rysselberge, and R. H. Martin, *Helv. Chim. Acta*, *38*, 2009 (1955).
27. G. Geuskens, G. Klopman, J. Nasielski, and R. H. Martin, *Helv. Chim. Acta*, *43*, 1927 (1960); *43*, 1934 (1960).
28. L. H. Sommer, *Stereochemistry, Mechanism and Silicon*, McGraw-Hill, New York, 1965, p. 146.
29. Reference 28, pp. 84–87.
30. G. W. Wheland, *Resonance in Organic Chemistry*, Wiley, New York, 1961, p. 99.
31. H. E. Simmons, private communication.
32. R. E. Rundle, *Record Chem. Progr. (Kresge-Hooker Sci. Lib.)*, *23*, 195 (1962).
33. H. A. Bent, *Organic Sulfur Compounds*, Vol. 2, Pergamon Press, London, 1966, chap. 1.

34. F. Sanger, *Nature*, *171*, 1025 (1953); A. M. Hutchison and S. Smiles, *J. Chem. Soc.*, *1912*, 570.
35. A. P. Lien, D. A. McCanlay, and W. A. Proell, *Am. Chem. Soc. Div. Petrol. Chem., Gen. Papers*, *28*, 169 (1952).
36. R. E. Benesch and R. Benesch, *J. Am. Chem. Soc.*, *80*, 1666 (1958).
37. A. Fava and G. Reichenbach, unpublished data.
38. J. L. Kice and G. Guaraldi, *J. Am. Chem. Soc.*, *88*, 5236 (1966).
39. J. L. Kice and E. H. Morkved, *J. Am. Chem. Soc.*, *86*, 2270 (1964).
40. W. A. Pryor, *Mechanisms of Sulfur Reactions*, McGraw-Hill, New York, 1962, p. 18.
41. D. Darwish and G. Tourigny, *J. Am. Chem. Soc.*, *88*, 4303 (1966).
42. H. Phillips, *J. Chem. Soc.*, *1925*, 2552.
43. C. A. Bunton, P. B. D. de la Mare, and J. G. Tillet, *J. Chem. Soc.*, *1958*, 4754.
44. R. E. Davis, *J. Am. Chem. Soc.*, *84*, 599 (1962).
45. N. J. Leonard and C. R. Johnson, *J. Am. Chem. Soc.*, *84*, 3701 (1962).
46. C. R. Johnson, *J. Am. Chem. Soc.*, *85*, 1020 (1963).
47. K. Mislow, M. M. Green, P. Laur, J. T. Melillo, T. Simmons, and A. L. Ternay, Jr., *J. Am. Chem. Soc.*, *87*, 1958 (1965).
48. P. Bickart, M. Axelrod, J. Jacobus, and K. Mislow, *J. Am. Chem. Soc.*, *89*, 697 (1967).
49. K. Mislow, T. Simmons, J. T. Melillo, and A. L. Ternay, Jr., *J. Am. Chem. Soc.*, *86*, 1452 (1964).
50. H. B. Henbest and S. A. Khan, *Proc. Chem. Soc.*, *1964*, 56.
51. C. R. Johnson and D. McCants, Jr., *J. Am. Chem. Soc.*, *86*, 2935 (1964).
52. D. R. Rayner, E. G. Miller, P. Bickart, A. J. Gordon, and K. Mislow, *J. Am. Chem. Soc.*, *88*, 3138 (1966).
53. J. L. Kice and G. Guaraldi, *Tetrahedron Letters*, *1966*, 6135.
54. K. Fries and W. Vogt, *Ann. Chem.*, *381*, 337 (1911).
55. E. Fromm, *Ann. Chem.*, *396*, 75 (1913).
56. K. Issleib and M. Tzschach, *Z. Anorg. Allgem. Chem.*, *305*, 198 (1960).
57. H. F. Herbrandson and R. T. Dickerson, Jr., *J. Am. Chem. Soc.*, *81*, 4102 (1959).
58. Ref. 47, footnote 20.
59. H. Kwart, private communication.
60. D. Landini, F. Montanari, H. Hogeveen, and G. Maccagnani, *Tetrahedron Letters*, *1964*, 2691.
61. G. Modena, G. Scorrano, D. Landini and F. Montanari, *Tetrahedron Letters*, *1966*, 3309.
62. (a) C. J. M. Stirling, *J. Chem. Soc.*, *1963*, 5741; (b) J. Krueger, private communication.
63. J. Day and D. J. Cram, *J. Am. Chem. Soc.*, *87*, 4398 (1965).
64. E. T. Kaiser, M. Panar, and F. H. Westheimer, *J. Am. Chem. Soc.*, *85*, 602 (1963).
65. B. D. Batts, *J. Chem. Soc.*, *1966*, 547, 551.
66. J. L. Kice, J. M. Anderson, *J. Am. Chem. Soc.*, *88*, 5242 (1966).
67. S. J. Benkovic, and P. A. Benkovic, *J. Am. Chem. Soc.*, *88*, 5504 (1966).
68. C. A. Bunton, P. B. D. de la Mare, A. Lennard, D. R. Llewellyn, R. B. Pearson, J. G. Pritchard, and J. G. Tillett, *J. Chem. Soc.*, *1958*, 4761.

69. D. R. Christman and S. Oae, *Chem. Ind.* (*London*), *1959*, 1251.
70. J. L. Kice and G. Guaraldi, *J. Org. Chem.*, *31*, 3568 (1966).
71. C. A. Bunton and Y. F. Frei, *J. Chem. Soc.*, *1951*, 1872.
72. J. F. Bunnett and J. Y. Bassett, Jr., *J. Org. Chem.*, *27*, 2345 (1962).
73. H. L. Goering, T. Rubin, and M. S. Newman, *J. Am. Chem. Soc.*, *76*, 787 (1954).
74. (a) G. Guaraldi, unpublished data; (b) D. Darwish and J. Noreyko, *Can. J. Chem.*, 1366 (1965).
75. R. T. van Aller, R. B. Scott, Jr., and E. L. Brockelbank, *J. Org. Chem.*, *31*, 2357 (1966) and references therein.
76. R. B. Scott, Jr., and M. S. Heller, *J. Org. Chem.*, *20*, 1159 (1955).
77. Ref. 40, pages 20–21.
78. O. Foss and O. Tjomsland, *Acta Chem. Scand.*, *12*, 1810 (1958).
79. S. Sunner, *Nature*, *176*, 217 (1955).
80. J. G. Tillett, *J. Chem. Soc.*, *1960*, 37.
81. P. D. Bartlett and E. S. Lewis, *J. Am. Chem. Soc.*, *72*, 1005 (1950).
82. P. B. D. de la Mare, J. G. Tillett, and H. F. van Woerden, *J. Chem. Soc.*, *1962*, 4888.
83. N. Pagdin, A. K. Pine, J. G. Tillett, and H. F. van Woerden, *J. Chem. Soc.*, *1962*, 3835.
84. E. T. Kaiser, I. R. Katz, and T. F. Wulfers, *J. Am. Chem. Soc.*, *87*, 3781 (1965).
85. E. T. Kaiser, M. Panar, and F. H. Westheimer, *J. Am. Chem. Soc.*, *85*, 602 (1963).
86. J. Kumamoto, J. R. Cox, Jr., and F. H. Westheimer, *J. Am. Chem. Soc.*, *78*, 4858 (1956).
87. O. R. Zaborsky and E. T. Kaiser, *J. Am. Chem. Soc.*, *88*, 3084 (1966).
88. E. T. Kaiser, K. Kudo, and O. R. Zaborsky, *J. Am. Chem. Soc.*, *89*, 1393 (1967).
89. E. B. Fleischer, E. T. Kaiser, P. Langford, S. Hawkinson, A. Stone, and R. Dewar, *Chem. Commun.*, *1967*, 197.
90. P. C. Haake and F. H. Westheimer, *J. Am. Chem. Soc.*, *83*, 1102 (1961).
91. E. A. Dennis and F. H. Westheimer, *J. Am. Chem. Soc.*, *88*, 3432 (1966).
92. E. A. Dennis and F. H. Westheimer, *J. Am. Chem. Soc.*, *88*, 3431 (1966).
93. O. R. Zaborsky, unpublished work.

Group Electronegativities

PETER R. WELLS*

Department of Chemistry, University of California, Irvine, California

CONTENTS

I. INTRODUCTION

The term electronegativity was introduced by Pauling (1) to describe "the power of an atom in a molecule to attract electrons to itself." It rapidly became a much-used, and sometimes ill-used, qualitative concept, and considerable effort has been expended in attempts to derive a universal quantitative set of electronegativity values.

In the approximation that permits polyatomic groups to be regarded as pseudo-atoms, the corresponding group electronegativities may be considered. These share the intrinsic problems associated with the measurement of electronegativities of atoms such as those discussed by Pritchard and Skinner (2), but are more useful in the understanding of effect of substituents on reactivity and physical properties of organic molecules.

Pauling's thermochemical method of measurement of electronegativities can be rationalized but it may be seriously criticized on many points. This scale, although not the individual values themselves, has become accepted so that the more satisfying electronegativities of Mulliken (3) and those from other sources are commonly expressed in these terms.

* On leave from University of Queensland, Brisbane, Australia.

In this chapter the view is developed that a mutually consistent set of electronegativity values can be derived for most of the commonly encountered substituent groups. The values are obtained from various sources relating to the Mulliken electronegativities of the univalent atoms F, Cl, Br, I, H, Li, and Na, expressed on the Pauling scale.

II. PAULING OR THERMOCHEMICAL ELECTRONEGATIVITIES

Starting from the observation that the energy of a bond A—B is generally larger than the mean of the A—A and B—B bond energies, Pauling (4,5) suggested that the energy enhancement could be used as a measure of the differences in electronegativity between A and B.

The expression employed

$$(x_A - x_B)^2 = \Delta_{AB}/23.05 \tag{1}$$

relates the difference in electronegativity between atoms A and B to the difference

$$\Delta_{AB} = D(A—B) - \frac{1}{2}[D(A—A) + D(B—B)] \tag{2}$$

between the bond dissociation energies of AB and the arithmetic mean of A_2 and B_2. These quantities are customarily given in kilocalories per mole and are converted to electron volts by the factor 23.05 so that the units of electronegativity superficially are volts$^{1/2}$.

For some molecules, e.g., the alkali metal hydrides, Δ_{AB} is found to be negative, although the geometric mean method

$$\Delta'_{AB} = D(A—B) - [D(A—A) \cdot D(B—B)]^{1/2} \tag{3}$$

gives positive differences. In most cases, however, Δ_{AB} and Δ'_{AB} are sufficiently similar not to change significantly most of the electronegativity values. The geometric mean method has not been generally employed.

A somewhat naïve rationalization of the arithmetic mean approach may be obtained from a consideration of the energy of the bonding molecular orbital, ψ_{AB}, constructed by a linear combination of atomic orbitals on atoms A and B, ϕ_A and ϕ_B, respectively,

$$\psi_{AB} = a\phi_A + b\phi_B \tag{4}$$
$$(a^2 + b^2 + 2abS = 1)$$

Energy minimization yields

$$E_+ = \tfrac{1}{2}(\alpha_A + \alpha_B) - \frac{S\gamma}{(1 - S^2)} + \frac{[\gamma^2 + \delta^2]^{\frac{1}{2}}}{(1 - S^2)} \tag{5}$$

In eq. (5), α_A and α_B are the Coulombic integrals for atoms A and B which are closely related to electronegativities of these atoms. S is the overlap integral in the bond A—B, γ is a modified resonance integral, $\gamma = \beta - \tfrac{1}{2}S(\alpha_A + \alpha_B)$, and $\delta = \tfrac{1}{2}(\alpha_A - \alpha_B)$ depends upon the electronegativity difference $(x_A - x_B)$.

Equation (5) may be approximated to yield the energy of the doubly occupied ψ_{AB} as

$$2E_+(AB) = \alpha_A + \alpha_B + 2\gamma/(1 + S) + \delta^2/\gamma(1 - S^2) + \cdots \tag{6}$$

Thus

$$2E_+(AB) - \tfrac{1}{2}[2E_+(A_2) + 2E_+(B_2)] \simeq \Delta_{AB}/23.05 \simeq \delta^2/\gamma(1 - S^2) \tag{7}$$

if $\gamma = \gamma_{AB} \simeq \tfrac{1}{2}(\gamma_{AA} + \gamma_{BB})$ and $S = S_{AB} \simeq \tfrac{1}{2}(S_{AA} + S_{BB})$.

Refinements and extensions of Pauling's original tabulation of electronegativity values, which are scaled upon $x_H = 2.1$, have been made by Huggins (6) and by Häissinsky (7) largely on the basis of new and improved thermal data. Brown (8) has drawn attention to an inadequacy in the procedure arising from the presence of contributions to bond energy due to nonbonded electron repulsions. This has the effect, for example, of reducing the bond strength in F_2 relative to HF and H_2, thus exaggerating Δ_{HF}. Brown's empirically corrected values, x_B, are included in Table I with the Pauling, x_P, and Huggins, x_H, values for univalent atoms.

TABLE I
Thermochemical Electronegativities (Univalent Atoms)

Atom	x_P	x_H	x_B
H	2.1	2.20	2.14
F	4.0	3.95	4.00
Cl	3.0	3.15	3.29
Br	2.8	2.95	3.13
I	2.4	2.65	2.84
Li	1.0	—	—
Na	0.9	—	—

A similar procedure can be applied to the derivation of group electronegativities, although the available thermochemical data are somewhat limited. For this purpose Constantinides (9) has suggested

$$x_R = x_H \pm 0.183\{D(R\!-\!H) - [D(R\!-\!R)\cdot D(H\!-\!H)]^{1/2}\}^{1/2} \qquad (8)$$

in which the geometric mean approach has been adopted.

A rather fuller use of the thermochemical data can be obtained by examining both

$$\Delta_{RH}^{1/2} = \{D(R\!-\!H) - [D(R\!-\!R)\cdot D(H\!-\!H)]^{1/2}\}^{1/2}$$

and

$$\Delta_{RCH_3}^{1/2} = \{D(R\!-\!CH_3) - [D(R\!-\!R)\cdot D(H_3C\!-\!CH_3)]^{1/2}\}^{1/2}$$

For R $=$ F, OH, Cl, Br, and I, these two quantities are linearly related with essentially unit slope although there are small deviations in the intercepts corresponding to R $=$ H and CH_3, and somewhat larger deviations when R $=$ NH_2, SH, and CN.

Using the Pauling values for F, Cl, Br, I, and H, it is found that

$$x_R = 0.36\Delta_{RH}^{1/2} + 1.55 \qquad \text{for } x_R > 2 \qquad (9a)$$
$$x_R = 0.385\Delta_{RCH_3}^{1/2} + 1.15 \quad \text{for } x_R > 2.5 \qquad (9b)$$

Correlation (9a) appears to be linear between x_F and x_I, but thereafter levels off to almost zero slope. Similarly correlation (9b) has a linear region between x_F and x_{Br} and runs almost parallel to correlation (9a).

These observations suggest that the thermochemical method is of dubious value; nevertheless, group electronegativities obtained in this way, x_C, and those reported by Yingst (10), x_Y, based upon the arith-

TABLE II
Thermochemical Electronegativities (Groups)

Group	x_C [a]	x_Y [b]	x_{MY}	Group	x_{MY}
OH	3.5	3.53	3.6	OC_6H_5	3.5
SH	$\simeq$2.3	—	3.2	$OCOCH_3$	3.6
NH_2	3.0–3.3	3.17	—	OH_2^+	3.8
NO_2	3.3–3.4	—	3.2	N_3	3.3
CH_3	2.2–2.4	2.64	—	NH_3^+	3.3
CF_3	2.8	3.10	—	SCN	3.1
CN	2.7–2.9	—	3.3	S^-	2.7

[a] $x_{SiH_3} \simeq x_H$.
[b] All alkyl groups $x_Y = 2.66 \pm 0.03$.

metic mean approach, are listed for comparison in Table II. In addition Table II contains some values derived by McDaniel and Yingst (11), x_{MY}, from eq. (10)

$$(x_R - x_H)^2 = 0.059 p K_{RH} - E°(R_2) \qquad (10)$$

in which pK_{RH} refers to the acid dissociation of RH in aqueous solution, and $E°(R_2)$ is the oxidation coupling potential for the reaction $2R^- \rightarrow R_2 + 2e$.

III. VALENCE-STATE ELECTRONEGATIVITIES

Mulliken (3) attempted to provide a theoretical basis for electronegativity and suggested the definition

$$\chi_X = \tfrac{1}{2}(I_X + E_X) \qquad (11)$$

where I_X and E_X are, respectively, the ionization potential and electron affinity of the atom, or presumably group, X. These electronegativity values, expressed in electron volts, i.e., in energy per electron units, and symbolized χ_X, are found to be approximately linearly related to the Pauling values, x_X.

Mulliken (12) and Moffitt (13) have discussed this definition and have drawn attention to the fact that valence-state ionization potentials, I_X^v, and electron affinities, E_X^v, are required to describe the character of an atom in a molecule. This concept has been taken up in detail by Hinze and Jaffé (14) who have deduced the energies required to promote the ground-state atom, P_X^0, cation, P_X^+, and anion, P_X^-, to the appropriate valence state, and hence to valence-state quantities

$$I_X^v = I_X + P_X^+ - P_X^0 \qquad (12a)$$
$$E_X^v = E_X + P_X^0 - P_X^- \qquad (12b)$$

required to obtain valence-state electronegativities

$$\chi_X^v = \tfrac{1}{2}(I_X^v + E_X^v) \qquad (13)$$

Their results for the monovalent atoms are given in Table III as χ^v, together with the values x' expressed on the Pauling scale. The latter were obtained employing the approximate empirical relationship illustrated in Figure 1,

$$x_X = 0.325 \chi_X - 0.05 \qquad (14)$$

based upon the Pauling values for F, Cl, Br, I, H, Li, and Na.

A more complete tabulation is given in Table IV.

TABLE III

Valence-State Electronegativities (Univalent Atoms)

Atom	Valence state [a]	I^v, eV	E^v, eV	χ^v, eV	x'
F	$2s^2p^4\mathbf{p}$	20.86	3.50	12.18	3.90
Cl	$3s^2p^4\mathbf{p}$	15.03	3.73	9.38	3.00
Br	$4s^2p^4\mathbf{p}$	13.10	3.70	8.40	2.68
I	$5s^2p^4\mathbf{p}$	12.67	3.52	8.20	2.60
H	1s	13.60	0.75	7.17	2.28
Li	2s	5.39	0.82	3.11	0.95
Na	3s	5.14	0.47	2.81	0.85

[a] The bond-forming orbital to which the electronegativity refers is given in bold face.

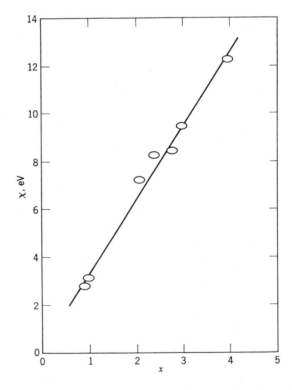

Fig. 1. Relationship between Mulliken and Pauling electronegativities.

The theoretical derivation of eq. (11) is discussed by Pritchard and Skinner (2) and a theoretical–empirical treatment is given by Iczkowski and Margrave (15). The latter draw attention to the fact that the energies of positive and negative ions relative to the neutral atom can be expressed to a good approximation as a power series in N, the excess over the nuclear charge of electrons in these species. In all those cases examined, eq. (15)

$$E(N) = aN + bN^2 \tag{15}$$

proved to be a very good approximation. Thus the ionization potential is given by

$$I = E(-1) - E(0) = -a + b \tag{16a}$$

and the electron affinity by

$$E = E(0) - E(+1) = -a - b \tag{16b}$$

Hence

$$a = -\tfrac{1}{2}(I + E) \tag{17a}$$
$$b = \tfrac{1}{2}(I - E) \tag{17b}$$

A suitable definition of the electronegativity of the neutral atom is

$$\chi(0) = -(\partial E/\partial N)_{N=0} = -a \tag{18}$$

which is precisely the definition suggested by Mulliken and employed by Hinze and Jaffé.

The added value of this approach arises from the possibility of considering valence-state electronegativities in situations where $N \neq 0$ through eq. (19)

$$\chi(N) = -a - 2bN \tag{19}$$

Hinze, Whitehead, and Jaffé (16) have discussed in detail the concept of electronegativity changes associated with fractional charges arising from the distribution of the electron pair in a covalent bond and suggest that at equilibrium for the A—B bond

$$(\partial E_A/\partial N)_{N=N_A} = (\partial E_B/\partial N)_{N=N_B} \tag{20}$$

i.e., there is bond electronegativity equalization.

Since $N_A = -N_B$, it follows that

$$N_A = -(a_A - a_B)/2(b_A + b_B) \tag{21}$$

and that

$$\chi_A(N_A) = \chi_B(N_B) = -(a_A b_B + a_B b_A)/(b_A + b_B) \tag{22}$$

Some values of the a and b parameters are given in Table V.

TABLE IV

Valence-State Electronegativities (Atoms)

	Valence state [a]	I^v, eV	E^v, eV	χ^v, eV	x'
Be	$2di^2$	8.58	0.99	4.79	1.51
	$2di^2$ vacancy	—	—	0.62	−0.03
B	$2tr^3$	11.29	1.38	6.34	2.01
	$2tr^3$ vacancy	—	—	4.26	1.27
C	$2te^4$	14.61	1.34	7.98	2.54
	$2tr^3\pi$	15.62	1.95	8.79	2.81
	$2tr^3\pi$	11.16	0.03	5.60	1.75
	$2di^2\pi^2$	17.42	3.34	10.39	3.33
	$2di^2\pi^2$	11.19	0.10	5.65	1.79
N	$2s^2p^3$	13.94	0.84	7.39	2.35
	$23\%s$ [b]	—	—	11.21	3.59
	$2tr^3\pi^2$	19.72	4.92	12.32	3.95
	$2tr^3\pi^2$ (lone pair)	—	—	3.57	1.11
	$2di^2\pi^3$	22.10	6.84	14.47	4.65
	$2di^2\pi^3$	14.11	2.14	8.13	2.55
N^+	$2te^4$	33.29	14.14	23.72	7.66
	$2tr^3\pi$	34.62	15.09	24.86	8.03
	$2tr^3\pi$	28.71	11.95	20.33	6.56
O	$2s^2p^4$	17.28	2.01	9.65	3.09
	$20\%s$ [b]	—	—	14.39	4.63
	$2di^2\pi^4$	28.71	9.51	19.11	6.16
	$2di^2\pi^4$ (lone pair)	—	—	6.34	2.01
O^+	$2tr^3\pi^2$	41.39	19.64	30.52	9.87
F	$2s^2p^5$ (lone pair)	—	—	7.35	2.34
Mg	$3di^2$	7.10	1.08	4.09	1.28
	$3di^2$ vacancy	—	—	1.22	0.35
Al	$3tr^3$	8.83	2.11	5.47	1.73
	$3tr^3$ vacancy	—	—	3.69	1.15
Si	$3te^4$	11.82	2.78	7.30	2.32
P	$3s^2p^3$	10.73	1.42	6.08	1.93
	$3tr^3\pi^2$	15.18	3.76	9.47	3.03
	$3tr^3\pi^2$ (lone pair)	—	—	5.06	1.59
P^+	$3te^4$	24.10	12.09	18.10	5.83
S	$3s^2p^4$	12.39	2.38	7.39	2.35
	$3di^2\pi^4$	17.42	6.80	12.11	3.89
	$3di^2\pi^4$ (lone pair)	—	—	5.86	1.85
S^+	$3tr^3\pi^2$	28.51	14.33	21.42	6.91
Cl	$3s^2p^5$ (lone pair)	—	—	6.58	2.09

[a] di, tr, and te signify sp, sp^2, and sp^3 hybrid orbitals.
[b] Estimated for partially hybridized valence states.

TABLE V

Parameters for Eq. (15)

	Valence state	$-a$, eV	$-b$, eV
H	1s	7.17	6.43
Li	2s	3.11	2.29
Be	$2di^2$	4.79	3.79
B	$2tr^3$	6.34	4.96
C	$2te^4$	7.98	6.64
	$2tr^3\pi$	8.79	6.84
	$2tr^3\pmb{\pi}$	5.60	5.57
	$2di^2\pi^2$	10.39	7.03
	$2di^2\pmb{\pi}^2$	5.65	5.55
N	$2s^2p^3$	7.39	6.55
	$2tr^3\pi^2$	12.32	7.40
	$2di^2\pi^3$	14.47	7.63
	$2di^2\pmb{\pi}^3$	8.13	5.99
N$^+$	$2te^4$	23.72	9.58
	$2tr^3\pi$	24.86	9.77
	$2tr^3\pmb{\pi}$	20.33	8.38
O	$2s^2p^4$	9.65	9.64
	$2di^2\pi^4$	19.11	9.60
O$^+$	$2tr^3\pi^2$	30.52	10.88
F	$2s^2p^4p$	12.18	8.68
Na	3s	2.81	2.34
Mg	$3di^2$	4.09	3.01
Al	$3tr^3$	5.47	3.36
Si	$3te^4$	7.30	4.52
P	$3s^2p^3$	6.08	4.66
P$^+$	$3te^4$	18.10	6.01
S	$3s^2p^4$	7.39	5.01
S$^+$	$3tr^3\pi^2$	21.42	7.09
Cl	$3s^2p^4p$	9.38	5.65
Br	$4s^2p^4p$	8.40	4.70
I	$5s^2p^4p$	8.20	4.58

It is of interest to compare the covalent bond charge distribution given by eq. (21) with that associated with the LCAO–MO description, eq. (4). An equal division of the overlap charge distribution gives

$$N_A = 2(a^2 + abS) - 1 \qquad (23a)$$
$$N_B = 2(b^2 + abS) - 1 \qquad (23b)$$
$$\therefore N_A = a^2 - b^2 \qquad (24)$$

If the energy of the occupied MO given in eq. (5) is introduced into the secular equation for the coefficients a and b one obtains

$$a^2 - b^2 = (\delta/\gamma)[1 - (1 - S^2)\delta^2/\gamma^2 + \cdots] \tag{25}$$

$$\therefore\ N_A \simeq \delta/\gamma \tag{26}$$

Equation (26) relates N_A and hence the ionic character of a bond to the difference, δ, between the Coulombic integrals of the bonding partners while eq. (21) relates N_A to the differences $(a_A - a_B)$ in valence-state electronegativities. Equation (7) was seen to relate similarly the Pauling electronegativities to δ.

Group electronegativities can, in principle, be derived from a consideration of the effect of the atoms associated with the bonding atom in the group. Thus, for the molecule X_nA—B, account would be taken of the effect of the X atoms on the electronegativity of A in its bond with B. Hinze, Whitehead, and Jaffé (16) have carried out a number of calculations of this type, employing an iterative method starting from a reasonable assumed value for the ionic character of the X—A bonds.

Whitehead and Jaffé (17) have employed the simple method of correcting the electronegativity of the bonding atom A by one-sixth of the difference between this quantity and the electronegativities of the attached groups, X. Thus for $R = X_nA$

$$\chi_R(0) = \chi_A(0) + (n/6)[\chi_X(0) - \chi_A(0)]$$
$$= [(6 - n)/6]\chi_A(0) + (n/6)\chi_X(0) \tag{27}$$

Employing the principle of orbital electronegativity equalization, eq. (20), Huheey (18) has calculated electronegativity values for a large number of groups. In essence, the method involves the calculation of N_A in the group X_nA as radical, cation, and anion from the relationships

$$-\chi_A(N_A) = a_A + 2b_A N_A = -\chi_X(N_X) = a_X + 2b_X N_X \tag{28}$$

with

$$N_A + nN_X = 0 \qquad \text{for } X_nA(N_R = 0) \tag{29a}$$
$$N_A + nN_X = -1 \quad \text{for } X_nA^+(N_R = -1) \tag{29b}$$
$$N_A + nN_X = 1 \qquad \text{for } X_nA^-(N_R = +1) \tag{29c}$$

Yielding for $R = X_nA$

$$-\chi_R(N_R) = a_R + 2b_R N_R \tag{30}$$

where

$$a_R = q_R a_A + (1 - q_R)a_X \tag{31a}$$
$$b_R = q_R b_A \tag{31b}$$
$$q_R = (1 + b_A/b_X)^{-1} \tag{32}$$

Equation (27) is clearly a simplified form in which $b_X/b_A = 6 - n$. Unfortunately this simple procedure leads to obviously erroneous results, since in complex groups complete orbital electronegativity equalization in which the $\chi_X(N_X)$ for all atoms are the same will lead to equal group electronegativities for all isomeric forms of the group as for example in the cases of $FCH_2CH_2CH_2$—, CH_3CHFCH_2—, CH_3CH_2CHF—, $FCH_2(CH_3)CH$—, and $(CH_3)_2CF$—. This point was recognized by Huheey (18), who suggested that incomplete equalization in the sense discussed by Pritchard (19) might correct the anomaly.

Pritchard (19) has drawn attention to the fact that bond electronegativity equalization corresponds to the neglect of overlap approximation. When these considerations are introduced there may remain a difference, ϵ, between resultant electronegativities approximately equal to one-tenth of the original atomic electronegativity difference. With this modification eq. (31a) will become

$$a_R \simeq q_R a_A + (1 - q_R)a_X[(1 - \epsilon) + \epsilon a_A/a_X] \tag{33}$$

and eq. (31b) becomes $\epsilon q_R b_A$. An overlap factor of this magnitude alone appears insufficient to account for the anomalies.

Further progress along these lines has been reported by Whitehead, Baird, and Kaplansky (20). The starting point is the work of Hinze (21), who examined the variation of the a and b parameters for a tetrahedral orbital of carbon by computing I^v and E^v for various values of the total charge N_T on carbon, arising from the occupancy of the other atomic orbitals. It is found that

$$I^v = \alpha + \beta N_T + \gamma N_T^2 \tag{34a}$$

and

$$E^v = \delta + \epsilon N_T + \zeta N_T^2 \tag{34b}$$

for several atoms and from these relationships it is possible to express χ_c as a function of N_T. In a particular molecule a reasonable estimate of charge distribution is made and applied by way of eqs. (34) to the estimation of the χ's. An iterative procedure based upon electronegativity equalization is followed to obtain the equilibrium charge distribution and electronegativities.

Some of the group electronegativities obtained in the treatments above are collected in Table XI.

IV. COVALENT BOUNDARY POTENTIALS

Several attempts have been made to express electronegativity in terms of the energy of a valence-state electron arising from its interaction with the screened nucleus at a distance corresponding to covalent-bond formation.

Gordy (22) suggests that values for atoms on the Pauling scale may be related to the number of valence electrons n and the covalent radius r (in Å) by

$$x_G = 0.31(n + 1)/r + 0.50 \qquad (35)$$

This arises from the interpretation of electronegativity as the potential due to the effective nuclear charge, Z^*, at the covalent boundary, i.e., Z^*e/r, employing

$$Z^* = n - 0.5(n - 1) = 0.5(n + 1) \qquad (36)$$

which assumes all electrons in closed shells below the valence shell exert a full screening effect, while the screening constant for one valence shell electron on another is 0.5.

An obvious refinement of this method is to employ the Slater effective nuclear charge, replacing eq. (36) by

$$
\begin{aligned}
\text{Hydrogen} \quad & Z^* = 1 \\
\text{1st row atoms} \quad & = 0.65(n + 1) \\
\text{2nd row atoms} \quad & = 0.65(n + 1) + 0.90 \\
\text{3rd and 4th row } d^0 \text{ atoms} \quad & = 0.65(n + 1) + 0.90 \\
\text{3rd and 4th row } d^{10} \text{ atoms} \quad & = 0.65(n + 1) + 1.20
\end{aligned}
\qquad (37)
$$

Pritchard and Skinner (2) report the following:

$$
\begin{aligned}
\text{1st row atoms} \quad x_{PS} & = 0.48Z^*/r + 0.50 \\
\text{2nd row atoms} \quad & = 0.44Z^*/r + 0.28 \\
\text{3rd row atoms} \quad & = 0.42Z^*/r - 0.07 \\
\text{4th row atoms} \quad & = 0.46Z^*/r - 0.12
\end{aligned}
\qquad (38)
$$

The appropriate combinations of eqs. (37) and (38) approximately satisfy eq. (35).

Allred and Rochow (23) consider that electronegativity should be measured not by a potential but by a force at the covalent boundary and should be proportional to Z^*/r^2. Employing an effective nuclear charge, Z_{eff} calculated by Slater's rules for *all* the electrons, not all except the one under examination as is usual, they find

$$x_{AR} = 0.36Z_{eff/r^2} + 0.74 \qquad (39)$$

expressed on the Pauling scale.

Electronegativities obtained through eqs. (35), (38), and (39) for univalent atoms are summarized in Table VI.

TABLE VI
Covalent Boundary Potentials (Univalent Atoms)

Atom	x_G	x_{PS}	x_{AR}
H	2.17	—	—
F	3.94	4.00	4.10
Cl	3.00	2.98	2.83
Br	2.68	2.73	2.74
I	2.36	2.50	2.21
Li	0.96	0.97	0.97
Na	0.90	0.91	1.01

The Allred-Rochow values are unusual in some respects. The usual $x_{Li} > x_{Na}$ order is inverted and x_I appears to be abnormally low. Drago (24) has pointed out that the x_{AR} values for the group IV elements do not appear to be reasonable and has commented critically on the method and its application.

A modification of Gordy's method has been applied by Wilmshurst (25) to the derivation of group electronegativities. For a group the covalent radius employed is that of bonding atom, but a modified valence electron number n^* is obtained as the sum of (1) the nonbonded electrons of the bonding atom, which is the value n of the free atom less the sum of the valences used in the group bonding; (2) the bonded electrons, according to the fraction considered associated with the bonding atom; and (3) the "resonance" electrons or those delocalized to or from (negative) the bonding atom according to s, the number of $\overset{-}{A}\overset{+}{X}$ or $\overset{+}{A}\overset{-}{X}$ contributing structures in the resonance hybrid picture.

Thus for X_mA—

$$n^* = (n - m) + 2mx_A/(x_A + x_X) + sx_A/(x_A + x_X) \qquad (40)$$

Table VII lists the values obtained as x_W and the atomic values, x_G, from which they were derived.

TABLE VII

Covalent Boundary Potentials (Groups)

Group	x_W	Group	x_W	Atom	x_G
CH_3	2.63	CO_2H	2.84	H	2.17
$CH{=}CH_2$	3.08	CO_2^-	2.92	C	2.52
$C{\equiv}CH$	3.29	OH	3.89	N	3.01
C_6H_5	3.13	NH_2	3.40	O	3.47
CH_2Cl	2.74	NC	3.49	F	3.94
$CHCl_2$	2.88	NO_2	3.45	Si	1.82
CCl_3	3.03	SiF_3	2.15	P	2.19
CN	3.17	PH_2	2.20	S	2.58
		SH	2.61	Cl	3.00

V. ELECTRONEGATIVITIES FROM BOND VIBRATIONS

Gordy (26) has observed that eq. (41) holds with some accuracy for many diatomic and simple polyatomic molecules.

$$k = 1.67B(x_A x_B/r_{AB}^2)^{3/4} + 0.30 \tag{41}$$

where k is the bond stretching force constant (dyne cm^{-1} $\times$ 10^5) of the A—B bond, B is the bond order, r_{AB} is the bond length (Å) and x_A and x_B are the electronegativities on the Pauling scale. For hydrides the coefficient 1.67 and the intercept 0.30 need to be modified.

TABLE VIII

Electronegativities from Bond Vibrations (ν_{HX})[a]

Group	x_W'	Group	x_W'
$CH{=}CH_2$	2.97	OH	3.72
$C{\equiv}CH$	3.29	OCH_3	3.70
C_6H_5	3.01	OCHO	3.58
CCl_3	2.99	SH	2.54
CO_2H	2.88	SeH	2.33
CO_2^-	2.81	NH_2	3.36
CHO	2.78	PH_2	2.29
CN	3.27	AsH_2	2.08

[a] Based upon $x_F = 3.94$, $x_{Cl} = 3.00$, $x_{Br} = 2.68$, $x_I = 2.36$.

A considerably simpler expression has been suggested by Wilms-hurst (27) in terms of the stretching frequencies, ν_{HX}, of hydrides which depend upon $k_{HX}^{\frac{1}{2}}$. An allowance for the reduced mass is included to give

$$x_X = 1.1 \times 10^{-4}(1 + M_H/M_X)^{-\frac{1}{2}}\nu_{HX} - 0.24 \tag{42}$$

Some of the values, x'_W, obtained are given in Table VIII.

Direct relationships between stretching frequencies and electro-negativities are yet simpler and these appear to have been observed in compounds of the type R_1R_2CO, ν_{CO}, by Kagarise (28) and $R_1R_2R_3PO$, ν_{PO}, by Bell, Heisler, Tannenbaum, and Goldstein (29).

The data reported by Kagarise can be examined first using $x_F = 3.95$, $x_{Cl} = 3.04$, and $x_H = 2.20$ to yield the rather precise relationship covering five compounds

$$\nu_{CO} = 62 \sum x + 1451 \quad (\text{cm}^{-1}) \tag{43}$$

By means of eq. (43), atom and group electronegativities can be derived from the remaining data and those reported by Lagowski (30).

Similarly the data for $R_1R_2R_3PO$ yield

$$\nu_{PO} = 9.93 - 0.24 \sum x \quad (10^{-4}\mu) \tag{44}$$

from which a large number of group electronegativities can be obtained.

With the availability of a large body of data it is now reasonable to indicate the mean deviations and the number of estimates of each

TABLE IX

Electronegativities from Bond Vibrations (ν_{CO})

Initial values		x_X	
F	3.95	CH_3	2.34 ± 0.17 (3)
Cl	3.04	CH_2Br	2.64 (1)
H	2.20	CH_2Cl	2.73 (1)
"Best" values		$CHCl_2$	2.76 ± 0.07 (2)
		CCl_3	2.95 (1)
F	3.90 ± 0.03 (3)	CF_3	3.4 ± 0.2 (4)
Cl	2.99 ± 0.05 (3)	I	2.72 (1)
Br	2.86 ± 0.07 (2)		
H	2.28 ± 0.08 (3)		

TABLE X

Electronegativities from Bond Vibrations (ν_{PO})

Initial values		x_X	
F	3.95	CH_3	2.06 ± 0.13 (3)
Cl	3.04	C_6H_5	2.43 ± 0.09 (3)
Br	2.79	CF_3	3.31 (1)
		CN	3.34 ± 0.05 (2)
"Best" values		OCH_3	2.89 ± 0.02 (2)
F	3.95 ± 0.02 (6)	OC_2H_5	2.91 ± 0.07 (6)
Cl	3.06 ± 0.03 (6)	OC_6H_5	3.20 ± 0.09 (5)
Br	2.81 ± 0.02 (6)	SC_2H_5	2.83 ± 0.05 (2)
H	2.55 ± 0.12 (3)	SC_6H_5	2.56 (1)
		NH_2	2.3 ± 0.2 (3)
		$N(CH_3)_2$	2.96 ± 0.17 (3)
		NHC_6H_5	3.13 ± 0.17 (3)

electronegativity value. These are given in Tables IX and X. Also included are the initial electronegativities employed to set up eqs. (42)–(44) and the "best" values appropriate to these atoms.

VI. MUTUALLY CONSISTENT GROUP ELECTRONEGATIVITIES

A survey of the empirical methods of obtaining group electronegativities leads to the set of values given in the second column of Table XI. The first six values appropriate to the univalent atoms may be considered known to within 0.05 Pauling units and constitute the "master set" from which all others have been obtained. The value for fluorine is essentially a defined value setting the scale. The remaining values are perhaps known to within 0.1 units although those given in parentheses may be considerably less certain.

Three sets of calculated group electronegativities also appear in Table XI for comparison. Set I is based upon the covalent boundary potential method of Gordy (22) as modified by Wilmshurst (25); set II are the values derived by Hinze, Whitehead, and Jaffé (16); and set III are those given by Huheey (18). The latter two sets are based upon Hinze and Jaffé's (14) orbital electronegativities of atoms.

The empirical values arise largely from bond vibrational data and to a lesser extent from thermochemical data. The values listed for groups marked with an asterisk are those which may act as donors when bonded to the PO group and are therefore based on ν_{HX}. The values

TABLE XI

Mutually Consistent Group Electronegativities

	Empirical values	Calculated values		
		I	II	III
F	3.95	3.94	3.90	
Cl	3.03	3.00	3.00	
Br	2.80	2.68	2.68	
I	2.47	2.36	2.60	
H	2.28	2.17	2.28	
Li	0.97	0.96	0.95	
CH_3	2.3	2.63	2.30	2.27
CH_2Cl	(2.75)	2.74	2.47	2.47
$CHCl_2$	2.8	2.88	2.63	2.66
CCl_3	3.0	3.03	2.79	2.84
CF_3	3.35	—	3.29	3.46
C_6H_5	3.0*	3.13	—	—
$CH{=}CH_2$	(3.0)	3.08	—	—
$C{\equiv}CH$	(3.3)	3.29	—	—
CN	3.3	3.17	—	—
CO_2H	(2.85)	2.84	—	—
$CO_2{}^-$	(2.85)	2.92	—	—
SiH_3	(2.2)	—	—	2.21
SiF_3	—	2.15	—	3.35
NH_2	3.35	3.40	2.82	2.61
$N(CH_3)_2$	(3.0)	—	—	2.40
NF_2	(3.25)	3.25	—	3.64
NO_2	3.4	3.45	—	—
PH_2	(2.3)	2.20	2.06	2.13
OH	3.7*	3.89	3.53	3.51
OCH_3	(3.7)*	—	—	2.68
SH	(2.8)	2.61	2.35	2.32
BH_2	—	—	—	2.09
$NH_3{}^+$	(3.8)	—	—	3.65
$OH_2{}^+$	(3.3)	—	—	4.51

obtained from ν_{PO} (see Table X) are smaller, presumably since donor action reduces the apparent electronegativity with respect to the PO group.

Some measure of agreement between the empirical and the calculated electronegativities suggests that the empirical values may be described as "mutually consistent electronegativities." This suggestion will be borne out or repudiated by the usefulness of these numbers in the correlation of other data, some of which are described below.

The Gordy-Wilmshurst method of calculation appears from Table X to be distinctly superior to methods II and III, although one suspects this is fortuitous. The test of method II is rather limited, but method III clearly overestimates the effect of the atoms attached to the bonding atom of the group. The deficiencies of method III, and presumably of method II, arise from the requirement of complete electronegativity equalization. This leads to equal group electronegativities for isomeric structures, a probable overestimation of the influence of attached atoms in the bonding atom and a large dilution of the effect of geminal atoms in large groups.

A more reasonable approach would appear to be one involving electronegativity equalization within each bond but only a partial transfer of this changed electronegativity to other orbitals. This is expressed by eq. (33) for groups of the type X_nA. Unfortunately, the group electronegativities are rather insensitive to small changes in ϵ and the empirical data are too imprecise and insufficient to establish the most suitable ϵ value.

The original and the most obvious use of electronegativity is in the estimation of ionic character. Gordy (31) has chosen to *define* ionic character, i.e., for a bond A—B as $\frac{1}{2}|x_A - x_B|$ in terms of Pauling scale electronegativities, but, more reasonably, ionic character corresponds to an unbalanced bonding electron pair distribution which following eqs. (21) and (14) would be given by

$$i = |N_A| = |(a_A - a_B)/2(b_A + b_B)|$$
$$= |(x_A - x_B)/0.65(b_A + b_B)| \qquad (45)$$

Equation (45) will clearly lead to much smaller estimates of ionic character by a factor $[0.325(b_A + b_B)]^{-1}$. This point was recognized by Whitehead and Jaffé (17). Nevertheless they have not considered the change significant and have preferred the i_G values.

Baird and Whitehead (32) have instead applied eq. (45) to obtain the results given in Table XII.

The direct measurement of bonding charge distribution and ionic character is not possible. Although electric dipole moments are consistent with electronegativity considerations, it is well known that these quantities arise as a complicated vector sum of various charge distributions.

The theory of nuclear quadrupole coupling constants (cf. ref. 33) suggests that this field may provide one of the least obscured approaches to ionic character. The nuclear quadrupole coupling energy, eQq,

TABLE XII
Ionic Character of Diatomic Molecules

	i_G	i		i_G	i
LiH	0.685	0.234	HBr	0.305	0.082
NaH	0.735	0.248	HI	0.255	0.069
HF	0.845	0.166	FCl	0.475	0.099
HCl	0.370	0.089	BrCl	0.065	0.019

depends upon the electronic charge, e, upon the electric quadrupole moment, Q, of the nucleus under examination, and upon the field gradient, q, at the nucleus. In free atoms the field gradient due to complete electronic shells and due to valence shell s electrons is zero unless these are polarized significantly away from spherical symmetry. Small field gradients may exist arising from unpaired d and to a lesser extent unpaired f electrons, but the main contributor to q will be the unbalanced distribution to p electrons, U_p. Townes and Dailey (34,35) have examined the nature of the field gradient at a particular nucleus in molecules in terms of (1) the valence electrons associated with chemical bonding of the atom, (2) lone-pair electrons on this atom (or orbital vacancies), (3) electrons and nuclei in the rest of the molecule, and (4) polarization of inner shell electrons. It is believed that factors 3 and 4 may be neglected and that the quadrupole coupling constant of an atom in a molecule, $[eQq]_{mol}$, may be related to that of the free atom, $[eQq]_{atom}$, presumed to be in pure p state, by

$$\rho = [eQq]_{mol}/[eQq]_{atom} = -U_p \tag{46}$$

Some results are given in Table XIII obtained from refs. 17 and 33.

Part of the unbalanced p-electron distribution arises from ionic character and part arises from hybridization. Thus for a molecule in which the atom employs a hybrid atomic orbital given by

$$\phi = \sqrt{s} \cdot \phi_s + \sqrt{p} \cdot \phi_p + \sqrt{d} \cdot \phi_d \tag{47}$$

one expects that

$$\rho = [1 - s + d - i(1 - s - d)] - \pi \tag{48}$$

where i is the ionic character of the bond, s is the fraction of s character, d the fraction of d character in the orbital employed, and π allows for any π bonding.

TABLE XIII
Unbalanced p-Electron Distribution from NQR
$$\rho = -U_p \text{ for X}$$

Group	X = ^{35}Cl	^{79}Br	^{127}I
X·	1.00	1.00	1.00
LiX	—	0.05	0.09
KX	0.00	0.01	0.03
XF	1.33	1.41	—
XCl	0.99	1.14	1.28
XBr	0.945	—	—
XI	0.75	—	—
XH	0.49	—	—
XD	0.50	0.69	0.80
XCF$_3$	0.695	—	—
XCX$_3$	0.745	0.83	0.93
XSiF$_3$	0.40	0.57	—
XSiX$_3$	0.37	0.46	0.58
XCH$_3$	0.62	0.75	0.84
XCN	0.60	0.74	1.055
XBX$_2$	0.395	—	—
XPX$_2$	0.48	—	—

Some rather drastic assumptions are required (cf. refs. 17 and 33), such as $d = 0$ and $\pi = 0$, in the use of eq. (48) which contains four unknowns. The main ambiguity is the extent to which the atomic orbitals employed by the halogens contain s and/or d character. Nevertheless with reasonable assumptions about hybridization the results given in Table XIII can be satisfactorily accounted for in terms of electronegativity differences. This field of investigation certainly promises to be most fruitful.

The store of nuclear magnetic resonance data continues to build up at a tremendous rate, and it has always been evident that much valuable information could be obtained from this source. If the diamagnetic shielding due to other atoms and groups is a major factor determining chemical shifts, then correlations with electronegativity are to be inspected. Allred and Rochow (36) report that the relative proton chemical shifts for infinitely dilute carbon tetrachloride solutions of the methyl halides are linearly related to halogen electronegativities. However, using the group electronegativities one finds that the results for other methyl compounds are not accommodated by this line (see Fig. 2). Essentially the same behavior is observed for gas-phase data on methyl derivatives determined by Spiesecke and Schneider (37). Most of the groups examined are derivatives containing carbon, or some other first row element, as the central atom. The relative chemical shifts of these plotted against electronegativity lie close to a line encompassing F, H,

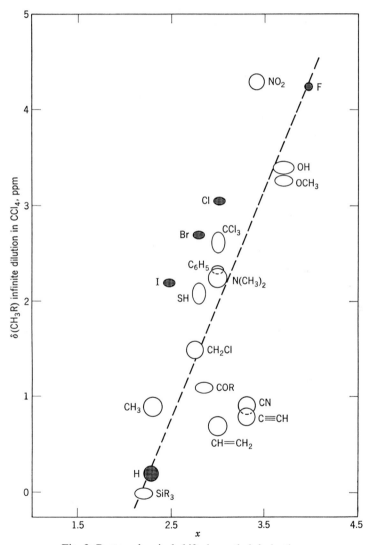

Fig. 2. Proton chemical shifts in methyl derivatives.

and Li. The halogens, the nitro group, and perhaps also sulfur-containing groups deviate from this rough relationship in the sense of yielding larger diamagnetic shifts than expected in the order NO_2, $I > Br > Cl > SR$. This is the expected order of contributions to shielding arising from anisotropic effects. To some extent it might be hoped that the internal chemical shift ($\delta_{CH_3} - \delta_{CH_2}$) will be free from anisotropic effects; an electronegativity scale based upon these values has been suggested by Dailey and Shoolery (38). Anisotropic effects are still evident, however, as indicated by the results summarized in Table

TABLE XIV
Proton Chemical Shifts

R	x_R	$-\delta(CH_3R)$ [a] ppm	$-\delta(CH_3R)$ [b] ppm	$(\delta_{CH_3} - \delta_{CH_2})$ [c] ppm	$-\delta(^{13}CH_3R)$ [d] ppm
F	3.95	4.26	4.00	3.09	77.5
OH	3.7	3.40	—	(2.4)	50.0
OR [e]	3.7	3.24	3.10	2.28	61.5
NO_2	3.4	4.31	3.91	(2.8)	59.4
CN	3.3	1.90	1.53	0.88	7.0
Cl	3.03	3.05	2.71	1.93	49.6
NR_2 [e]	(3.0)	(2.25)	2.03	1.49	27.0
C_6H_5	3.0	2.32	—	(1.4)	24.6
COR [e]	2.85	2.10	—	(1.2)	30.0
Br	2.80	2.69	2.32	1.65	11.5
SR [e]	(2.8)	(2.1)	1.90	1.28	21.6
CH_2Cl	(2.75)	(1.5)	1.29	—	20.9
I	2.47	2.19	1.85	1.31	−20.2
CH_3 [f]	2.3	(0.9)	0.75	0.40	33.7
H	2.28	(0.2)	0.00	0.00	0.00
Li	0.97	—	—	(−2.2)	—

[a] Chemical shifts relative to $(CH_3)_4Si$ for "infinite dilution" in CCl_4 (36). Values in parentheses are for 5% solutions in $CDCl_3$.

[b] Chemical shifts relative to CH_4 for gas-phase samples (37).

[c] Chemical shift of CH_3 relative to CH_2 in CH_3CH_2R for gas-phase sample (37). Values in parentheses from other sources.

[d] Chemical shift of $^{13}CH_3$ relative to $^{13}CH_4$ (37,39,40).

[e] R = H or alkyl group.

[f] And all alkyl groups.

XIV and illustrated in Figure 3. On the other hand, the remaining results do tend to substantiate the electronegativity values for groups containing a first row central atom. These conclusions are seen to apply less well to ^{13}C chemical shifts.

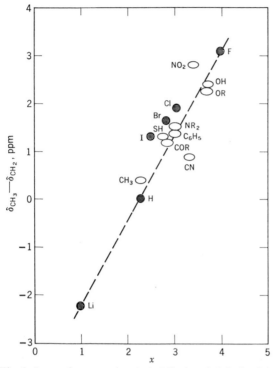

Fig. 3. Internal proton chemical shifts in ethyl derivatives.

TABLE XV

Coupling Constants in Vinyl Derivatives

R	x_R	$J(gem)$, cps	$J(cis)$, cps	$J(trans)$, cps
F	3.95	−3.2	4.65	12.75
OR	3.7	−1.9	6.7	14.2
NO₂	3.4	−2.0	7.6	15.0
CN	3.3	1.3	11.3	18.2
Cl	3.03	−1.4	7.3	14.6
NR₂	(3.0)	0	9.4	16.1
C₆H₅	3.0	1.3	11.0	18.0
COR	(2.85)	1.6	10.3	17.3
Br	2.80	−1.8	7.1	15.2
CH₃	2.3	1.6	10.3	17.3
H [a]	2.28	2.5	11.7	19.1
Li	0.97	7.1	19.3	23.9

[a] From ref. 42.

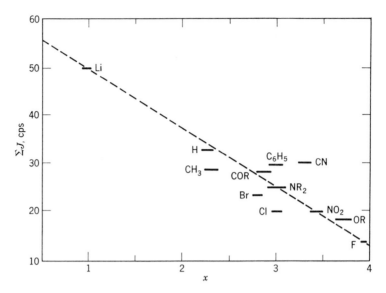

Fig. 4. Proton–proton coupling constants in vinyl derivatives.
$\Sigma J = J(gem) + J(cis) + J(trans)$.

Empirical correlations between proton–proton coupling constants and electronegativities have been observed. Perhaps the best of these is reported by Schaefer (41) applying to the coupling constants in vinyl derivatives, $RCH = CH_2$. Table XV illustrates the correspondence between electronegativity and the various J values. Figure 4 indicates the relationship between the sum of these coupling constants and electronegativity.

With respect to the coupling constants between directly bonded protons and ^{13}C (J_{CH}) a situation exists which is similar to that prevailing for the NQR data. The magnitude of J_{CH} is markedly dependent upon the hybridization state of carbon but is also affected by the nature of the attached groups. Shoolery (43) and Muller and Pritchard (44) have shown that J_{CH} in methane (125 cps), olefins (156 cps), and acetylenes (249 cps) can be precisely accounted for in terms of the s character of the bonding carbon atomic orbital by $J_{CH} = 500s$. The extrapolated value of J_{CH} for a pure p orbital is zero as expected if the Fermi contact coupling mechanism predominates. Superimposed upon these variations is an apparent dependence of J_{CH} on group electronegativity and the principal quantum number of the central atom. The results for sp^3 hybridized carbon (45–49), collected in Table XVI, can be fitted to a

TABLE XVI

^{13}C–H Coupling Constants in CH_3X

X	x_X	J_{CH}, cps	X	x_X	J_{CH}, cps
First row series			Second row series		
F	3.95	149	Cl	3.03	150
NR_3^+	(3.8)	146	SR_2^+	—	146
OH	3.7	141	PR_3^+	—	134
OR	3.7	139.5	SR	(2.8)	137.5
NO_2	3.4	147	SiR_3	—	119
NH_2	(3.35)	133	Al_2R_5	—	113
CN	3.3	136	MgR	—	108
C≡CH	(3.3)	132	Third row series		
NR_2	(3.0)	131			
CCl_3	3.0	134	Br	2.80	152
CR_2^+	—	130	SeR	—	140
$CH=CH_2$	(3.0)	126	GeR_3	—	124
C_6H_5	3.0	126	GaR_2	—	122
COR	(2.85)	126–130	Fourth row series		
CR_3	2.3	120–126			
H	2.28	125	I	2.47	151
Li	0.97	98	TeR	—	140.5
			SnR_3	—	128
			InR_2	—	126
			CdR	—	126

satisfactory linear relationship covering the groups with a first row central atom, and it appears that the second, third, and fourth row series may be fitted by separate relationships (see Fig. 5).

Similar behavior (49) is observed for ^{13}C–H coupling for a limited series of sp^2-hybridized carbon in $H \cdot COX$ (Table XVII).

There is again ambiguity in the interpretation of these data since electronegativity and hybridization effects are not readily disentangled. It is not yet clear whether the relationship between electronegativity and coupling constants is more apparent than real. Current theoretical treatments imply a rather complex situation. Thus Grant and Litchman (50) suggest for the Fermi contact contribution,

$$\frac{J(RCH_3)}{J(CH_4)} \simeq \left(\frac{\Delta_{CH_4}}{\Delta_{RCH_3}}\right)\left(\frac{N_{RCH_3}}{N_{CH_4}}\right)^2\left(\frac{s_{RCH_3}}{s_{CH_4}}\right)\left(\frac{Z_{RCH_3}}{Z_{CH_4}}\right)^3 \tag{49}$$

in which Δ is the average excitation energy arising from the perturbation treatment employed, N is a bond normalization constant, s is the fraction of s character and Z is the effective nuclear charge appearing

TABLE XVII
^{13}C–H Coupling Constants in H·COX

X	x_X	J_{CH}, cps
F	3.95	267
OR	3.7	226
CCl$_3$	3.0	207
NR$_2$	(3.0)	192
C$_6$H$_5$	3.0	174
CH$_3$	2.3	173

in the radical function of a carbon 2s orbital. Nevertheless it is to be hoped that the further accumulation of data, particularly those for other nuclei, will be rewarding.

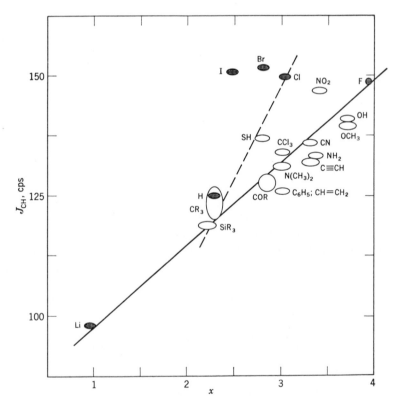

Fig. 5. ^{13}C—^{1}H coupling constants in methyl derivatives.

One of the most encouraging applications of electronegativities is in the estimation of Coulombic integrals (α) for use in Hückel molecular orbital (HMO) calculations as described in eq. (5). In π-electron calculations it is customary to express the Coulombic integral for an atom X in terms of a standard integral, α_C, for a $2p_\pi$ orbital of carbon and the resonance integral, β, for a $2p_\pi$–$2p_\pi$ carbon–carbon bond, i.e.,

$$\alpha_X = \alpha_C + h_X\beta \qquad (50)$$

Baird and Whitehead (51) have derived a set of h_X values obtained directly from the valence state electronegativities of Hinze and Jaffé (14) by setting $\alpha_X = -\chi_X$ with due allowance for number, N, of $2p$ electrons supplied by the atom. Comparison in Table XVIII with the "best" estimates of these parameters given by Streitwieser (52) shows excellent agreement. In addition eqs. (51a)–(51c) are suggested to make allowance for the excess charge, Q, on the atom, arising from σ-bond formation.

$$h_C = 0.00 + 2.28Q + 0.43Q^2 \qquad (51a)$$
$$h_N = 0.56 + 2.43Q + 0.49Q^2 \qquad (51b)$$
$$h_O = 0.96 + 3.17Q + 0.32Q^2 \qquad (51c)$$

TABLE XVIII

Coulomb Parameters for HMO Calculations

X	N_X	$(h_X)_{BW}$ [a]	$(h_X)_s$ [b]	X	N_X	$(h_X)_{BW}$ [a]
Be	0	−1.28		Mg	0	−1.28
B	0	−1.07	−1.0	Al	0	−0.94
C	1	0.00	0.0	Si	1	0.00
N	1	0.56	0.5	P	1	0.21
N	2	1.50	1.5	P	2	1.10
O	1	0.96	1.0	S	1	0.43
O	2	2.13	2.0	S	2	1.29
F	2	2.96	3.0	Cl	2	1.84

[a] Calculated values of reference 51.
[b] Recommended values of reference 52.

From these equations and Streitwieser's recommended values of $h_{N^+} = 2.0$ and $h_{O^+} = 2.5$, it appears that $Q \simeq 0.5$ in both cases.

The HMO method with these parameters can be employed to account for the stabilization of carbonium ions and aromatic radical cations by π-delocalization effects. More general effects due to electronegativity are also evident in the "stabilization energies" reported by

TABLE XIX

Stabilization Energies of Cations

X	x_X	$A(XCH_2{}^+)$, eV	$\Delta I(C_6H_5X)$, eV
F	3.95	−1.1	−0.05
OH	3.7	−2.6	−0.745
OCH$_3$	3.7	−3.0	−1.025
NH$_2$	3.35	−4.1	−1.545
N(CH$_3$)$_2$	(3.0)	−4.6	—
C$_6$H$_5$	3.0	−2.4	−0.975
CH$_3$	2.3	−1.5	−0.425
H	2.28	0.0	0.00
Cl	3.03	−1.4	−0.175
SH	(2.8)	−2.8	−0.915
SCH$_3$	(2.8)	−3.2	—
P(CH$_3$)$_2$	(2.3)	−3.4	—
Br	2.80	−2.2	−0.265
I	2.47	−2.3	−0.514
NO$_2$	3.4	—	+0.675
CF$_3$	3.35	—	+0.435
CN	3.3	+0.4	+0.46

Taft, Martin, and Lampe (53) as the relative appearance potentials of substituted methyl cations

$$\Delta A(XCH_2{}^+) = A(XCH_2{}^+) - \Delta(CH_3{}^+) \tag{52}$$

for

$$XCH_3 + e \longrightarrow XCH_2{}^+ + 2e + H$$

are listed in Table XIX. The first five entries corresponding to first row donors clearly stabilize the cation by π-bonding delocalization mechanism (resonance effect) that is resisted by the electronegativity of X. The second row and higher row groups, though less effective donors, have to contend with a reduced electronegativity. The final entry presumably shows a simple electronegativity effect (see also Fig. 6). Similar results are observed for the relative ionization potentials

$$\Delta I(C_6H_5X) = I(C_6H_5X) - I(C_6H_6) \tag{53}$$

of substituted benzenes also listed in Table XIX (see also Fig. 7).

Finally one may look for a relationship between group electronegativities and the substituent parameters of the Hammett and Taft relationships or parameters derived therefrom (cf. ref. 54). Taft (53) has examined the data in Table XIX from the viewpoint of correlations with the σ_I and σ^+ parameters. With the marked exceptions of H, CH_3, C_6H_5, and CN there appears to be a satisfactory linear relationship between σ_I and ΔA. Furthermore for the first row donor groups there is an excellent relationship between σ^+ and ΔA. A second parallel line accommodates the second row donor atoms. The σ^+ correlations agree

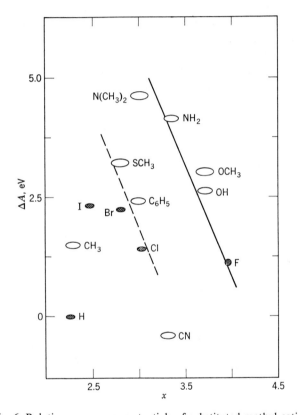

Fig. 6. Relative appearance potentials of substituted methyl cations.

well with the view that this parameter is sensitive to donor character in situations of high electron demand. The ΔI values for all the donor groups are linearly related to the σ_I parameters, but there appears to be no relationship between ΔI and σ_I for nondonors. Taft (55) has also employed Coulomb integrals in HMO calculations justifying the use of ^{19}F chemical shift data for substituted fluorobenzenes as a measure of

σ_R values (resonance effect). The Coulomb integrals employed, though not the same as those given in Table XVIII, are proportional to them. It seems that the group electronegativities and the σ_I values show some parallels in as far as large values of one correspond to large values of the other (see Fig. 8). Both quantities parallel σ_R and are in accord with the view that this parameter measures π-delocalization effects which are proportional to electronegativity within a given periodic row (see Fig. 9). Electronegativities are compared in Table XX with σ_R and σ_I values of donor groups (56) and with the relative strengths, ΔpK, of substituted acetic acids. The electronegativities give a rather poor account of relative acid strength since this is probably largely determined not by the classical inductive effect, which should parallel ionic character and hence

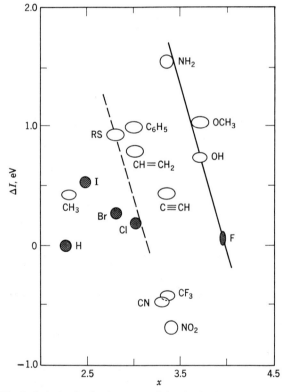

Fig. 7. Relative ionization potentials of substituted benzenes.

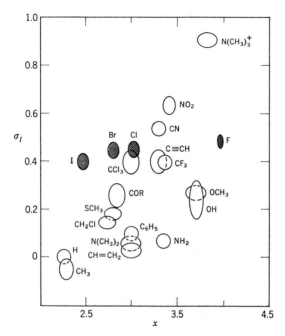

Fig. 8. Substituent inductive parameters (σ_I).

electronegativity closely, but by the electrostatic field at the reaction site due to the substituent dipole. This depends on other factors besides ionic character (cf. ref. 57).

E. D. Jenson, L. D. McKeefer, and R. W. Taft have recently described emf-cell measurements of the standard free energy changes in the reactions

$$R^+ + \tfrac{1}{2}R_0 - R_0 \rightleftharpoons R_0{}^+ + \tfrac{1}{2}R - R \qquad \Delta F^\circ(R^+)$$
$$R^- + \tfrac{1}{2}R_0 - R_0 \rightleftharpoons R_0 + \tfrac{1}{2}R - R \qquad \Delta F^\circ(R^-)$$

for $R_0 =$ triphenylmethyl and $R =$ substituted triphenylmethyl. Combining these two quantities obtained under identical reaction conditions yields

$$\Delta\Delta F^\circ(R) = \Delta F^\circ(R^+) - \Delta F^\circ(R^-)$$

which is the standard free-energy change for the reaction

$$R^+ + R_0{}^- \rightleftharpoons R_0{}^+ + R^-$$

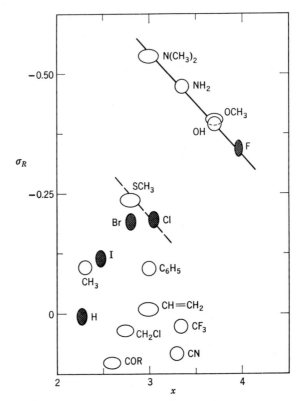

Fig. 9. Substituent resonance parameters (σ_R).

TABLE XX

Substituent Parameters

Group	x	σ_R	σ_I		x	ΔpK
F	3.95	−0.46	0.52	F	3.95	2.10
OH	3.7	−0.67	0.30	$NR_3{}^+$	(3.8)	2.93
OCH_3	3.7	−0.57	0.30	OCH_3	3.7	1.23
NH_2	3.35	−0.79	0.13	NO_2	3.4	3.08
CH_3	2.3	−0.12	−0.05	CF_3	3.35	1.69
H	2.28	0.00	0.00	CN	3.3	2.33
Cl	3.03	−0.24	0.47	Cl	3.03	1.90
SH	(2.8)	−0.08	0.23	Br	2.80	1.90
SR	(2.8)	−0.19	0.19	SH	(2.8)	1.09
Br	2.80	−0.22	0.45	I	2.47	1.64
I	2.47	−0.22	0.40	CH_3	2.3	−0.12

But, if entropy effects can be neglected

$$\Delta\Delta F^0(R) = \Delta I + \Delta E = 2\Delta x_R$$

where ΔI and ΔE are the relative ionization potentials and electron affinities, i.e.

$$R\cdot + R_0{}^+ \rightleftharpoons R^+ + R_0\cdot \qquad \Delta I$$
$$R\cdot + R_0{}^- \rightleftharpoons R^- + R_0\cdot \qquad \Delta E$$

The $\Delta\Delta F^0(R)$ values given below are taken from ref. 59 and have been converted by means of eq. (14) to relative electronegativities on the Pauling scale.

Radical	$\Delta\Delta F^0(R_1)$, kcal/mole^{-1}	$\Delta\chi$
$(p\text{-}Me_2NC_6H_4)_3C$	-20.2	-0.14
$(p\text{-}MeOC_6H_4)_3C$	-9.8	-0.07
$(p\text{-}CH_3C_6H_4)_3C$	-3.1	-0.02
$(C_6H_5)_3C$	0.0	0.00
$(p\text{-}FC_6H_4)_3C$	-0.6	~ 0.00
$(p\text{-}F_3CC_6H_4)_3C$	$+7.3$	0.05
$(m\text{-}O_2NC_6H_4)_3C$	$+11.3$	0.08
$(p\text{-}O_2NC_6H_4)_3C$	$+21.4$	0.15

V. CONCLUSIONS

Several viewpoints may be taken in regard to the concept of electronegativity. It is certainly an approximation with which one hopes to be able to reduce complex molecular problems to the level of diatomics and treat individual atoms and groups more or less independently of their detailed surroundings. For some the approximation is too crude but the alternative of treating each molecule individually is rather terrifying and often unrewarding in its yield of general principles. Refinement of calculated electronegativities is possible by the methods indicated above and will undoubtedly be continued by these and other approaches. The empirical values given in Table XI are considered to be the most reasonable presently available, but are by no means to be regarded as a finalized set. The author is encouraged to believe that further empirical studies will be pursued which are directed towards expanding and refining the tabulation and testing the usefulness or otherwise of this concept.

References

1. L. Pauling and D. M. Yost, *Proc. Natl. Acad. Sci. U.S.*, *14*, 414 (1932).
2. H. O. Pritchard and H. A. Skinner, *Chem. Rev.*, *55*, 745 (1955).
3. R. S. Mulliken, *J. Chem. Phys.*, *2*, 782 (1934).
4. L. Pauling, *J. Am. Chem. Soc.*, *54*, 3570 (1932).
5. L. Pauling, *The Nature of the Chemical Bond*, 3rd Ed., Cornell University Press, Ithaca, N.Y., 1960.
6. M. L. Huggins, *J. Am. Chem. Soc.*, *75*, 4123 (1953).
7. M. Haïssinsky, *J. Phys. Radium*, *7*, 7 (1946); *J. Chim. Phys.*, *46*, 298 (1949).
8. R. F. Brown, *J. Am. Chem. Soc.*, *83*, 36 (1961).
9. E. Constantinides, *Proc. Chem. Soc.*, *1964*, 290.
10. A. Yingst, *Chem. Commun.*, *1965*, 480.
11. D. H. McDaniel and A. Yingst, *J. Am. Chem. Soc.*, *86*, 1334 (1964).
12. R. S. Mulliken, *J. Chim. Phys.*, *46*, 497 (1949).
13. W. Moffitt, *Proc. Roy. Soc. (London) Ser. A*, *202*, 548 (1950).
14. J. Hinze and H. H. Jaffé, *J. Am. Chem. Soc.*, *84*, 540 (1962); *J. Chem. Phys.*, *67*, 1501 (1963).
15. R. P. Iczkowski and J. L. Margrave, *J. Am. Chem. Soc.*, *83*, 3547 (1961).
16. J. Hinze, M. A. Whitehead, and H. H. Jaffé, *J. Am. Chem. Soc.*, *85*, 148 (1963).
17. M. A. Whitehead and H. H. Jaffé, *Theoret. Chim. Acta*, *1*, 209 (1963).
18. J. E. Huheey, *J. Phys. Chem.*, *69*, 3284 (1965).
19. H. O. Pritchard, *J. Am. Chem. Soc.*, *85*, 1876 (1963).
20. M. A. Whitehead, N. C. Baird, and M. Kaplansky, *Theoret. Chim. Acta*, *3*, 135 (1965).
21. J. Hinze reported in ref. 20.
22. W. Gordy, *Phys. Rev.*, *69*, 604 (1946).
23. A. L. Allred and E. G. Rochow, *J. Inorg. Nucl. Chem.*, *5*, 264 (1958).
24. R. S. Drago, *J. Inorg. Nucl. Chem.*, *15*, 237 (1960).
25. J. K. Wilmshurst, *J. Chem. Phys.*, *27*, 1129 (1957).
26. W. Gordy, *J. Chem. Phys.*, *14*, 304 (1946).
27. J. K. Wilmshurst, *J. Chem. Phys.*, *28*, 733 (1957).
28. R. E. Kagarise, *J. Am. Chem. Soc.*, *77*, 1377 (1955).
29. J. V. Bell, J. Heisler, H. Tannenbaum, and J. Goldstein, *J. Am. Chem. Soc.*, *76*, 5185 (1954).
30. J. J. Lagowski, *Quart. Rev. (London)*, *13*, 233 (1959).
31. W. Gordy, W. V. Smith, and R. F. Trambaruto, *Microwave Spectroscopy*, Wiley, New York, 1953.
32. N. C. Baird and M. A. Whitehead, *Theoret. Chem. Acta*, *2*, 264 (1964).
33. W. J. Orville-Thomas, *Quart. Revs. (London)*, *11*, 162 (1957).
34. C. H. Townes and D. P. Dailey, *J. Chem. Phys.*, *17*, 782 (1949).
35. D. P. Dailey and C. H. Townes, *J. Chem. Phys.*, *23*, 118 (1955).
36. A. L. Allred and E. G. Rochow, *J. Am. Chem. Soc.*, *79*, 5361 (1957).
37. H. Spiesecke and W. G. Schneider, *J. Chem. Phys.*, *35*, 722 (1961).
38. D. P. Dailey and J. N. Shoolery, *J. Am. Chem. Soc.*, *77*, 3977 (1955).
39. C. H. Holm, *J. Chem. Phys.*, *26*, 707 (1957).
40. P. C. Lauterbur, *Ann. N.Y. Acad. Sci.*, *70*, 841 (1958).
41. T. Schaefer, *Can. J. Chem.*, *40*, 1 (1962).

42. N. Sheppard and R. M. Lynden-Bell, *Proc. Roy. Soc. (London) Ser. A, 269,* 385 (1962).
43. J. N. Shoolery, *J. Chem. Phys., 31,* 1427 (1959).
44. N. Muller and D. E. Pritchard, *J. Chem. Phys., 31,* 768 (1959).
45. P. C. Lauterbur, *J. Chem. Phys., 26,* 217 (1957).
46. N. Muller and D. E. Pritchard, *J. Chem. Phys., 31,* 1471 (1959).
47. A. W. Douglass, *J. Chem. Phys., 40,* 2413 (1964).
48. R. S. Drago and N. A. Matwiyoff, *J. Organometal. Chem. (Amsterdam), 3,* 62 (1965).
49. N. A. Matwiyoff and R. W. Taft, unpublished results.
50. D. M. Grant and W. M. Litchman, *J. Am. Chem. Soc., 87,* 3994 (1965).
51. N. C. Baird and M. A. Whitehead, *Can. J. Chem., 44,* 1933 (1966).
52. A. Streitwieser, *Molecular Orbital Theory for Organic Chemists,* Wiley, New York, 1961.
53. R. W. Taft, R. H. Martin, and F. W. Lampe, *J. Am. Chem. Soc., 87,* 2490 (1965).
54. P. R. Wells, *Chem. Rev., 63,* 171 (1963).
55. R. W. Taft, F. P. Prosser, L. Goodman, and G. T. Davis, *J. Chem. Phys., 38,* 380 (1963).
56. C. D. Ritchie and W. F. Sager, *Progress in Physical Organic Chemistry,* Vol. 2, Interscience, New York, 1964, p. 323.
57. P. R. Wells, *Linear Free Energy Relationships,* Academic Press, in press.
58. E. D. Jenson, L. D. McKeefer, and R. W. Taft, *J. Am. Chem. Soc., 86,* 116 (1964); *ibid., 88,* 4544 (1966).
59. L. D. McKeefer, Ph.D. Thesis, University of California, Irvine, 1966.

Substituent Effects in the Naphthalene Series. An Analysis of Polar and Pi Delocalization Effects*

P. R. WELLS

Department of Chemistry, University of California, Irvine, California

S. EHRENSON

Department of Chemistry, Brookhaven National Laboratory, Upton, New York

R. W. TAFT

Department of Chemistry, University of California, Irvine, California

I. INTRODUCTION†

In previous publications (1), it has been established that the substituent parameters, σ, of the Hammett equation can be usefully separated to components, σ_I, associated with polar or "inductive" effects, and σ_R, associated with pi electron delocalization or "resonance"

* This work was supported in part by the National Science Foundation (UCI) and the U.S. Atomic Energy Commission (Brookhaven).
† References for Sections I-V will be found on p. 186.

effects. Despite a sensitivity to solvent in some cases (3), the σ_I parameters are essentially invariant through a wide variety of reaction series so that it is possible to derive a unique parameter for each substituent. The σ_R parameters, on the other hand, are generally dependent upon both the substituent and the nature of the reaction processes (4). However, in the so-called "σ^0-reactivities," where the reaction takes place at some site effectively insulated from the pi electron framework of the benzene ring (5), a unique set of $_R\sigma^0$ parameters has been obtained. To a limited extent, useful "apparent" $\bar{\sigma}_R$ parameters can be found for most substituents for other types of reaction series involving "characteristic" changes in pi electron delocalization.

In the use of σ_I and σ_R^0 it is necessary to bear in mind the origin of these sets of characterizing parameters. The σ_I parameter is defined (1) by a proportionality to the aliphatic series polar effect parameter, σ^*. The σ_I values have been strongly supported as having applicability to polar effects in the benzene series by generalized correlations of the fluorine nuclear magnetic resonance (F NMR) shielding effects of *meta* substituents in fluorobenzenes (3) and of the IR intensities of the in-plane C—H stretching vibration in monosubstituted benzenes (1c). The σ_R^0 parameters are based upon the benzene series, with $\sigma_R^0 \equiv \sigma_{(p)}^0 - \sigma_I$. The σ_R^0 parameters are strongly supported by generalized correlations of the F NMR shifts of *para*- relative to *meta*-substituted fluorobenzenes (5), and of the IR intensities for the ν_{16} benzene skeletal mode of monosubstituted benzenes (1d). Consequently, the application of σ_I and σ_R^0 values to aromatic systems other than benzene involves the use of parameters in no way determined by observations in such aromatic systems.

The extension of this analytical approach beyond the original benzene system accomplishes two objectives. First, a more critical test is provided of the σ_I and σ_R^0 parameters. If these parameters are meaningful quantities characteristic of the substituent, they quite clearly should find application in other aromatic systems, independent of reaction or position of substitution. Secondly, the successful bisection of the total substituent effect into components from more than two aromatic positions (a frequent limitation in the benzene series) assists in understanding the mode of operation of substituent effects. Substituent effects in the naphthalene series are analyzed in these terms in this chapter. For the benefit of the reader unfamiliar with the earlier literature, five papers which summarize much of the background material relevant to this chapter have been reproduced in Appendixes 2–6.

II. THE BASIC EQUATION

The applicability of σ_I and σ_R^0 to aromatic systems other than benzene may be examined by the use of least squares computer fitting of the data to the dual substituent parameter equation (6):

$$\log (K/K_0)_i \quad \text{or other substituent property} \quad P^i = \sigma_I \rho_I^i + \sigma_R^0 \rho_R^i \quad (1)$$

where the susceptibility parameters ρ_I^i and ρ_R^i depend upon the position of the substituent (indicated by the index, i) with respect to the reaction (or detector) center, the nature of the measurement at this center, and the conditions of solvent and temperature. Equation (1) attributes the observed substituent effect to an additive blend of polar and pi delocalization effects which are scaled to corresponding effects in the benzene series; ρ_I^i and ρ_R^i may be regarded as the respective blending coefficients of these effects. It is useful to refer to the blending factor, λ, defined as $\rho_R/\rho_I \equiv \lambda$.

The successful application of eq. (1) to a few positions has been strikingly demonstrated by Bryson for ionization of naphthols and naphthylamines (7). In the present analysis, the treatment is extended to additional reactivities and to F NMR shielding effects, including all positions of naphthalene which are not subject to serious steric effects.

The following reaction series have been analyzed and are labeled according to the scheme:

A. Acidity of naphthoic acids in 50% aqueous ethanol at 25°C [Dewar and Grisdale (8), Wells and Adcock (9)]

B. Saponification of ethyl naphthoates in 85% ethanol at 50°C [Vaughan (10)]

C. Saponification of methyl naphthoates in 70% aqueous dioxane at 25°C [Wells and Adcock (11)]

D. Acidity of naphthylammonium cations in water at 25°C. [Bryson (7a)]

E. Acidity of naphthols in water at 25°C [Bryson (7b)]

F. Acidity of pyridinium, quinolinium, and isoquinolinium cations in water at 25°C [Perrin tables (12)]

G. Substituent F NMR chemical shifts in DMF solution [Adcock and Dewar (13)]

and according to the relative dispositions of substituent and reaction site, thus:

1. *meta* and *para* for benzene derivatives (for comparison with naphthalene results)
2. 3α for a 3-R-substituted α-naphthalene derivative,

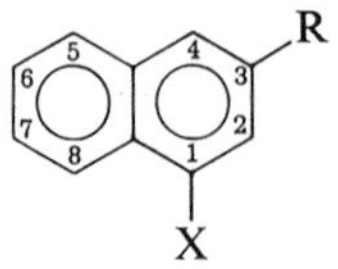

3. 4β for a 4-R-substituted β-naphthalene derivative,

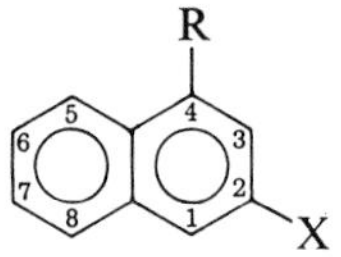

etc.

Table I lists the substituents used in this analysis and gives a summary of the σ_I and σ_R^0 values. In a few instances the values given have been slightly modified from the original values listed (5). The modifications are based upon a reexamination of presently available σ^0 reactivity and other data used in the original definition of σ_R^0 values. The reexamination of substituent effects in the benzene series will be reported elsewhere (14).

TABLE I

Substituent Parameters

Subst.	σ_I	σ_R^0	σ^0	Subst.	σ_I	σ_R^0	σ^0
NMe$_2$	0.05	−0.52	−0.47	I	0.39	−0.12	0.27
NH$_2$	0.10	−0.48	−0.38	CH$_3$	−0.05	−0.10	−0.15
NHAc	0.26	−0.22	+0.04	SMe	0.19	−0.17	0.02
OH	0.27	−0.44	−0.17	CF$_3$	0.41	0.13	0.54
OCH$_3$	0.26	−0.41	−0.15	MeCO	0.28	0.19	0.47
F	0.51	−0.34	0.17	CO$_2$R	0.31	0.15	0.46
Cl	0.47	−0.20	0.27	CN	0.52	0.14	0.66
Br	0.45	−0.16	0.29	NO$_2$	0.64	0.19	0.83

The substituents listed in Table I are a basic set from the standpoint of frequent use and the feature essential to the present application, namely, a pronounced nonlinearity between the σ_I and σ_R^0 parameters. A plot of these σ_I vs. σ_R^0 values (Fig. 1) affords what appears to be almost a random scatter diagram. Further, upon analysis in terms of the equation $\sigma_I = p + q\sigma_R^0$, we have found the standard deviation from the least squares line to be one-half the root mean square of the σ_I and

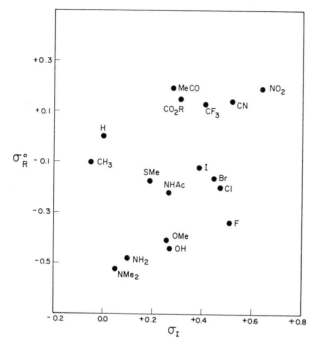

Fig. 1. σ_R^0 vs. σ_I for a basic set of substituent parameters.

σ_R values of Table I. Such poor precision is certainly indicative of nonlinearity between these parameter sets. Although the number of substituents in many of the reaction series is distinctly limited, the only series included in the present analysis are those for which σ_I and σ_R^0 for the available substituents are distinctly nonlinear. The data on which the present paper is based are summarized in Appendix 1 (Tables AI.I and AI.II). The need for additional data in many series is emphasized by the missing entries in this data summary.

The first observation to be made is that from a judgement based on the standard deviation (SD)—the naphthalene data are, with few exceptions, correlated at least as well as benzene data. A generally good correlation is one for which SD/rms ≤ 0.100, where rms is the root-mean-square value of the data. The notable exceptions to such generally good correlations appear for 4α-series data. The 4α position is formally similar to the *para* position of benzene where similar exceptions are frequently noted (4,5). With exclusion of the 4α data, the reactivity data (183 points) give an overall SD of 0.097, which is 10% of the rms. The F NMR shielding effects, including the 4α data, are fit with an overall SD of 0.370, which is 9% of the rms.

These precision measures may be compared with the results (14) of a similar analysis of 23 reaction series of the σ^0 type. A total of 265 *meta* and *para* log (k/k_0) points fit with a SD of 0.051, which is 12% of

TABLE II

Analysis of Substituent Effects on Reactivities in the
Naphthalene Series by eq. (1)

Reaction series	n	"Free-fitting" results			General K_I constraint results			General K_I and K_R constraint results		
		ρ_I	ρ_R	SD	ρ_I	ρ_R	SD	ρ_I	ρ_R	SD
A 4β	9	1.23	0.75	0.060	1.36	0.80	0.080	1.36	0.75	0.080
A 5β	5	0.83	0.58	0.027	0.76	0.57	0.040	0.76	0.56	0.041
A 6β	9	0.83	0.88	0.025	0.76	0.85	0.035	0.76	0.98	0.050
A 7β	11	0.79	0.58	0.046	0.76	0.56	0.047	0.76	0.56	0.047
A 8β	9	0.50	0.73	0.048	0.57	0.78	0.057	0.57	0.60	0.074
A 3α	6	1.36	0.65	0.058	1.36	0.65	0.058	—	—	—
A 5α	8	1.11	0.69	0.072	0.86	0.61	0.119	0.86	0.60	0.120
A 6α	7	0.79	0.69	0.025	0.76	0.68	0.026	0.76	0.56	0.041
A 7α	6	0.64	0.75	0.029	0.76	0.78	0.059	0.76	0.90	0.067
B 3α	4	2.20	0.86	0.035	2.20	0.86	0.035	—	—	—
C 6β	7	1.74	1.93	0.033	1.58	1.88	0.073	1.58	1.90	0.073
C 7β	5	1.42	1.11	0.071	1.58	1.14	0.105	1.58	1.10	0.106
D 4β	7	2.30	1.30	0.075	2.58	1.36	0.146	2.40	1.24	0.091
D 7β	4	1.36	1.00	0.016	1.44	1.03	0.040	1.35	0.93	0.026
D 3α	9	2.86	0.24	0.078	2.58	0.14	0.136	—	—	—
D 5α	3	1.59	1.01	0.027	1.62	1.02	0.032	1.51	0.99	0.043
D 6α	4	1.34	0.88	0.022	1.44	0.93	0.048	1.35	0.93	0.026
D 7α	4	1.66	1.56	0.039	1.44	1.47	0.097	1.35	1.49	0.136
E 4β	8	2.06	1.44	0.144	2.19	1.46	0.189	2.06	1.44	0.144
E 3α	7	2.35	0.27	0.083	2.19	0.21	0.104	—	—	—
F 4β	4	5.54	2.54	0.150	5.26	2.47	0.189	5.16	2.56	0.219
F 5β	3	2.60	1.21	0.124	2.94	1.24	0.188	2.89	1.92	0.318
F 6β	3	2.30	3.30	0.335	2.94	3.36	0.426	2.89	3.33	0.413
F 7β	3	2.28	2.08	0.016	2.94	2.14	0.268	2.89	1.92	0.262
F 8β	3	2.34	1.90	0.031	2.21	1.89	0.062	2.17	2.05	0.100
F 3α	8	5.44	1.46	0.151	5.26	1.35	0.163	—	—	—
F 5α	9	3.24	1.97	0.167	3.31	2.01	0.169	3.25	2.05	0.168
F 6α	10	2.77	2.10	0.052	2.94	2.19	0.078	2.89	1.92	0.097
F 7α	8	3.28	3.44	0.189	2.94	3.27	0.230	2.89	3.07	0.248

the rms, using the constraints: $\rho_I{}^m = \rho_I{}^p$, $\rho_R{}^m = 0.50 \, \rho_I{}^m$ and $\rho_R{}^m = 0.50 \, \rho_R{}^p$. Relaxation of all or part of these ρ constraints, of course, somewhat reduces the SD (in the limit to 10% of the rms).

Table II lists the individual ρ_I and ρ_R values according to reaction and position obtained by "free-fitting" to eq. (1), together with the number of data points and the SD. Also included in Table II are similar quantities obtained by placing additional constraints on the data, as will be discussed later. Table III gives similar results for "free-fitting" of the 4α data to eq. (1) and, for comparison, the results of a similar analysis for the same reactions of *meta-* to *para-*substituted benzene derivatives. It will be noted that the *meta-*benzene derivative fittings are closely similar in precision to those of the naphthalene data given in Table II. As noted above, the fits of the *para-*benzene and the 4α-naphthalene data are of comparably poorer precision.

TABLE III
Analysis of Substituent Effects on Reactivities in the
meta and *para* Benzene Positions and the 4α
Naphthalene Position by eq. (1)

	n	ρ_I	ρ_R	SD
A m	11	1.51	0.51	0.05
A p	12	1.46	1.77	0.11
A 4α	9	1.58	2.38	0.14
B m	8	2.47	1.32	0.06
B p	10	2.60	3.25	0.17
B 4α	6	2.04	2.39	0.11
D m	15	2.89	0.88	0.13
D p	15	3.85	4.47	0.36
E m	15	2.26	0.39	0.12
E ρ	15	3.01	2.22	0.42
F m	14	5.88	2.58	0.14
F p	13	4.27	6.38	0.54
F 4α	5	6.73	9.33	0.42

Table IV lists the results of analysis of substituent F NMR shielding effects by eq. (1). The values listed in Table IV are based upon the results of Adcock and Dewar in DMF [the only solvent common to all positions (13)]. For comparison, the results in methanol solution for the *meta* and *para* positions of benzene (3.15) are also listed in Table IV. It is to be noted that although these shielding data involve substantially different λ blending factors than are found for the corresponding

TABLE IV

Analysis of F NMR Substituent Shielding Effects in DMF

	n	rms	"Free-fitting" to eq. (1)			Fitting to eq. (1) with σ_R freedom[b]		
			ρ_I	ρ_R	SD	ρ_I	ρ_R	SD
3α	5	3.13	-6.84	-2.81	0.32	-6.85	-2.68	0.31
4α	6	10.01	-12.10	-32.12	0.69	-12.17	-30.44	0.51
6α	5	1.26	-2.20	-2.79	0.26	-2.21	-2.64	0.26
7α	8	1.89	-3.38	-5.15	0.29	-3.39	-4.90	0.28
4β	5	0.84	-1.78	$+1.04$	0.17	-1.77	$+0.96$	0.18
6β	8	4.77	-6.39	-14.74	0.49	-6.42	-13.96	0.44
8β	8	3.90	-8.87	-6.27	0.31	-8.88	-5.95	0.27
m	14	2.00	-6.78	-1.36[a]	0.50	—	—	—
p	14	8.76	-7.80	-32.10[a]	0.90	—	—	—

[a] In methanol.

[b] The resulting best values of $\bar{\sigma}_R$ were: NH_2, -0.52; OH, -0.49; and OCH_3, -0.41 (based upon data in all solvents).

positions in reactivities, they are fitted by a precision comparable to that for the naphthalene and *meta*-benzene reactivities, i.e., SD/rms $\sim$ 0.10. This fit is also somewhat better than that achieved for the methanol F NMR shifts in the *meta*- and *para*-benzene series.

III. ESSENTIAL FEATURES OF EQ. (1)

For subsequent discussion of the results, the positional and reaction series dependencies of ρ_I and ρ_R values will be considered separately. At this point we wish to draw the readers attention to the following points regarding eq. (1). This equation specifically resembles the original Hammett equation for *meta*- and *para*-substituted derivatives of benzene in that it involves two substituent parameters. The Hammett equation utilizes one set for the *meta* position and the other for the *para* position. In contrast, eq. (1) utilizes one set for the polar effect and one set for the resonance effect, but there is *no positional dependence* of either parameter. Equation (1) utilizes additional "reaction" parameters.

It is of further interest to compare eq. (1) with eq. (2) of Yukawa and Tsuno (16), which has been applied to electrophilic substitution in the benzene series (16).

$$\log (k/k_0) = \rho[\sigma + r(\sigma^+ - \sigma)] \qquad (2a)$$

$$\log (k/k_0) = \rho[\sigma^0 + r(\sigma^+ - \sigma^0)] \qquad (2b)$$

Formally, these equations have the form of eq. (1), i.e., there are two substituent (σ^0, σ^+) and two reaction parameters (r, ρ). However, the Yukawa and Tsuno equations, like the Hammett equation, places the positional dependence in the substituent rather than in the reaction parameter. Thus, in the benzene series this treatment actually utilizes an additional set (3 altogether) of substituent parameters: a set of $\sigma_{(m)}$, a set of $\sigma_{(p)}$ [or of $\sigma_{(p)}^0$], and a set of $\sigma_{(p)}^+$.

In most proposed applications of the Hammett equation (presumably also the Yukawa and Tsuno equation) to other aromatic systems, a new set of σ values is (or may be) required for every ring position. Thus, it is highly probable that eq. (1) affords in such applications a substantial reduction in the number of parameters. Equation (1), of course, reduces to a single substituent parameter treatment at a given position if appropriate constraints are applicable to the reaction series constants, ρ_I and ρ_R. Thus, for σ^0 reactivities, for example, the constraints $\rho_I^m/\rho_I^p = 1.00$, $\rho_R^m/\rho_I^m = 0.50$, and $\rho_R^m/\rho_R^p = 0.50$ give (1):

$$\log (k/k_0)_m = \rho_I^m(0.50\sigma_R^0 + \sigma_I) = \rho\sigma_{(m)}^0$$

$$\log (k/k_0)_p = \rho_I^m(\sigma_R^0 + \sigma_I) = \rho\sigma_{(p)}^0$$

Although the constraints in this specific instance are not very serious (cf. above discussion), in general, such a series of highly restrictive constraints may be unrealistic.

In the general application of eq. (1), the substituents used must not be limited to a series of structurally similar substituents. The linear relationships between σ_R^0 and σ_I, which are known for narrowly defined families of structurally similar substituents (15), preclude meaningful analysis by eq. (1). As noted above, the full series of substituents listed in Table I, for example, avoids this difficulty. A minimal recommended set of seven substituents from Table I is CH_3; any two of NMe_2, NH_2, OH, and OMe; any two of the halogens (except both Cl and Br); and any two of the +R substituents.

Finally, it must be appreciated that the precision of fit achieved by eq. (1) is also applicable to the equivalent relationship (3):

$$\log (k/k_0)_i \quad \text{or} \quad P^i = \alpha\sigma_I + \beta\sigma^0 \qquad (3)$$

Equation (3) is obtained by substituting into eq. (1) the defining (2) eq. (4):

$$\sigma_R^0 \equiv \sigma_{(p)}^0 - \sigma_I \qquad (4)$$

from which it follows that $\alpha = \rho_I - \rho_R$ and $\beta = \rho_R$. Consequently, the generality and precision of eq. (1) in itself does not provide evidence for the correctness of the proposed bisection of σ^0 (eq. 4).

It might be proposed instead of the special case ($\gamma = 1.0$) represented by eq. (4) that $\sigma_R' \equiv \sigma_{(p)}^0 - \gamma\sigma_I$, where γ is any constant and σ_R' is a more correctly defined pi-delocalization effect parameter than σ_R^0. If, then (5) $\log (k/k_0) = \sigma_I\rho_I' + \sigma_R'\rho_R'$, it follows that $\rho_R' = \beta = \rho_R$ and $\rho_I' = \rho_I + (\gamma - 1)\rho_R$. Thus the same resonance effect rho value is obtained whether eqs. (1) or (3) or the family of eqs. (5) is utilized. Therefore, the adherence of the positional dependence of this reaction parameter to a unique scheme expected by a theoretical model also does not provide evidence for the correct bisection of σ^0. As indicated, this result may be achieved without separation of σ^0!

A decision as to the correct γ value can be based upon the finding of a unique set of ρ_I or ρ_I' values which display no (or a minimum) dependence upon ρ_R (or ρ_R') and show a positional pattern dependence which is expected by theoretical models for the polar effect. Fortunately, chemical considerations confine the acceptable value of γ at least within the range $+0.5$ to $+1.3$. That is, $\gamma > 1.3$ gives, for example, the unacceptable result that σ_R' values for NO_2 and CN are negative quantities (net pi donor) and $-\sigma_R'$ values for F are equal or greater than those for OCH_3 and NH_2. For $\gamma \leq 0.5$, σ_R' values for the halogens take on unacceptable positive values (net pi acceptor).

Although it is difficult to eliminate with certainty the possibility that γ may differ from unity within ± 0.2, the results discussed in the following sections based upon $\gamma = 1.0$ show that little or no apparent correlation exists between the positional dependencies of ρ_I and ρ_R values. Using $\gamma \neq 1.0$, we have been unable to find any evidence that this situation may be improved or that better agreement of ρ_I' rather than ρ_I values can be achieved with the presently available theoretical models.

As intimated here and detailed below (cf. Table V), simple theoretical models are adopted in an attempt to explain the total observed effects of substituents in terms of four distinctly segmented interaction models. Because of inherent complexities in the actual systems, a degree of arbitrariness in the separations is inevitable. Dewar and Grisdale

TABLE V

Simplified Component Interaction Models for Propagation and Distribution of Substituent Electronic Effects

Interaction type[a]	Symbol[b]	Models for interaction	Atoms in π system[c]	Type effect[d]
1. Classical inductive		Direct **R** with X[e] a) By field effect: $D^{-1}r^{-2}\cos\theta$ or $(Dr)^{-1}$ b) By sigma bond transmission	6	I
2. Pi-inductive inducto-electromeric		R to C_1 by σ bond and/or field; C_1 to other ring sites by π-delocalization; from ring sites to X by field, σ-bonds or π-delocalization	7	I
3. Resonance or mesomeric polar		R to ring C's by π-delocalization; from these sites to X by field and/or σ-bonds	7	R
4. Direct conjugation or electromeric		Direct R with X by π-delocalization	8	R

[a] Second-order effects, for example such as are expected for the σ-bond framework in interaction type 1 caused by the presence of π-electrons, and X-induced ring—π-charge field interactions, as in interaction type 2, with R are ignored in this presentation.

[b] Analogous descriptions are applicable to *ortho* and *meta* derivatives.

[c] R and X are considered as united atoms.

[d] $I = \rho_I\sigma_I$; $R = \rho_R\sigma_R$.

[e] Pi-electrons omitted since in field models these can only exert an effect through the effective dielectric, D.

have attempted to avoid the complexities of deriving the total substituent effect from the models of Table V by combining the σ values of
Hammett for the benzene system with physical quantities based upon
simplified theoretical considerations (8). In this manner σ values are
derived which are presumably applicable to naphthalene and other
aromatic systems.

The line of investigation followed by Dewar and Grisdale has
been avoided here. While such hybrid treatments are of potential
correlational and conceptual use when data are quantitatively fitted,
the treatment provides little of use to aid in understanding of the
frequently observed deviations. Our approach has been to maintain a
separation between the statistical fittings and the theoretical results so
that the correlational use of the former is not demeaned by the at
present inevitable deficiences in the latter.

IV. DISCUSSION

A. Polar Effects

The reaction constants, ρ_I (of Table II), obtained by "free-fitting"
to eq. (1) for a given reactivity series, tend to closely similar values at
certain positions. This result is summarized in Table VI, which lists
position-averaged ρ_I values ($<\rho_I>$) for various reaction series. Using
the 3α, 4β average as standard, average-relative ρ_I values,

$$<K_I> \; \equiv \; <\rho_I>/<\rho_I(3\alpha, 4\beta)>,$$

are also listed in Table VI. Mean deviations from $\langle\rho_I\rangle$ and $\langle K_I\rangle$ values
are also listed in Table VI.

Although the deviations of individual ρ_I from $\langle\rho_I\rangle$ for the positions
averaged in Table VI do not appear to be significant, there is a substantial presumption from the results that $\langle K_I\rangle$ has some dependence upon
reaction type. (Note in particular the variable $\langle K_I\rangle$ for the 6α, 5β, 6β,
and 7β positions between reactions A and F.) This conclusion is further
supported by the results of imposing on all the reactivities the still more
restrictive I-effect constraints (independent of reaction type): $K_I \equiv 1.00$
for 3α, 4β; $K_I = 0.56$ for 5β, 6β, 7β, 6α, and 7α; $K_I = 0.63$ for 5α; and
$K_I = 0.42$ for 8β. The results of these constraints are shown in Table II
under the heading "General K_I constraint." It will be noted that, in
general, the K_I constraint has relatively little effect on ρ_R values, but
the overall SD is significantly increased from 10 to 13% of the rms. The

TABLE VI

Average Values of ρ_I from "Free-Fitting" of eq. (1)

Reaction series	Positions averaged	$\langle \rho_I \rangle$	$\langle K_I \rangle$
A	m, p	1.49 ± 0.03	1.15 ± 0.10
A	$3\alpha, 4\beta$	1.29 ± 0.07	(1.00)
A	$6\alpha, 5\beta, 6\beta, 7\beta$	0.81 ± 0.02	0.63 ± 0.05
D	$3\alpha, 4\beta$	2.58 ± 0.28	(1.00)
D	$6\alpha, 7\beta(5\beta, 6\beta?)$	1.35 ± 0.01	0.52 ± 0.07
D	$5\alpha, 7\alpha$	1.63 ± 0.03	0.63 ± 0.09
E	$3\alpha, 4\beta$	2.21 ± 0.16	(1.00)
F	$3\alpha, 4\beta$	5.49 ± 0.05	(1.00)
F	$6\alpha, 5\beta, 6\beta, 7\beta$	2.51 ± 0.20	0.46 ± 0.04
F	$5\alpha, 7\alpha$	3.36 ± 0.02	0.59 ± 0.01

generally increased SD for individual reactions and positions may be seen by inspection of Table II. Nevertheless, it must be agreed that the overall SD for the "general K_I constraint" is such that this treatment does give a reasonable first-approximation account of the naphthalene reactivities.

There seems to be good grounds for suggesting that the $\sigma_I \rho_I$ term in the naphthalene series, especially for substituents positioned in the second ring, arises from two major effects: the field and pi inductive effects (cf. Table V). Dewar and Grisdale (8) have argued that the field effect, to good approximation, may be represented as a charge–charge interaction involving only that end of the dipole closest to the aromatic ring. The relative effect is then expected to vary with position according to $1/r_{rX}$, where r_{rX} is the distance between the ring atom, r, to which the substituent is attached, and the charge site, X, of the reacting group. Alternatively, this effect may be approximated, following the Kirkwood-Westheimer model (17), in terms of the interaction of X and a point dipole at the substituent R by $\cos \theta_{RX}/r_{RX}^2$, where r_{RX} is the distance between the point dipole and X, and θ_{RX} is the angle between the dipole vector and the vector $\mathbf{r}_{RX}$. Table VII summarizes the relative magnitudes expected for the direct electrostatic interaction, assuming a common dielectric constant. Calculations were made for various values of $\gamma = r_{C-X}/r_{C-C}$ and various point dipole positions. Reaction series F corresponds to $\gamma = 0$; series D and E to $\gamma \simeq 1$; and series A, B, and C to $\gamma \simeq 1.5$, if solvation does not seriously displace the effective centers of charge.

TABLE VII
Relative Field Effects

Position	Relative $1/r_{rx}$				Relative $\cos \theta_{RX}/r^2_{RX}$			
	$\gamma = 0$	$\gamma = 1$	$\gamma = 1.5$	Average	$\gamma = 0$	$\gamma = 1$	$\gamma = 1.5$	Average
m, 3α, 4β	1.00	1.00	1.00	1.00	1.00	1.00	1.00	1.00
p, 4α	0.86	0.87	0.91	0.88 ± 0.02	0.83 ± 0.00	0.87 ± 0.03	0.89 ± 0.02	0.86 ± 0.02
6α	0.57	0.74	0.78	0.70 ± 0.08	0.46 ± 0.03	0.64 ± 0.00	0.71 ± 0.02	0.60 ± 0.10
5β	0.57	0.66	0.70	0.64 ± 0.05	0.38 ± 0.07	0.43 ± 0.04	0.44 ± 0.06	0.42 ± 0.02
7β	0.50	0.61	0.66	0.59 ± 0.06	0.36 ± 0.04	0.44 ± 0.02	0.47 ± 0.04	0.42 ± 0.04
5α	0.66	0.76	0.81	0.74 ± 0.06	0.53 ± 0.05	0.44 ± 0.02	0.71 ± 0.03	0.57 ± 0.10
7α	0.66	0.87	0.94	0.82 ± 0.11	0.53 ± 0.05	0.6–0.8	0.6–0.85	Variable high
6β	0.48	0.58	0.63	0.56 ± 0.06	0.34 ± 0.03	0.44 ± 0.01	0.48 ± 0.03	0.43 ± 0.06
8β	0.66	0.87	0.94	0.82 ± 0.11	0.2–0.6	0.25–0.5	0–0.4	Variable low

The relative merits of these two approximations to the expected relative field effects have already been discussed [Wells and Adcock (9)]. Either approach roughly gives agreement with observed $\langle K_I \rangle$ pattern (cf. Table VI) for 6α, 5β, 6β, and 7β positions, and indicates that a somewhat larger $\langle K_I \rangle$ is expected for the A than the F series reactions. The difference in the two approximations mainly rests upon the predicted 7α and 8β effects, which in the $1/r$ treatment are large (relative to the other effects predicted for the naphthalene 6α, 5β, 7β, 5α, and 6β positions), but which in the $\cos\theta/r^2$ treatment are very sensitive to the relative orientation of dipole and reaction site. Also taking into account the fact that a higher effective dielectric constant may be more appropriate in these two cases than in all other cases, the $\cos\theta/r^2$ approach suggests the 7α effect may not be large compared with the others, while the 8β effect is probably quite small. This is indeed observed in the pattern of relative ρ_I values (cf. "free-fitting" results of Tables II and VI).

The pi inductive effect (Table V) is the second major effect which may be included in the $\sigma_I \rho_I$ term. An attempted assessment of the relative magnitudes of this effect are reported in Table VIII for cases

TABLE VIII
Relative Pi Inductive Effects

Position	π_{rx}	Relative $\sum \pi_{\mathrm{ri}}/r_{\mathrm{ix}}$			Average
		$\gamma = 0.5$	$\gamma = 1$	$\gamma = 1.5$	
		Nonconjugative R			
m	-0.009	1.00	1.00	1.00	1.00
3α	-0.018	-0.19	0.03	0.15	
4β	-0.018	1.25	1.37	1.41	1.34 ± 0.06
p	0.102	3.25	2.03	1.59	
4α	0.140	4.47	2.76	2.26	
6α	-0.006	0.51	0.45	0.56	0.51 ± 0.04
5β	-0.006	0.58	0.68	0.82	0.7 ± 0.1
7β	0.000	0.92	1.08	1.22	1.1 ± 0.1
5α	0.023	1.32	1.32	1.33	1.32 ± 0.00
7α	0.032	1.60	1.26	1.22	1.35 ± 0.15
6β	0.033	1.34	0.87	0.78	1.0 ± 0.2
8β	0.032	0.40	-0.36	-0.74	

where the reaction site is insulated from conjugation with the aromatic system. The simplest approach is to suppose that this effect arises from changes at the atom to which the reacting side chain is attached. For a nonconjugating substituent this may be approximated by the HMO atom–atom polarizabilities of the parent hydrocarbon, π_{rx}, and, for conjugating substituents (represented as a pseudoatom, R), by the atom–atom polarizabilities, π'_{rx}, corresponding to the aryl methyl radical (18). Granting this, one would expect that all the framework atoms would undergo pi-electron density perturbations (polarizations) from the inductive effect exercised by the substituent at its point of attachment. This distortion may reach the reaction site by a means which can be represented as direct electrostatic interactions. Effectively, the modified pi electron distribution is regarded as a multipole interacting with the reaction site. The relative effects may then be approximated by summations over all ring atoms, i, of π_{ri}/r_{iX} for nonconjugating substituents and π'_{ri}/r_{iX} for conjugating substituents. That is, the multipole is approximated as six or seven point charges which are proportional to atom–atom polarizabilities. It should be noted that transmission by ring σ-bond polarization to the reaction center, involving the familiar bond falloff model, is not invoked in Table VIII. Nor has this effect been included in the estimation of the classical inductive effect shown in Table VII.

The pi-inductive mechanism has been semiquantitatively treated in a slightly different manner using a perturbation treatment in a modified Hückel framework with σ-bond falloff to evaluate pi-inductive energy effects in benzene reactivities (cf. Appendix 3). This mechanism of transmission is more important in *para*-type (ring positions capable of direct conjugation with one another) than in *meta*-type substitutions, since direct modification of the ring charge by this mechanism at the position bearing the reaction group is possible in the former, but not in the latter, position.

The magnitudes of the estimated relative pi inductive effects given in Table VIII (as well as those estimated for conjugating substituents) show little correspondence to the observed relative ρ_I pattern for naphthalene reactivities. This result may be due to either the inadequacy of the pi inductive effect calculations or to the fact that the pi inductive contributions to the $\sigma_I \rho_I$ term are of secondary importance relative to the direct field effects (19). Unfortunately, a decision between these alternatives is not presently possible. What is of consequence, however, is the fact that the naphthalene series provides a sufficient number of positional dependencies that a decision may ultimately be possible.

A third effect on $\sigma_I \rho_I$ (of a somewhat indirect nature) may be expected in the substituted naphthalenes. In the cases where the substitution is in the 4, 5, and 8 positions, the perihydrogen exerts a twisting effect on those substituents having planarity demands with the ring system. This effect would tend to decrease the effective σ_R value for these substituents. However, because there are substitution positions which are not so influenced in the entire set of naphthalene reactivity data, a counter tendency exists toward retention of normal σ_R values. Hence, it is reasonable to expect some of the effect of twisting to be numerically shifted off to the ρ_I's, especially if groups at one end of the σ_R scale are affected more strongly than at the other. The $+R$ groups, such as NO_2, CO_2R and $COMe$, would be influenced by this steric inhibition of resonance factor as well as the $-R$ substituent, $N(CH_3)_2$. However, no reactivity data are available for the latter substituent in the indicated positions. Consequently, we would expect that in the usual cases where ρ_I and ρ_R are of the same sign, this steric inhibition of resonance effect would be reflected in an increase in the magnitude of ρ_I, concomitant with a partial decrease in ρ_R. This would have the effect of making the 4α, 5α, and 8β ρ_I proportionality factors appear somewhat greater than they would in the absence of such a steric effect, as well as con- tributing to somewhat decreased values of ρ_R. Such an influence appears to exist for the ρ values for the 4α and 8β cases.

The comparison of reactivity and F NMR shielding ρ_I values as a function of position provides striking confirmation of the utility of eq. (1) in uncovering what must surely be regarded as two different polar or inductive interaction mechanisms. Table IX compares the ρ_I values from "free-fitting" to eq. (1) for these two different sources (Tables II and IV), using the A series reactivity data as approximately typical.

Values of $K_I \equiv \rho_I/\rho_I^{4\beta}$ are included in Table IX. From these values it is immediately apparent that the positional dependencies of the two sets of data have little or nothing in common. Although there are some medium effects which should be taken into account in the comparison, it is very probable that these are too small to affect the main conclusion. The conclusion is also unaffected if one allows that the ρ_I for F NMR shielding effects at the 4β position may be anomalous. The order of increasing ρ_I values for F NMR is 4β, $6\alpha < 7\alpha < 6\beta$, 3α, meta, para $< 8\beta < 4\alpha$. The order for A series values has virtually nothing in common with the latter: $8\beta < 6\alpha$, 7α, 5β, 6β, $7\beta < 5\alpha < 4\beta$, $3\alpha <$ meta, para, 4α.

Adcock and Dewar have made the important suggestion that the substituent effect in F NMR shielding is due to a polarization of the electrons forming the C—F bond so that the field effect is dependent only (in contrast to reactivity field effects) upon the field along the bond

TABLE IX

Comparison of ρ_I Values for F NMR Substituent Shielding
Effects and A Series Reactivity Effects

Position	F NMR ρ_I	F NMR K_I	Aρ_I	AK_I
meta	−6.78	3.8	1.51	1.2
para	−7.80	4.3	1.46	1.2
3α	−6.84	3.8	1.36	1.1
4α	−12.10	6.7	1.58	1.3
5α	—	—	1.11	0.9
6α	−2.20	1.2	0.79	0.6
7α	−3.38	1.9	0.64	0.5
4β	−1.78	(1.00)	1.23	(1.00)
5β	—	—	0.83	0.7
6β	−6.39	3.6	0.83	0.7
7β	—	—	0.79	0.6
8β	−8.87	4.9	0.50	0.4

axis (15). Using the point charge model, these authors consider the important variable to be cos θ, where theta is the angle defined by a line from the midpoint of the C—F bond to the polarizing group and the line of the C—F bond. This consideration does explain the relatively high F NMR K_I values in Table IX for the 3α, 4α, *meta*, *para*, 6β, and 8β positions and the relatively low values for the 4β, 6α, and 7α positions. However, the higher K_I for 7α (1.9) than for 6α (1.2), for example, cannot be accounted for by this consideration. Clearly a satisfactory quantitative interpretation of the F NMR ρ_I values is needed but is not available at the present time.

Unfortunately, Adcock and Dewar have detracted from their interpretation by attempting to correlate the observed substituent effects in the F NMR shieldings with substituent effects on the naphthoic acid ionizations (A series reactivities), i.e., $\int (F) = a(pK_a) + b$. As indicated by the present analysis, this correlation is of poor quality, although the quoted correlation coefficients are misleading in this regard.

Table X compares the precision measures for Dewar's correlation of the F NMR substituent shielding effects and those obtained by the application of eq. (1). Since the data available for fitting to the two relationships is not identical, the number of points and the rms of the data are given for both correlations. Attention is called in particular to the poor measure of precision of fit provided by the correlation

TABLE X

Comparative Precision Measures for Correlation of F NMR
Substituent Shielding Effects in the Naphthalene Series

Position	rms	n	SD	Correlation coefficient
Correlation by eq. (1): $\int (F) = \sigma_I \rho_I + \sigma_R^{\,0} \rho_R$				
3α	3.13	5	0.32	
4α	10.01	6	0.69	
6α	1.26	5	0.26	
7α	1.89	8	0.29	
4β	0.84	5	0.17	
6β	4.77	8	0.49	
8β	3.64	8	0.31	
Correlation by $\int (F) = a(pK_a) + b$				
3α	3.45	4	0.50	0.96
4α	10.13	5	1.47	0.99
6α	1.29	4	0.15	0.98
7α	1.80	6	0.23	0.99
4β	0.72	5	0.28	0.63
6β	4.19	5	0.69	0.98
8β	3.80	6	0.87	0.96

coefficient. Thus, for example, the correlation coefficient of 0.96 for Dewar's relationship applied to the 8β data corresponds to a SD of 23% of the rms. In contrast, the 8β data are fitted by eq. (1) to a SD of 9% of the rms.

It is appropriate at this point to draw attention to the inherent reliability of the SD/rms ratio as a statistical measure reflecting success of fitting, in contrast to the often used and occasionally abused correlation coefficient measure. Since in least squares procedures the SD is the quantity explicitly minimized, comparison among series requires normalization which is provided by the rms for the series. In comparison of cases where the number of parameters to be determined are grossly different, the measure GOF/rms is recommended. The goodness-of-fit (GOF) is the sum of the squares of individual deviations divided by the quantity given by the number of data minus the number of parameters determined (6c).

B. Pi Electron Delocalization Effects

The reaction constants, ρ_R, of Table II obtained by "free-fitting" to eq. (1) are both strongly reaction and position dependent. There is only a limited tendency for ρ_R values to define average values such as those listed in Table VI for ρ_I. In Table XI are given relative ρ_R values

TABLE XI

The Positional Dependencies of ρ_R Values from Eq. (1)

Reaction	Position	ρ_R	K_R
A	3α	0.65	0.86
D	3α	0.24	0.18
E	3α	0.27	0.19
F	3α	1.46	0.57
A	4β	0.75	(1.00)
D	4β	1.30	(1.00)
E	4β	1.44	(1.00)
F	4β	2.54	(1.00)
A	5β	0.58	0.77
F	5β	1.21	0.48
A	7α	0.75	1.00
D	7α	1.56	1.20
F	7α	3.44	1.35
A	7β	0.58	0.77
D	7β	1.00	0.77
F	7β	2.08	0.82
A	8β	0.73	0.97
F	8β	1.90	0.75

defined as $K_R \equiv \rho_R/\rho_R^{4\beta}$, which illustrate this result. Values of K_R vary from as small as 0.18 for D 3α to as large as 1.35 for F 7α or 3.68 for F 4α. At the 3α position, K_R varies from 0.18 to 0.86, at 5β from 0.48 to 0.77, and at 8β from 0.75 to 0.97. Only the 7β position gives essentially equal K_R values for all of the available reaction series. Different orders of ρ_R are obtained for each kind of reaction. For reaction series A, the order is 5β, $7\beta < 3\alpha < 5\alpha$, 6α, 4β, $8\beta < 7\alpha < 6\beta$; for reaction series F, $5\beta < 3\alpha < 5\alpha$, 6α, 7β, $8\beta < 4\beta < 7\alpha$, 6β; for reaction series D, $3\alpha < 6\alpha < 5\alpha$, $7\beta < 4\beta < 7\alpha$. It is also noteworthy that none of these progressions has any correspondence to the approximately generalized order of increasing reactivity ρ_I values, $8\beta < 5\beta$, 6β, 7β, 6α, $7\alpha < 5\alpha < 3\alpha$, 4β.

In the far right-hand columns of Table II are shown the results of placing the following K_I and K_R constraints on all of the naphthalene reactivity data:

	$K_I \equiv \rho_I/\rho_I^{4\beta}$	$K_R = \rho_R/\rho_R^{4\beta}$
3α	1.00	Variable
5α	0.63	0.80
5β	0.56	0.75
6α	0.56	0.75
6β	0.56	1.30
7α	0.56	1.20
7β	0.56	0.75
8β	0.42	0.80

The resulting ρ_I, ρ_R, and SD are listed in Table II under the heading "General K_I and K_R constraint." The overall SD is increased from 10% of the rms for "free-fitting" to 16% of rms, and, in general, individual SD are increased over that for "free-fitting" or for only the K_I constraint. The increased SD, we believe, is significant and supports the conclusion that K_R is both position and reaction dependent.

It is clear that none of the reaction series A–F belong to the σ^0 type (1b) so that σ_R^0 values may, in fact, be inappropriate parameters for the pi delocalization effects in the naphthalene series. For example, *para*-benzoic reactivities can be correlated with much improved precision by replacing σ_R^0 by a set of σ_R values. The σ_R set differs primarily in that the values of strong $-R$ substituents are substantially more negative (1). In Table XII are given the results obtained when the naphthalene reactivity data are processed iteratively only under the above K_I constraints to obtain the best set of effective $\bar{\sigma}_R$ values for the strong $-R$ substituents: NMe_2, NH_2, OH, and OMe. For comparison, Table XI also contains the results for "free-fitting" and for fitting with the K_I constraint, but without σ_R freedom. The overall SD is reduced from 13 to 12% of rms upon freeing the σ_R values for the strong $-R$ substituents. The "best values" of the resulting $\bar{\sigma}_R$ are: NMe_2, -0.55; NH_2, -0.67; OH, -0.55; and OCH, -0.45. The value obtained for NMe_2 is probably not very reliable since it is based upon only 4 points (cf. Appendix, Table AI.I).

Because of the somewhat reduced SD and increased negative $\bar{\sigma}_R$ values for each of the strong $-R$ substituents, there is the strong presumption that at least several of the naphthalene positions involve the effects of direct delocalization interaction between substituent and reaction center. It appears significant and expected that the most conjugative positions, 7α and 6β, show the most marked reduction of SD (in F reaction series) resulting from relaxing the σ_R^0 constraints for the strong $-R$ substituents.

TABLE XII
Results of Giving σ_R Freedom to Strong $-R$ Substituents

		"Free-fitting" to eq. (1)		Fitting with general K_I constraint without σ_R freedom		Fitting with general K_I constraint with σ_R freedom	
		ρ_R	SD	ρ_R	SD	ρ_R	SD
A 4β	9	0.75	0.060	0.80	0.080	0.64	0.100
A 5α	5	0.58	0.027	0.57	0.040	0.49	0.045
A 6β	9	0.88	0.025	0.85	0.035	0.82	0.032
A 7β	11	0.58	0.046	0.56	0.047	0.49	0.037
A 8β	9	0.73	0.048	0.78	0.057	0.71	0.053
A 3α	6	0.65	0.058	0.65	0.058	0.56	0.062
A 5α	8	0.69	0.072	0.61	0.119	0.50	0.118
A 6α	7	0.69	0.025	0.68	0.026	0.60	0.020
A 7α	6	0.75	0.029	0.78	0.059	0.68	0.060
B 3α	4	0.86	0.035	Not affected		Not affected	
C 6β	7	1.93	0.033	1.88	0.073	1.80	0.077
C 7β	5	1.11	0.071	1.14	0.105	1.11	0.109
D 4β	7	1.30	0.075	1.36	0.146	1.29	0.150
D 7β	4	1.00	0.016	1.03	0.040	0.90	0.041
D 3α	9	0.24	0.078	0.14	0.136	0.13	0.133
D 5α	3	1.01	0.027	1.02	0.032	0.87	0.012
D 6α	4	0.88	0.022	0.93	0.048	0.81	0.047
D 7α	4	1.56	0.039	1.47	0.097	1.27	0.105
E 4β	8	1.44	0.144	1.46	0.153	1.25	0.105
E 3α	7	0.27	0.083	0.21	0.104	0.19	0.100
F 4β	4	2.54	0.150	2.47	0.189	1.90	0.185
F 5β	3	1.21	0.124	1.24	0.188	0.94	0.194
F 6β	3	3.30	0.335	3.36	0.426	2.62	0.309
F 7β	3	2.08	0.016	2.14	0.268	1.66	0.224
F 8β	3	1.90	0.031	1.89	0.062	1.42	0.125
F 3α	8	1.46	0.151	1.35	0.163	1.10	0.174
F 5α	9	1.97	0.167	2.01	0.169	1.65	0.175
F 6α	10	2.10	0.052	2.19	0.078	1.82	0.120
F 7α	8	3.44	0.189	3.27	0.230	2.72	0.138

This point is further illustrated in Table XIII which gives results for the F series reaction fitted alone under three conditions: (a) "free-fitting" to eq. (1); (b) with the "averaged" I constraint, $K_I = 0.49 = \rho_I/\rho_I^{4\beta}$; and (c) with $K_I = 0.49$ constraint but σ_R freedom for NMe_2, NH_2, OH, and OMe substituents. Only the 7α and 6β positions give a reduction in SD for condition c as compared with that for either a or b.

TABLE XIII
ρ_R Values for the Ionization of Quinolinium and
Isoquinolinium Ions (F Series)[a]

		(a) "Free-fitting"		(b) $K_I = 0.49$ and $\sigma_R{}^0$		(c) $K_I = 0.49$ and σ_R freedom	
		ρ_R	SD	ρ_R	SD	ρ_R	SD
4β	4	2.54	0.150	2.58	0.167	1.84	0.151
5α	9	1.97	0.167	1.75	0.232	1.35	0.222
6α	10	2.10	0.052	2.11	0.053	1.61	0.116
7α	8	3.44	0.189	3.20	0.265	2.50	0.169
5β	3	1.21	0.124	1.23	0.150	0.85	0.175
6β	3	3.30	0.335	3.34	0.393	2.39	0.252
7β	3	2.08	0.016	2.13	0.211	1.51	0.167
8β	3	1.90	0.031	1.94	0.188	1.37	0.168

[a] Values of $\rho_I(4\beta)$ are: (a) 5.54; (b) 5.72; (c) 5.76.

The F NMR substituent shielding effects consist of distinctly different blends of polar and pi delocalization as compared to reactivity substituent effects (Table XIX summarizes values of the blending factor, λ). As discussed in the previous section, the positional dependencies of ρ_I values differ drastically. In striking contrast, the positional dependencies of ρ_R values appear to display essentially identical patterns for the appropriate comparison between these two kinds of measurements.

Unquestionably, among the several reaction series, the A series, involving a change from CO_2H to $CO_2{}^-$ at the reaction center, provides the minimum perturbation of the reaction center on the naphthalene system. It seems therefore appropriate to compare ρ_R's for the A reaction series and the F NMR shielding series. This comparison is made in Table XIV, which lists (when available) the F NMR ρ_R values in both cyclohexane and methanol solutions (3,15,17). Considering the relatively small variation with solvent, the order of increasing ρ_R values, meta $< 3\alpha$, $6\alpha < 7\alpha$, $8\beta < 6\beta < para < 4\alpha$, is the same for both reactivity and F NMR shielding. The F NMR ρ_R value for the 4β position ($+1.04$) is anomalous and is not included. It is probably significant, however, that this ρ_R is relatively small (approaching zero), which is in keeping with the results at the similar meta and 3α positions.

For the reasons indicated, the results in Table XIV provide impressive evidence that the $\sigma_R{}^0\rho_R$ terms for reactivity and F NMR shielding effects have a common origin in the effects of substituent pi electron delocalization. Support is also provided for the presumption

TABLE XIV

Comparison of F NMR and Naphthoic Acid Ionization
ρ_R Values

	F NMR $-\rho_R$		A Series ρ_R
Position	Cyclohexane	MeOH	50% aq. EtOH
meta	0.62	1.36[a]	0.51
3α	—	2.81[a]	0.65
6α	—	2.79[a]	0.69
7α	5.46	4.81	0.75
8β	4.92	5.74	0.73
6β	12.62	12.45	0.88
para	30.38	32.10	1.80
4α	—	32.12[a]	2.38

[a] In DMF.

that pi electron delocalization effects are separated from polar effects by eq. (1).

F NMR substituent shielding effects also appear to support the conclusion that the positional dependence of pi electron effects has an appreciable dependence upon reaction type. In Table XV are given

TABLE XV

Comparison of F NMR Shielding Changes and ρ_R Values for
the Ionization of Quinolinium and Isoquinolinium Ions

		Free-fitting	$K_I = 0.49$
Position	$-\Delta$, ppm[a]	ρ_R	ρ_R
5β	3.30	1.21	1.23
5α	4.56	1.97	1.75
7β	5.96	2.08	2.13
8β	6.45	1.90	1.94
6α	6.63	2.10	2.11
7α	10.95	3.44	3.20
6β	11.13	3.30	3.34

[a] The position of the heterocylic N is given by α or β.

results obtained by Takayama (21) for the F NMR shielding change, Δ, which takes place on protonation of quinolines or isoquinolines in 75% aqueous (vol.) MeOH for F substituted in the nonheterocyclic ring positions. In this instance, the F atom may be regarded as probing the effect which the change at the reaction site produces at the various ring positions where substituents are introduced in reaction series F. Values of ρ_R for the F series reaction corresponding to Δ are given in Table XV. Essentially the same order of increasing ρ_R and $-\Delta$ values is found: $5\beta < 5\alpha < 8\beta, 6\alpha < 7\alpha, 6\beta$. This order is in noteworthy contrast to that shown in Table XIV. The comparison of Δ with F series ρ_R values does assume that the polar effect contribution to Δ in the nonheterocyclic ring positions is small or nearly constant, compared to the variable (or predominant) pi electron effect contribution. Further work is in progress to test this assumption.

According to Hückel MO theory, the positions at which direct conjugation between a substituent and a reaction center occurs in the α series are, 4α, 5α, and 7α; and in the β series, 6β and 8β. The excess pi electron charge densities (Δq_π) in the nonheterocyclic ring positions of quinoline and isoquinoline obtained by Adam and Grimison (22) from Hoffmann's extended Hückel theory are typical. The values of Δq_π are compared in Table XVI with ρ_R values for A and F series reactivities.

TABLE XVI

Comparison of Excess Pi Electron Charge Densities in Quinoline and Isoquinoline with ρ_R Values

Position	Δq_π	Aρ_R	Fρ_R
5β	-0.02	0.58	1.21
6α	-0.01	0.69	2.10
7β	-0.01	0.58	2.08
5α	$+0.02$	0.69	1.97
8β	$+0.03$	0.73	1.90
7α	$+0.03$	0.75	3.44
6β	$+0.04$	0.88	3.30

The orders of increasing Δq_π and Aρ_R values reflect very much the same pattern. Although there is some grossly similar trend for Fρ_R values, it is noteworthy that the 5α and 8β positions show no enhancement in Fρ_R values over that for the 6α and 7β positions, and Fρ_R is essentially the same for 7α and 6β positions. It is therefore clear that Δq_π values for only one state of a reaction series are inadequate for interpretation of ρ_R values. The usefulness of HMO or other theoretical calculations in a truly satisfactory correlation or prediction of ρ_R values must await the results of calculations of the change in Δq_π at the position of substitution (and probably at adjacent positions as well) which correspond to the reaction process.

The ρ_R value is not thought to depend only upon the existence of direct conjugative effects (cf. Table V). The accumulation of excess pi charge density at the carbon positions adjacent to that bearing the reaction center may result in polar effects on reactivity. The effect which originates from the delocalization of pi charge between substituent and aromatic ring, but which is "inductively" transmitted from ring to the reaction center, has been called the resonance polar effect (1). One approach is to estimate relative resonance polar effects at the various naphthalene ring positions from HMO calculations of the bond–atom polarizabilities ($\pi'_{Rr,i}$) (in a manner similar to the calculation of relative pi inductive effects, see Table VIII). The results of such a calculation for $\gamma = 1.5$ (corresponding to A reaction series) are given in Table XVII, together with the K_R values for the A reaction series.

TABLE XVII
Relative Resonance Polar Effects

	$\pi'_{Rr,x}$	Relative $\sum \dfrac{\pi'_{Rr,i}}{r_{ix}}$ $(\gamma = 1.5)$	AK_R
meta	0.0	0.90	0.68
para	0.163	1.06	2.35
3α	0.0	0.82	0.87
4β	0.0	(1.00)	(1.00)
4α	0.180	1.11	3.1
6α	0.0	0.70	0.92
5β	0.0	0.72	0.77
7β	0.0	0.80	0.77
5α	0.045	0.92	0.92
7α	0.062	1.03	1.00
6β	0.062	0.74	1.17
8β	0.045	0.69	0.97

The relative values of $\sum(\pi'_{Rr,i}/r_{ix})$ give a reasonably good account of the observed K_R values for those positions where $\pi'_{Rr,x} = 0$ (Table XVII). When $\pi'_{Rr,x} \neq 0$, the observed K_R value is generally greater than the calculated value (the 5α and 7α positions are exceptional in that both quantities are essentially equal). These results appear to support the view that pi delocalization effects of substituents introduced in the 5–8 positions of the naphthoic acids are largely those of the "indirect" resonance polar effect. However, the 6β and, of course, 4α and *para* results differ sufficiently from the calculated K_R value to indicate appreciable contributions to the ρ_R value from direct conjugative effects.

The only reaction series for which "free-fitting" to eq. (1) could be carried out and for which direct conjugative effects of $+R$ substituents are probably involved is D 7α. The ρ_I value for this reaction series obtained by "free-fitting" to eq. (1) (Table II) is probably too large since $\sigma_R{}^0$ for the $(+R)NO_2$ group should be replaced by an appropriately enhanced $\bar{\sigma}_R$ value. If one applies the general constraints $K_I = 0.56$ and $K_R = 0.75$ for the nonconjugating 5β, 6α, and 7β positions, the observed log (K/K_0) values (7) are in excellent agreement with the values calculated from eq. (1).

Position	log $(K/K_0)_{calc}$	log $(K/K_0)_{obs}$
5β	$+1.11$	$+1.15$
6α	$+1.11$	$+1.03$
7β	$+1.11$	$+1.06$
		(av. $= 1.08 \pm 0.05$)

The observed log (K/K_0) for the NO_2 group at the conjugating positions are all larger than the average, $+1.08 \pm 0.05$, for the nonconjugating positions, increasing in a pattern already discussed: 5α, $+1.19 < 7\alpha$, $+1.37$, $8\beta + 1.43 < 6\beta + 1.54$. The result for the 6β position with, for example, $K_I = 0.65$ and $K_R = 1.30$, gives for NO_2 the $\bar{\sigma}_R$ value of $+0.36$. Data also available (7) for the OCH_3 in the D 6β reaction give in a similar fashion the $\bar{\sigma}_R$ value of -0.48.

Subsequent to obtaining the fitting results listed in Table II, the results of Kontyu, Petrov, and Gerasimova (23) have come to our attention on the effects of the SO_2CH_3 substituent in the D reaction series. Consequently, their results may be compared with those predicted by eq. (1), using ρ_I and ρ_R values of Table II (for "free-fitting") and the $\sigma_I = +0.62$ and $\sigma_R{}^0 = +0.10$ values available (3,24) for the SO_2CH_3 substituent. Comparison may also be made with the less precise predictions of the Dewar F—M procedure (25). These comparisons are given in Table XVIII.

Exner has argued (26) that $+R$ groups, such as CN and NO_2, do not give rise to electron-withdrawing pi delocalization effects. The observed effects of all "$+R$" type substituents, except carbonyl derivatives, are considered by Exner to be equal to the polar effect $\sigma_I\rho_I$.

We have examined this assumption by applying the data for the series A–G to eq. (1), using the following three sets of "Exner-type"

TABLE XVIII

Comparison of Observed and Calculated Values of log (K/K_0) for Reaction Series
D for the SO_2CH_3 Substituent

Position	log $(K/K_0)_{obs}$	log (K/K_0) eq. (1)	log (K/K_0) F—M
3α	$+1.85$	$+1.80$	$+1.73$
4β	1.55	1.56	1.73
5α	1.12	1.09	1.33
5β	1.09	0.99	1.01
6α	0.92	0.92	1.01
6β	1.39	—[a]	1.07
7α	1.18	1.19	1.33
7β	1.04	0.94	0.87

[a] Using $\rho_I = 1.44$ and $\rho_R = 0.98$ (corresponding to $K_I = 0.56$ and $K_R = 1.30$) the observed effect gives the enhanced value of $\bar{\sigma}_R = +0.30$, as expected by the above discussion for the NO_2 group.

σ_I' and σ_R' values. Set I utilizes the σ_I and σ_R values listed by Exner (based upon "ordinary" $\sigma_{(m)}$ and $\sigma_{(p)}$ values). Set II utilizes the σ_I and σ_R values listed by Exner for the $+R$ type. For the $-R$ type, σ_R' values are calculated for the σ^0 values of Table I, using the Exner relationship: $\sigma_R' = \sigma_{(p)}^0 - 1.14\sigma_I$ (σ_I as listed by Exner). Set III utilizes the σ_I and σ_R values given by Exner for the $+R$ type and σ_I and σ_R^0 values given in Table I for the $-R$ type.

For all data, reaction series A–F (excluding 4α data) and F NMR series G (in DMF solution), i.e., 183 reactivity points and 44 NMR points, the rms is 2.25. With the σ_I and σ_R^0 values of Table I, SD $= 0.19$; with the three "Exner-type" sets of sigma values, SD $= 0.23$–0.25. The Exner-type set which gives the best fit is set III. The σ_R and σ_I values for set III are specifically: CF_3, 0.00, $+0.46$; CO_2R, $+0.03$, $+0.35$; CN, 0.00, $+0.61$; NO_2, 0.00, 0.70, respectively. For the reactivity data alone, rms $= 0.96$; SD $= 0.097$ for the σ_I and σ_R^0 values of Table I, and SD $= 0.087$ for the "best" Exner-type set III (cf. Table XIX for comparison of individual reaction series results). For the F NMR data alone, rms $= 4.72$, and SD $= 0.38$ and 0.50, respectively.

These results appear to be quite revealing. The reactivity data have blending factors (λ) from 0.08 to 1.4, with most values less than unity (cf. Table XX). On the contrary, for the F NMR shifts (omitting the spurious 4β result) λ is generally greater than unity, ranging from 0.71 to 2.65. Series data with small λ are characterized by relative insensitivity

TABLE XIX
Comparison with Exner Treatment

Reaction series	n	σ_R^0 and σ_I values of Table I			"Exner-type" set III σ_R' and σ_I' values		
		ρ_I	ρ_R	SD	ρ_I	ρ_R	SD
A 4β	9	1.23	0.75	0.060	1.26	0.75	0.060
A 5β	5	0.83	0.58	0.027	0.88	0.62	0.018
A 6β	9	0.83	0.88	0.025	0.91	0.96	0.028
A 7β	11	0.79	0.58	0.046	0.83	0.60	0.042
A 8β	9	0.50	0.73	0.048	0.59	0.85	0.045
A 3α	6	1.36	0.65	0.058	1.37	0.65	0.043
A 5α	8	1.11	0.69	0.072	1.15	0.73	0.068
A 6α	7	0.79	0.69	0.025	0.86	0.76	0.024
A 7α	6	0.64	0.75	0.029	0.74	0.82	0.042
B 3α	4	2.20	0.86	0.035	2.23	0.87	0.040
C 6β	7	1.74	1.93	0.033	1.96	2.08	0.067
C 7β	5	1.42	1.11	0.071	1.51	1.15	0.073
D 4β	7	2.30	1.30	0.075	2.37	1.37	0.066
D 7β	4	1.36	1.00	0.016	1.48	1.09	0.032
D 3α	9	2.86	0.24	0.078	2.66	0.06	0.070
D 5α	3	1.59	1.01	0.027	1.70	1.13	0.004
D 6α	4	1.34	0.88	0.022	1.44	0.96	0.033
D 7α	4	1.66	1.56	0.039	1.88	1.74	0.067
E 4β	8	2.06	1.44	0.144	2.19	1.69	0.117
E 3α	7	2.35	0.27	0.083	2.27	0.24	0.100
F 4β	4	5.54	2.54	0.150	5.70	2.64	0.147
F 5β	3	2.60	1.21	0.124	2.70	1.23	0.132
F 6β	3	2.30	3.30	0.335	2.97	3.61	0.283
F 7β	3	2.28	2.08	0.016	2.64	2.22	0.011
F 8β	3	2.34	1.90	0.031	2.65	2.02	0.054
F 3α	8	5.44	1.46	0.151	5.39	1.39	0.147
F 5α	9	3.24	1.97	0.167	3.41	2.14	0.156
F 6α	10	2.77	2.10	0.052	2.99	2.28	0.063
F 7α	8	3.28	3.44	0.189	3.73	3.89	0.112

to the relatively small σ_R^0 values for $+R$ substituents. That is, in small λ series the small $\sigma_R^0 \rho_R$ terms for $+R$ substituents can be readily incorporated in the $\sigma_I \rho_I$ term without being distinguished. For this reason, together with the steric inhibition of resonance influence discussed on p. 163, it is a significant fact that σ_R^0 and σ_I values of Table I give the best overall fit only when series data with large λ values are included. This conclusion is also supported by the analysis of the reaction data (A–F) alone, as given in Table XIX. If one considers the difference in

SD's of the two treatments for each reaction series to be of probable significance only when it exceeds 10% of the larger SD (which also exceeds 0.030), there are 15 such cases represented in Table XIX. For series with $\lambda > 0.70$, the fit using the σ_R^0 and σ_I values of Table I is better in 6 out of the 8 cases, whereas for series with $\lambda \leq 0.70$ the fit using the Exner sigma set is better in 4 out of the 7 cases. Thus the statistics support the conclusion that the $+R$ substituents give rise to relatively small but definite pi delocalization effects. It should be noted, however, that the decision must be based upon series data (λ large) which are discriminating in this regard.

C. Conclusions

Equation (1) has generally been found to fit substituent effects in the naphthalene series to a precision of SD/rms ≤ 0.10. No single substituent parameter treatment in which positional dependence of a reaction-independent σ value is allowed will provide an equivalent fit of the data. In order for eq. (1) to be simplified to a single substituent constant treatment at a given position, the ratio $\rho_R/\rho_I \equiv \lambda$ must be a constant, independent of reaction type, and condition of measurement. The wide variation of λ at a given position for the data analyzed in the present treatment is shown in Table XX. The success of eq. (1) in

TABLE XX
λ Values

Reaction series	Position									
	3α	4α	4β	5α	5β	6α	6β	7α	7β	8β
A	0.47	1.51	0.61	0.62	0.70	0.87	1.06	1.17	0.73	1.46
B	0.39	1.17	—	—	—	—	—	—	—	—
C	—	—	—	—	—	—	1.11	—	0.78	—
D	0.08	—	0.57	0.64	—	0.66	—	0.94	0.74	—
E	0.11	—	0.70	—	—	—	—	—	—	—
F	0.27	1.39	0.46	0.61	0.47	0.76	1.43	1.05	0.91	0.81
G	0.41	2.65	−0.58	—	—	1.27	2.31	1.53	—	0.71

discerning the different characteristic patterns of positional (and reaction series) dependencies of the ρ_R and ρ_I parameters, which generally appear to conform to the expectations of theoretical models, offers further evidence that substituent effects in aromatic systems may be treated as additive blends of polar and pi delocalization effects. Both of these effects in the treatment are characterized by the substituent (σ_I and σ_R^0 values), but allowance for enhanced $\bar{\sigma}_R$ values must be made (as discussed earlier) for the effects of direct conjugation. Applicability of eq. (1) to aromatic systems other than benzene and naphthalene may be anticipated from the success of the treatment in the naphthalene series.

V. A GRAPHICAL PROCEDURE FOR APPLICATION OF EQ. (1)

Equation (1) is not accompanied by the simple graphical linear relationship afforded by a single substituent parameter treatment. The data must be machine-analyzed to obtain the fit and best values of ρ_I and ρ_R (cf. Appendix 1.B). However, it is possible to derive a useful apparent single substituent parameter procedure from which the approximate evaluation of ρ_I and ρ_R may be made by the conventional graphical method.

Let

$$\log (k/k_0)_i, P_i = \bar{\sigma}\bar{\rho} = \sigma_I\rho_I + \sigma_R^0\rho_R$$

and

$$\bar{\sigma} \equiv \frac{\sigma_I + \lambda\sigma_R^0}{1 + |\lambda|}$$

Then,

$$\rho_I = \bar{\rho}/(1 + |\lambda|)$$

$$\rho_R = \frac{\bar{\rho}\lambda}{1 + |\lambda|}$$

$$\lambda \equiv \rho_R/\rho_I$$

The observed substituent effect may be plotted versus $\bar{\sigma}$ from the extended compilation of such values which are given in Table XXI. By trial and error, the set of $\bar{\sigma}$ values corresponding to a given λ value may be found which provides the best apparent fit of the data. From the slope, $\bar{\rho}$, of this plot and the λ value which defines the $\bar{\sigma}$ set, values of ρ_I and ρ_R are then obtained from the above equations. Figures 2–6 illustrate the application of $\bar{\sigma}$ values to typical data involving λ values varying from 0.08 to 5.2. The following limiting λ value relationships are to be noted:

$$\lambda = 0, \bar{\sigma} = \sigma_I; \quad \lambda = 1, 2\bar{\sigma} = \sigma^0; \quad \lambda = \pm\infty, \bar{\sigma} = \pm\sigma_R^0$$

The lower plot in Figure 2 shows the data for reaction series D 3α (ionization of 3-substituted 1-naphthylammonium ions, H_2O, 25°C) plotted versus the $\bar{\sigma}$ set (Table XXI) with $\lambda = 0.080$. This λ value corresponds closely to the results of "free-fitting" eq. (1) for this

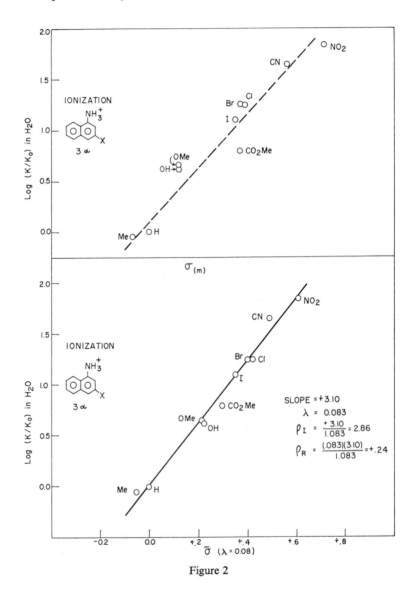

Figure 2

reaction series (Table II) which gives $\lambda = 0.083$. For comparison, the upper plot in Figure 2 shows the data plotted versus Hammett $\sigma_{(m)}$ values. Although the 3α position is formally analogous to the *meta* position in the benzene series, the $\sigma_{(m)}$ parameters correspond to a λ value (0.50) which is substantially different from the above figure for best fit of the D 3α results. It is noteworthy that both the D 3α and E 3α reaction series data (Table XX) have such small λ values that the observed substituent effects in these series (for most practical purposes) offer direct confirmation of the σ_I scale of polar effects. The low λ value is probably associated with the strong donor action of the functional groups, NH_2 and O^-, from the 1 position of naphthalene.

In Figure 3, the data for the G 8β series (F NMR shifts of 8-substituted-2-fluoronaphthalenes) is shown plotted (lower plot) versus the $\bar{\sigma}$ set (Table XXI) for which $\lambda = 0.80$ (corresponding to the results listed in Table XX). Since the 8β position corresponds to a "*para*-like" position, the comparison plot, shift versus $\sigma_{(p)}$, is shown in the upper plot of Figure 3. Figure 4 shows similar plots for the A 8β series data (ionization of 8-substituted-2-naphthoic acids).

Figure 5 illustrates the relatively poor precision of the Dewar correlation of the G 8β data plotted versus corresponding A 8β data (lower figure). The nonlinearity between the $\bar{\sigma}$ set with $\lambda = 0.80$ versus the corresponding $\bar{\sigma}$ set with $\lambda = 1.40$ is illustrated in the upper plot of Figure 5 for all of the substituents of Table XXI.

Figure 6 illustrates the correlations of the A 6β data (ionization of 6-substituted-2-naphthoic acids) and the G 6β data (F NMR shifts of 6-substituted-2-fluoronaphthalenes) with $\bar{\sigma}$ sets corresponding (approximately) to the λ values of Table XX, i.e., 1.10 and 2.33, respectively. These $\bar{\sigma}$ sets were obtained by interpolation from the sets given in Table XXI.

Figure 7 illustrates the correlation of the *para*-substituted fluoro-benzene F NMR shifts in methanol solution with the $\bar{\sigma}$ set for $\lambda = 5.2$ (lower figure). The plot of this data versus $\sigma_{(p)}$ is shown in the upper figure. The improved precision of fit with the appropriate $\bar{\sigma}$ value compared to σ values is apparent in Figures 2–4, and 7.

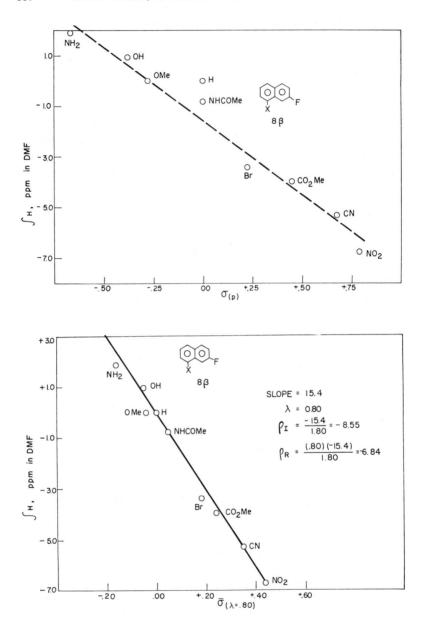

Figure 3

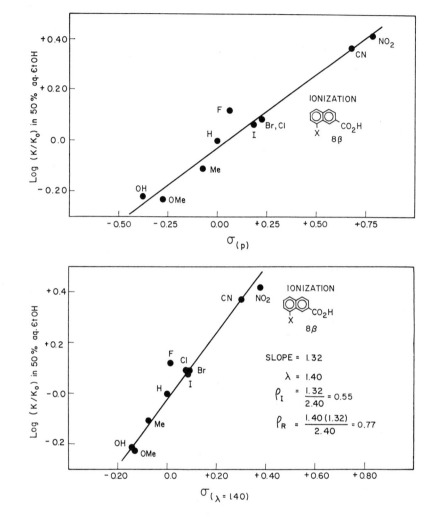

Figure 4

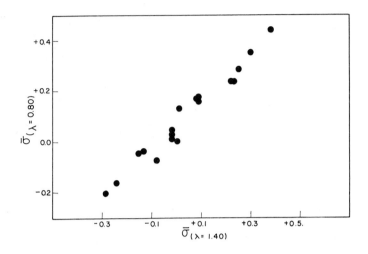

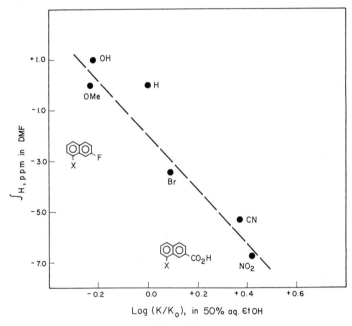

Figure 5

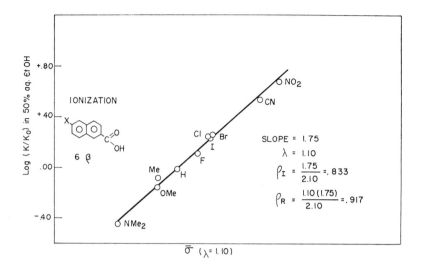

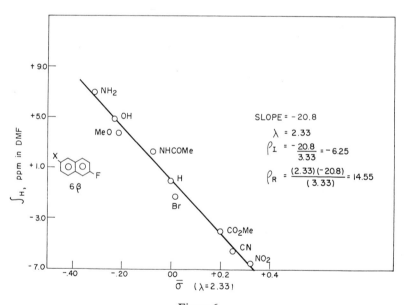

Figure 6

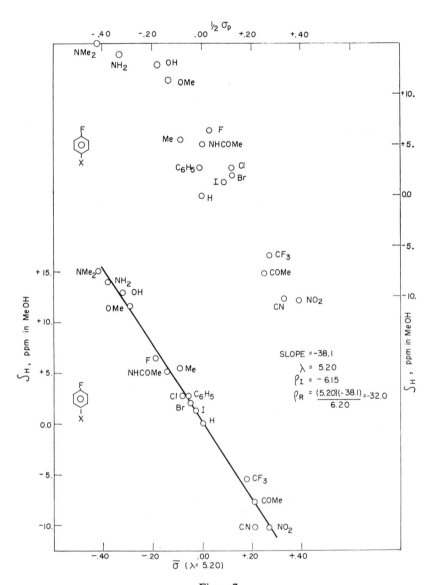

Figure 7

TABLE XXI

Blended $\bar{\sigma}$ Values $\left(\bar{\sigma} \equiv \dfrac{\sigma_I + \lambda\sigma_R^\circ}{1 + |\lambda|}\right)$

λ(LAMBDA) (+, ABOVE; −, BELOW)

SUBSTITUENT	σ_I	σ_R°	.0800	.1650	.2500	.4000	.6000	.8000	1.2500	1.7000	2.5000	4.0000	6.0000	12.0000
N(CH3)2	.05	−.52	.008	−.031	−.064	−.113	−.164	−.203	−.267	−.309	−.357	−.406	−.439	−.476
NH2	.10	−.48	.057	.018	−.016	−.066	−.117	−.158	−.222	−.265	−.314	−.364	−.397	−.435
NHCOCH3	.26	−.22	.224	.192	.164	.123	.080	.047	−.007	−.042	−.083	−.124	−.151	−.183
OH	.27	−.44	.217	.169	.128	.067	.004	−.046	−.124	−.177	−.237	−.298	−.339	−.385
OCH3	.26	−.41	.210	.165	.126	.069	.009	−.038	−.112	−.162	−.219	−.276	−.314	−.358
F	.51	−.34	.447	.390	.340	.267	.191	.132	.038	−.025	−.097	−.170	−.219	−.275
CL	.47	−.20	.420	.375	.336	.279	.219	.172	.098	.048	−.009	−.066	−.104	−.148
BR	.45	−.16	.405	.364	.328	.276	.221	.179	.111	.066	.014	−.038	−.073	−.113
I	.39	−.12	.352	.318	.288	.244	.199	.163	.107	.069	.026	−.018	−.047	−.081
CH3	−.05	−.10	−.054	−.057	−.060	−.064	−.069	−.072	−.078	−.081	−.086	−.090	−.093	−.096
C6H5	.10	−.10	.085	.072	.060	.043	.025	.011	−.011	−.026	−.043	−.060	−.071	−.085
SCH3	.19	−.17	.163	.139	.118	.087	.055	.030	−.010	−.037	−.067	−.098	−.119	−.142
CF3	.41	.13	.389	.370	.354	.330	.305	.286	.254	.234	.210	.186	.170	.152
CH3CC	.28	.19	.273	.267	.262	.254	.246	.240	.230	.223	.216	.208	.203	.197
CO2CH3	.31	.15	.298	.287	.278	.264	.250	.239	.221	.209	.196	.182	.173	.162
CN	.52	.14	.492	.466	.444	.411	.377	.351	.309	.281	.249	.216	.194	.169
NO2	.64	.19	.607	.576	.550	.511	.471	.440	.390	.357	.319	.280	.254	.225
N(CH3)2	.05	−.52	.085	.117	.144	.184	.226	.259	.311	.346	.386	.426	.453	.484
NH2	.10	−.48	.128	.154	.176	.209	.242	.269	.311	.339	.371	.404	.426	.451
NHCOCH3	.26	−.22	.257	.254	.252	.249	.245	.242	.238	.235	.231	.228	.226	.223
OH	.27	−.44	.283	.294	.304	.319	.334	.346	.364	.377	.391	.406	.416	.427
OCH3	.26	−.41	.271	.281	.290	.303	.316	.327	.343	.354	.367	.380	.389	.398
F	.51	−.34	.497	.486	.476	.461	.446	.434	.416	.403	.389	.374	.364	.353
CL	.47	−.20	.450	.432	.416	.393	.369	.350	.320	.300	.277	.254	.239	.221
BR	.45	−.16	.429	.409	.392	.367	.341	.321	.289	.267	.243	.218	.201	.182
I	.39	−.12	.370	.352	.336	.313	.289	.270	.240	.220	.197	.174	.159	.141
CH3	−.05	−.10	−.039	−.029	−.020	−.007	.006	.017	.033	.044	.057	.070	.079	.088
C6H5	.10	−.10	.100	.100	.100	.100	.100	.100	.100	.100	.100	.100	.100	.100
SCH3	.19	−.17	.189	.187	.186	.184	.182	.181	.179	.177	.176	.174	.173	.172
CF3	.41	.13	.370	.334	.302	.256	.207	.170	.110	.070	.024	−.022	−.053	−.088
CH3CC	.28	.19	.245	.213	.186	.146	.104	.071	.019	−.016	−.056	−.096	−.123	−.154
CO2CH3	.31	.15	.276	.245	.218	.179	.137	.106	.054	.020	−.019	−.058	−.084	−.115
CN	.52	.14	.471	.427	.388	.331	.272	.227	.153	.104	.049	−.008	−.046	−.089
NO2	.64	.19	.579	.522	.474	.403	.329	.271	.179	.117	.047	−.024	−.071	−.126

References

1. (a) R. W. Taft, in *Steric Effects in Organic Chemistry*, M. S. Newman, Ed., Wiley, New York, 1956, chapter 13; (b) R. W. Taft, *J. Phys. Chem.*, *64*, 1805 (1960) and references cited therein; (c) E. D. Schmid et al. *Spectrochim. Acta*, *22*, 1615, 1621, 1633, 1645, 1659 (1966); (d) R. T. C. Brownlee, A. R. Katritzky, and R. D. Topsom, *J. Am. Chem. Soc.*, *87*, 326 (1965); *88*, 1413 (1966); (e) M. Charton, *ibid.*, *86*, 2033 (1964).

3. R. W. Taft, E. Price, I. R. Fox, I. C. Lewis, K. K. Andersen, and G. T. Davis, *J. Am. Chem. Soc.*, *85*, 709 (1963).

4. R. W. Taft and I. C. Lewis, *J. Am. Chem. Soc.*, *81*, 5343 (1959).

5. R. W. Taft, S. Ehrenson, I. C. Lewis, and R. E. Glick, *J. Am. Chem. Soc.*, *81*, 5352 (1959); cf. also ref. 2(b).

6. (a) R. W. Taft, *J. Am. Chem. Soc.*, *79*, 1045 (1957); (b) I. R. Fox, P. L. Levins, and R. W. Taft, *Tetrahedron Letters*, *7*, 249 (1961); (c) S. Ehrenson, *ibid.*, *7*, 351 (1964); (d) S. Ehrenson, in *Progress in Physical Organic Chemistry*, Vol. 2, S. G. Cohen, A. Streitwiser, and R. W. Taft, Eds., Interscience, New York, 1964.

7. (a) A. Bryson, *J. Am. Chem. Soc.*, *82*, 4862 (1960); (b) A. Bryson and R. W. Matthews, *Australian J. Chem.*, *16*, 401 (1963).

8. M. J. S. Dewar and P. J. Grisdale, *J. Am. Chem. Soc.*, *84*, 3939 (1962).

9. P. R. Wells and W. Adcock, *Australian J. Chem.*, *18*, 1365 (1965).

10. A. Fischer, J. D. Murdoch, J. Packer, R. D. Topsom, and J. Vaughan, *J. Chem. Soc.*, *1957*, 4358.

11. P. R. Wells and W. Adcock, *Australian J. Chem.*, *19*, 221 (1966).

12. (a) D. D. Perrin, *Dissociation Constants of Organic Bases in Aqueous Solution* Butterworths, London, 1965; (b) A. Fischer, W. J. Galloway, and J. Vaughn, *J. Chem. Soc.*, *1964*, 3591.

13. W. Adcock and M. J. S. Dewar, *J. Am. Chem. Soc.*, *89*, 379 (1967).

14. S. Ehrenson and R. W. Taft, work in progress.

15. R. W. Taft, E. Price, I. R. Fox, I. C. Lewis, K. K. Andersen, and G. T. Davis, *J. Am. Chem. Soc.*, *85*, 3146 (1963).

16. Y. Yukawa and Y. Tsuno, *Bull. Chem. Soc., Japan*, *32*, 971 (1959); Y. Yukawa, Y. Tsuno, and M. Sawada, *ibid.*, *39*, 2274 (1966).

17. J. G. Kirkwood and F. H. Westheimer, *Bull. Chem. Soc., Japan*, *6*, 506 (1938).

18. C. A. Coulson, A. Streitwieser, Jr., and J. I. Brauman, *Dictionary of π-Electron Calculations*, Vols. 1 and 2, Pergamon, New York, 1965.

19. M. J. S. Dewar and Y. Takeuchi, *J. Am. Chem. Soc.*, *89*, 390 (1967).

20. Cf. also arguments presented by P. Wells, *Linear Free Energy Relationships*, Academic Press, in press.

21. H. Takayama and R. W. Taft, unpublished results.

22. W. Adam and A. Grimison, *Tetrahedron*, *21*, 3417 (1965).

23. V. A. Kontyu, U. P. Petrov, and T. N. Gerasimova, *J. Org. Chem. USSR*, *1*, 912 (1965).

24. D. H. McDaniel and H. C. Brown, *J. Org. Chem.*, *23*, 420 (1958).

25. M. J. S. Dewar and P. J. Grisdale, *J. Am. Chem. Soc.*, *84*, 3548 (1962).

26. O. Exner, *Collection Czech. Chem. Commun.*, *31*, 65 (1966).

A. Experimental Data

The data on which the present analysis is based are summarized in Tables A1.I–A1.III.

Table A1.I log (K/K_0) Values

Subst.	A 3α	B 3α	D 3α	E 3α	F 3α	A 4α	B 4α	F 4α	A 5α
1. $N(CH_3)_2$									
2. NH_2				+0.04	−0.01	−1.09		−4.23	−0.20
3. $NHCOCH_3$				+0.53					
4. OH	+0.09		+0.62		+0.65	−0.79			−0.10
5. OCH_3			+0.66			−0.55	−0.60	−1.56	−0.02
6. F					+2.17	+0.08	+0.27		
7. Cl	+0.46	+0.85	+1.26	+1.00	+2.27	+0.39	+0.57	+1.22	+0.44
8. Br	+0.52	+0.85	+1.25	+1.06	+2.17	+0.45	+0.60		+0.45
9. I			+1.10	+1.04					
10. CH_3	−0.08	−0.13	−0.04		−0.24	−0.21	−0.33	−0.66	+0.02
11. C_6H_5									
12. $C_6H_5CH_2$									
13. SCH_3					+1.06			−0.95	
14. CF_3									
15. CH_3CO									
16. C_6H_5CO									
17. CO_2CH_3			+0.80	+0.65					
18. CN	+0.90		+1.66			+1.20			+0.70
19. NO_2	+0.93	+1.59	+1.85	+1.54	+3.85	+1.25	+1.63		+0.79

Subst.	D 5α	F 5α	A 6α	D 6α	F 6α	A 7α	D 7α	F 7α
1. $N(CH_3)_2$								
2. NH_2		−0.52			−0.69			−1.71
3. $NHCOCH_3$								
4. OH	−0.04	−0.30	−0.12	−0.05	−0.27	−0.15	−0.28	−0.58
5. OCH_3			−0.09	+0.02	−0.13	−0.12	−0.15	
6. F		+0.96			+0.68			+0.65
7. Cl	+0.58	+1.25	+0.25	+0.44	+0.89		+0.44	+1.05
8. Br		+1.28	+0.28		+0.99	+0.11		+1.03
9. I								
10. CH_3		−0.32	−0.07		−0.34	−0.11		−0.44
11. C_6H_5								
12. $C_6H_5CH_2$								
13. SCH_3		+0.44			+0.19			
14. CF_3		+1.78						+1.72
15. CH_3CO								
16. C_6H_5CO								
17. CO_2CH_3					+1.08			
18. CN			+0.50				+0.47	
19. NO_2	+1.19	+2.23	+0.62	+1.03	+2.20	+0.55	+1.37	+2.52

(*continued*)

Table A.1. I (*continued*)

Subst.	A 4β	D 4β	E 4β	F 4β	A 5β	D m	A 6β	C 6β
1. $N(CH_3)_2$							−0.44	−0.94
2. NH_2	−0.13		−0.75	−0.86		−0.04		
3. $NHCOCH_3$			+0.28					
4. OH	−0.01			+0.60	−0.04	+0.41		
5. OCH_3	−0.02	+0.11			−0.01	+0.37	−0.16	−0.33
6. F						+1.01	+0.11	+0.20
7. Cl	+0.40	+0.78	+0.77			+1.09	+0.24	+0.48
8. Br	+0.38	+0.76	+0.85	+2.04	+0.27	+1.07	+0.26	
9. I	+0.34	+0.75	+0.78			+0.99	+0.23	
10. CH_3	−0.13					−0.13	−0.08	−0.22
11. C_6H_5						+0.44		
12. $C_6H_5CH_2$								
13. SCH_3						+0.57		
14. CF_3						+1.10		
15. CH_3CO						+1.03		
16. C_6H_5CO								
17. CO_2CH_3		+0.78	+0.71			+1.04		
18. CN	+0.86	+1.50	+1.18		+0.56	+1.86	+0.54	+1.16
19. NO_2	+0.92	+1.73	+1.50	+4.00	+0.61	+2.14	+0.68	+1.48

Subst.	F 6β	A 7β	C 7β	D 7β	F 7β	A 8β	F 8β	F m	F p
1. $N(CH_3)_2$		−0.20	−0.45						−4.38
2. NH_2	−1.76	−0.23			−0.79		−0.64	−0.83	−3.91
3. $NHCOCH_3$								+0.75	−0.66
4. OH	−0.43	−0.14		−0.09	−0.28	−0.22	−0.24	+0.42	−2.12
5. OCH_3		−0.01		−0.03		−0.23		+0.40	−1.37
6. F		+0.22	+0.32			+0.12		+2.24	
7. Cl		+0.27	+0.39	+0.43		+0.09		+2.40	+1.38
8. Br		+0.29				+0.09		+2.36	+1.46
9. I		+0.27				+0.07		+1.92	+1.20
10. CH_3		−0.07				−0.11		−0.46	−0.82
11. C_6H_5								+0.41	−0.14
12. $C_6H_5CH_2$									−0.38
13. SCH_3								+0.90	−0.62
14. CF_3									
15. CH_3CO								+1.99	+1.70
16. C_6H_5CO								+2.03	+1.86
17. CO_2CH_3								+2.12	+1.72
18. CN		+0.54	+1.02			+0.37		+3.86	+3.35
19. NO_2	+1.99	+0.56	+1.07	+1.06	+1.85	+0.42	+1.87	+4.03	+3.82

TABLE A1.II

F NMR Substituent Shielding Effects in Dimethyl Formamide Solutions
(in ppm)

Subst.	G 3α	G 4α	G 6α	G 7α	G 4β	G 6β	G 8β
1. N(CH$_3$)$_2$		(+6.68)[a]			(−1.01)[a]		(−0.63)[a]
2. NH$_2$	+0.81	+14.88	+1.05	+2.51	−0.47	+6.96	+1.88
3. NHCOCH$_3$	−0.85	+3.04	+0.44	+0.33	−0.96	+2.31	−0.80
4. OH				+1.34		+4.87	+0.97
5. OCH$_3$				+0.65		+3.72	0.00
6. F							
7. Cl							
8. Br	−3.20	−0.62	−0.84	−0.38	−1.05	−1.23	−3.40
9. I							
10. CH$_3$							
11. C$_6$H$_5$							
12. C$_6$H$_5$CH$_2$							
13. SCH$_3$							
14. CF$_3$							
15. CH$_3$CO							
16. C$_6$H$_5$CO							
17. CO$_2$CH$_3$		−8.89		−1.67		−4.08	−3.95
18. CN	−3.68	−11.34	−1.29	−2.35	−0.80	−5.45	−5.32
19. NO$_2$	−4.87	−12.77	−2.08	−3.39	−0.79	−6.54	−6.76

[a] Omitted from correlation because of apparent steric inhibition of resonance effect.

TABLE A1.III

F NMR Substituent Shielding Effects in Other Solvents (in ppm)

Solvent	G 7α				G 6β				G 8β			
	CH	B	A	MeOH	CH	B	A	MeOH	CH	B	A	MeOH
1. N(CH₃)₂									(−0.07)[a]	(−0.13)[a]	(−0.24)[a]	(−0.41)[a]
2. NH₂	+2.18	+2.07	+2.24	+1.98		+5.45	+6.47	+5.49	+0.69	+0.83	+1.43	+0.81
3. NHCOCH₃		+0.18	+1.23	+0.10								
4. OH		+0.99				+3.55	+4.43	+4.46		+0.07	+0.69	+1.01
5. OCH₃	+1.22	+1.00	+0.76	+0.84	+3.71	+3.18	+3.65	+3.60	+0.26	0.00	0.00	+0.20
8. Br	−0.39	−0.56	−0.34	−0.25	−1.31	−1.02	−1.27		−3.20	−3.16	−3.24	−3.21
17. CO₂CH₃	−2.14	−2.10	−1.76	−1.95	−3.63	−3.80	−4.05		−4.07	−4.22	−3.91	−4.04
18. CN	−2.13	−2.50	−2.36	−2.54			−5.56		−5.49	−5.42	−5.35	
19. NO₂	−3.63	−3.62	−3.33			−6.39	−6.64			−7.09	−6.84	

[a] Omitted from correlation because of apparent steric inhibition of resonance effect. CH = cyclohexane, B = benzene, A = acetone, and MeOH = methanol.

B. Least Squares Procedure

The least squares relationships used in parameter optimization for eq. (1), rewritten with explicit dependence upon the substituent indexed by j,

$$\log (K_j/K_0)_i \equiv P^{ij} = \rho_I{}^i \sigma_I{}^j + \rho_R{}^i \sigma_R{}^j$$

are generally nonlinear when both *sigmas* and *rhos* are to be determined simultaneously or when such interpositional constraints as K_I or K_R are employed. The nonlinear methods used are of the Gauss-Newton type and have been discussed in detail previously (6c, 6d). For the special cases where the sigmas are assumed known and are not varied and where all ρ^i are independent variables, the least squares equations are bilinear and have the solutions,

$$\rho_{I,R}^i = \frac{(\sum P^{ij}\sigma_I{}^i)(\sum \sigma_R{}^j\sigma_{R,I}^j) - (\sum P^{ij}\sigma_R{}^j)(\sum \sigma_I{}^j\sigma_{R,I}^j)}{\sum (\sigma_I{}^j)^2 \sum (\sigma_R{}^j)^2 - (\sum \sigma_I{}^j\sigma_R{}^j)^2}$$

All sums are over the index j.

C. Background Reprints

For the reader unfamiliar with the background material on which this chapter is based, we have reproduced five papers which summarize such material (Appendixes 2–6). Reproduction seems further justified by the limited availability of much of this material.

Equation (1), in terms of σ_R rather than $\sigma_R{}^0$ parameters, was first proposed by Taft [*J. Am. Chem. Soc.*, 79, 1045 (1957)]. In that paper it was pointed out that ρ_I and ρ_R need not be identical for a given reaction series [cf. also discussion of Taft and Lewis, *J. Am. Chem. Soc.*, 2436 (1958)]. The first application of eq. (1) to both reactivity data and F NMR shift data was made by Fox, Levins, and Taft [*Tetrahedron Letters*, 7, 249 (1961)] in considering the conjugative acceptor capacities of benzoyl and protonated benzoyl derivatives. The condition that $\lambda \cong 0.50$ for $\sigma_{(m)}$ values was established by Taft, Ehrenson, Lewis, and Glick [*J. Am. Chem. Soc.*, 81, 5352 (1959)]. However, the first recognition and discussion of placing the positional dependence in ρ_I and ρ_R

in a generalized treatment according to eq. (1) was made by Ehrenson [*Tetrahedron Letters*, *7*, 351 (1964)].

The σ^0-type reactivity was first distinguished from general $\sigma\rho$ reactivities by Taft, Ehrenson, Lewis, and Glick (*loc. cit.*). The σ_R^0 parameters were defined in this paper following the proposed *I–R* separation of σ values as developed by Taft (Newman, *Steric Effect in Organic Chemistry*, Chapter 13). An independent analysis in which σ^n values were proposed by Van Bekkum, Verkade, and Wepster [*Rec. Trav. Chim.*, *78*, 815 (1959)] provided similar although not entirely identical results.

These developments and related matters were summarized by Taft [*J. Phys. Chem.*, *64*, 1804 (1960)] and by Ehrenson (*Abstracts of Papers, Symposium on Linear Free Energy Correlation*, U.S. Army Research Office, Durham, October 19–21, 1964). Reproduction of these papers follows as Appendixes 2 and 3, respectively.

Evidence favoring the field effect model of Kirkwood and West-heimer (with a $\cos \theta / r^2$ dependence) over that of Dewar and Grisdale (with a $1/r$ dependence) was first obtained for the A.8β substituent effects by Wells and Adcock [*Australian J. Chem.*, *18*, 1365 (1965)]. This paper is reproduced as Appendix 4.

The *meta-* and *para*-substituent F NMR shielding effects played a dominant role in support of eq. (1) in the benzene series. The first measurements of these shielding effects were obtained by Gutowsky, McCall, McGarvey, and Meyer [*J. Am. Chem. Soc.*, *74*, 4809 (1952)], who demonstrated rough relationships with Hammett σ values. Much more extensive investigations, both of substituent effects and solvent effects on substituent effects, were subsequently reported by Taft, Price, Fox, Lewis, Andersen, and Davis [*J. Am. Chem. Soc.*, *85*, 709, 3146 (1963)]. A direct relationship was proposed between σ_R^0 values and excess pi-electron charge density produced by the substituent at the *para* carbon atom. These papers provide extensive tables of σ_I and σ_R^0 values, including solvent effect studies. Linear relationships between σ_I and σ_R^0 values for families of structurally related substituents are reported and discussed. These papers are reproduced as Appendixes 5 and 6.

VII. APPENDIX 2. SIGMA VALUES FROM REACTIVITIES(1)*†§

A. Introduction

In principle the effect of structure on reactivity may be quite complex. The effect of changing structure from R_0 (an arbitrary standard of comparison) to R (a general substituent) on a reaction equilibrium or rate is given by (2,3a)

For equilibria: $\Delta F^0 = RT \ln (K/K_0) = \Delta E_0{}^0 - RT \ln (\pi Q)$ (1a)

For rates: $\Delta F^{\ddagger} = RT \ln (k/k_0) = \Delta E_0{}^{\ddagger} - RT \ln (\pi Q^{\ddagger})$ (1b)

for the generalized process

$$R-Y + R_0-Y' \rightleftharpoons R_0-Y + R-Y' \qquad (2)$$

where Y and Y' are functional groups (reaction centers) in the reactant state and the product state (for equilibria) or the transition state (for rates), respectively.

The $\Delta E_0{}^0$ (or $\Delta E_0{}^{\ddagger}$) term measures the change in potential energy accompanying the reaction process, a quantity which in appropriate cases can presumably be discussed in terms of the effect of electronic distribution. The πQ term involves a quotient of partition functions which is determined by molecular motions. This term may be very complex, especially since most of the systematic reactivity studies have been carried out in solution and motions of the solvent are in some way included in this term. An exact analysis of the contribution of potential and kinetic energy terms to the measured free energy change is unknown. It is apparent, however, in the instance that $\pi Q \cong 1$, $\Delta F^0 \cong \Delta E_0{}^0$ (or $\Delta F^{\ddagger} \cong \Delta E_0{}^{\ddagger}$). This simple result can be expected to apply, approximately, in special circumstances (2,4).

In spite of the potential complexity of the free energy changes of the kind in question, it is possible experimentally to make such determinations with rather good precision. It is further possible to make extremely systematic investigations of the effect of structure on ΔF^0 (or $\Delta F^{\ddagger}$). A

* By Robert W. Taft, Jr., College of Chemistry and Physics, The Pennsylvania State University, University Park, Pennsylvania.

† Reprinted from the *Journal of Physical Chemistry*, **64**, 1805–1815 (1960). Copyright 1961 by the American Chemical Society and reprinted by permission of the copyright owner.

§ References and footnotes for this Appendix will be found on pp. 220–222.

very extensive literature of such determinations exists. For these reasons (and those given below), reactivity data in part militate against their potential complexity and provide an important source of information on electronic distributions in organic molecules.

It is a matter of empirical fact that in especially suited reactivity systems, the measured free energy changes are simply and systematically related. Relationships of this kind are known which have broad scopes of applicability. It is difficult to imagine how results of this character could be obtained unless the measured free energy changes are in fact direct measures of (equal or proportional to) the potential energy changes resulting from electronic distribution effects (2,3b). Hammett first applied this reasoning to the class of reactivities involving *meta*- and *para*-substituted side-chain derivatives of benzene (5,6). The present paper will be concerned principally with the knowledge of electronic distributions as deduced from reactivities of this type. Brief consideration will also be given to the empirical confirmation of these deductions based upon powerful new physical techniques for obtaining detailed information on electronic distributions.

B. The Hammett Linear Free Energy Relationship

The linear free energy relationship of Hammett is illustrated in Figure A2.1. The approximately quantitative linear relationship between the effects of corresponding *meta* and *para* substituents on the rates of saponification of ethyl benzoates and on the ionization of benzoic acids in water is shown. A σ value is defined as $\log (K/K_0)$ for the latter reaction, and is presumed to be a substituent parameter independent of reaction type or conditions. Hammett selected the benzoic acid ionization as the standard reaction series because of the availability of accurate data. The slope of the plot in Figure A2.1 is the reaction constant, ρ. It is a susceptibility factor reflecting the change in electron density at the reaction center (7b). A great many reactivities follow this relationship with rather comparable precision. Jaffé in 1953 summarized the fit of available reactivities to the relationship (7a).

C. Limited Linear Free Energy Relationships

1. Deviations from the Hammett Relationship

Certain reactivities do not correlate at all well with the benzoic acid ionization. These are reactivities aside from those of *ortho*-substituted

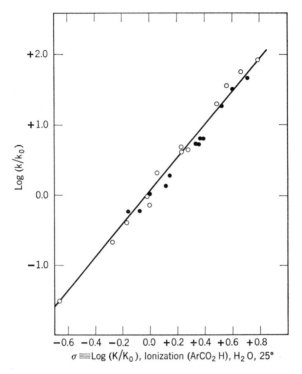

Fig. A2.1. Hammett linear free energy relationship, $\log (k/k_0) = \sigma\rho$, for the rates of saponification of ethyl benzoates, 60% aq. acetone, 25°. Cf. reaction A.15, ref.10. $\rho = + 2.33$ (○) *para* substituent, (●) *meta* substituent.

derivatives of benzene which frequently deviate because of steric effects (3,6). For example, Hammett noted that in reactions of the derivatives of phenols and anilines the p-NO_2 and p-CN substituents have correspondingly exalted effects. Recently, H. C. Brown has demonstrated the existence of linear free energy relationships for certain electrophilic reactivities, in which *para* substituents such as CH_3O and CH_3 show substantially exalted effects (8). Figure A2.2 illustrates the limited linear free energy relationship obtained for the ionization of phenols in water, 25°. Figure A2.3 illustrates the rather complete breakdown of the relationship for the rates of decomposition of diazonium salts in water, 29°. Deviations such as those illustrated in Figure A2.2 and A2.3 call for modification of the original Hammett linear free energy relationship, especially if precise correlations and predictions of reactivities are

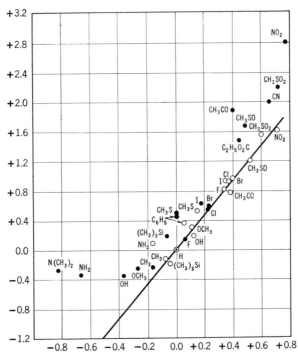

Fig. A2.2. Limited linear free enegy relationship for ionization of *meta-* and *para-*substituted phenols and benzoic acids, H_2O, $25°$: ordinate gives $\log (K/K_0)$ values for phenols and abscissa corresponding values for benzoic acids. (●) *para* substituents, (○) *meta* substituents.

to be achieved. The modifications, which still permit in similar terms certain precise descriptions of broad ranges of reactivity, are quite intelligible in terms of electronic distributions.

2. Formulation and Classification of the Problem

The reactivities of *meta-* and *para-*substituted derivatives of benzene may be considered from either of two points of view. Since both points of view may have distinct advantages depending upon the problem in question, it is essential to distinguish clearly the particular viewpoint. First, one may consider, in the generalized reaction 2, the substituent, R, as the *m-* or *p-*X-C_6H_4 group, the standard substituent R_0, as the unsubstituted phenyl group, and the side-chain reaction

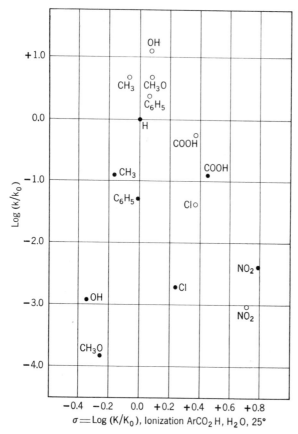

Fig. A2.3. Failure of linear free energy relationship for decomposition of benzenediazonium chlorides, H_2O, 20°, cf. reaction E.4, ref. 10. (●) *Para* substituent, (○) *meta* substituent.

centers as the functional groups, Y and Y'. Alternately, one may look upon the variation of structure as involving for the group, R, the *meta* or *para* substituent, X, with the standard substituent, R_0, the H atom. Let us, for the present, pursue the former viewpoint.

If, in reaction 2, there is a change in energy of a particular type of interaction between R and Y (of R–Y) and R and Y' (of R–Y') as compared with that between R_0 and Y' (of R_0–Y') and R_0 and Y (of R_0–Y), there will result an effect on reactivity attributable to this kind of interaction. Thus a change in resonance interactions is said to produce a resonance effect on ΔF^0 (or $\Delta F^{\ddagger}$) and so on (3). In the discussion which follows, the generally accepted assumption will be made that the reactivities [ΔF^0 or log (k/k_0) values] of *meta*- and *para*-substituted derivatives of benzene can be treated as the sum of inductive and resonance effects (3,6,7).

Quite apparently many of the deviations from precise Hammett linear free energy relationships (cf. Figs. A2.2 and A2.3) involve systems in which a change in resonance interactions in reaction 2 is expected for p-X-C_6H_4 substituents which are strongly conjugated with either Y or Y′ (hereafter referred to as an Ar–Y resonance effect) (9).

3. Precise Hammett Relationships for Select meta Substituents

Taft and Lewis recently have completed a critical examination of the data for eighty-eight reaction series in which substituents effects are relatively large and for which, in general, at least four *meta* substituents had been investigated (10). While this paper was in preparation, Professor B. M. Wepster kindly provided the author with a manuscript by himself and co-workers in which a similar examination was carried out along somewhat different lines. On nearly all major points, their conclusions are similar to those of Taft and Lewis (11).

A select group of m-X-C_6H_4 substituents exhibit precise linear free energy relationships with much more generality than for other substituents. The mean sigma values for this group are listed in Table A2.I and are designated as σ^0 values. In general, about 90% of the available data for each substituent follow the Hammett equation to a precision of a standard error of ± 0.03 in the sigma value (10). In no case does the mean value differ from that originally given by Hammett by more than this standard error. Within the standard error of ± 0.03 the linear free energy relationships for these substituents are independent of whether the process is rate or equilibrium, the solvents or the temperature. The solvent variation is from nonhydroxylic solvents such as benzene to aqueous solutions. However, for the several substituents indicated in Table A2.I, this degree of precision was achieved only by restriction to the indicated solvent classes. The maximum temperature variation involved is from 273 to 470°K. Each σ^0 value encompasses extremely wide variations in reactivity type.

In contrast to the results for the select group of *meta*-substituted phenyl groups, no *para*-substituted phenyl substituent was found which meets the same generality and precision criterion. Using the linear free energy relationships for the select group of *meta*-substituted phenyl

TABLE A2.I

Inductive Constants for the *meta*- and *para*-Substituted
Phenyl Substituents: σ^0 Values

Substituent	Position	
	meta[a]	*para*
—$C_6H_4N(CH_3)_2$	−0.15	−0.44
—$C_6H_4NH_2$	−0.14	−0.38
—$C_6H_4OCH_3$	+0.*13*[b] (+0.06)[c]	−0.12[b] (−0.16)[c]
—C_6H_4OH	—[d] (+0.04)[e]	—[d] (−0.13)[e]
—C_6H_4F	+0.*35*	+0.17
—C_6H_4Cl	+0.*37*	+0.27
—C_6H_4Br	+0.*38*	+0.26
—C_6H_4I	+0.*35*	+0.27
—$C_6H_4CH_3$	−0.07	−0.15
—C_6H_5	*0.00*	0.00
—C_6H_4CN	+0.62	+0.69[f] (+0.63)[e]
—$C_6H_4CO_2R$	+0.36	+0.46[f] (?)
—$C_6H_4COCH_3$	+0.*34*	+0.46[f] (+0.40)[e]
—$C_6H_4NO_2$	+0.*70*	+0.82[f] (+0.73)[e]

[a] Italicized values indicate the select group of *meta* substituents given in ref. 10.

[b] Value for pure aqueous solutions only.

[c] Value for nonhydroxylic media and most mixed aqueous organic solvents.

[d] Value strongly dependent upon hydroxylic solvent.

[e] Value for nonhydroxylic solvents only.

[f] Value for pure aqueous and most mixed aqueous organic solvents.

substituents to determine the value of ρ, the reaction constant, effective sigma values, $\bar{\sigma}$, may be obtained for comparison purposes, $\bar{\sigma} = 1/\rho$ [log (k/k_0)]. The values of $\bar{\sigma}$ obtained for *para*-substituted phenyl substituents cover quite substantial ranges, even for series of rather similar reactivities (including the ionization of benzoic acids in various media). In some cases $\bar{\sigma}$ values are actually of opposite sign (10).

4. Interpretation

The precise and general linear free energy relationships observed only for the select group of *meta*-substituted phenyl substituents are inherently reasonable in terms of the classical qualitative theory of electronic distribution in benzene derivatives, namely, that there is localization of π-electronic charge alternately about a benzene ring in the positions *ortho* and *para* to a resonating substituent, X (12). In fact, the relationships described above to some extent were anticipated on this basis by Branch and Calvin almost twenty years ago (9).

The precise and highly general linear free energy relationships for the select group of *meta*-substituted phenyl substituents are ascribed to the fact that in general substituents in the *meta* position are not directly conjugated with the side-chain reaction centers, Y and Y', and thus specific Ar–Y resonance effects do not contribute. The substituent effect measured is considered to result formally from the inductive transmission (by field or internal bond polarity effects) of charge across the Ar–Y bond. Consequently, if viewed at this position, the reactivity effects described by $\log (k/k_0) = \sigma^0 \rho$ are appropriately regarded as following an inductive effect relationship (hereafter referred to as Ar–Y inductive effects).

The σ^0 values represent inductive constants for the substituted phenyl groups (R) relative to the unsubstituted phenyl group (R_0). The σ^0 values include contributions, of course, from the resonance interaction of the *meta* substituent, X, with the ring (i.e., conjugative X–Ar interaction in the general system X–Ar–Y) which are considered in detail in Section E. It is important to note at this point, however, that substantial evidence is available (10) indicating that the quantitative inductive order for X-C_6H_4 groups can be altered appreciably (especially for the easily polarizable NH_2 and SCH_3 groups) by polarization induced by highly polarizing functional groups, Y and Y'. A formal charge on the first atom of the functional group appears especially effective in this regard. The polarization effect apparently results from the effect on the resonance interaction of the *meta* substituent, X, with the benzene ring (10).

The rather infrequent deviations from the relationship $\log (k/k_0) = \sigma^0 \rho$ for *meta*-substituted phenyl groups (those exceeding 0.1 sigma unit) are in part due to this cause, although in very special reactivities the effect of direct conjugation between the *meta* substituent, X, and the side-chain functional group, Y, can lead to deviations (13). Specific interactions (especially H-bonding) between solvent and the substituent, X, can also alter the Ar–Y inductive order (7,14). These solvent effects can be taken into account by assigning a σ^0-value characteristic of a particular solvent class (as given in Table A2.I for the several pertinent substituents).

5. Precise Linear Ar–Y Inductive Effect
Relationships for para Substituents

In order to obtain Ar–Y inductive constants (σ^0 values) for *para*-substituted phenyl substituents, Taft and co-workers selected reactivities in which a methylene group is interposed between the benzene ring and the reaction center (15). It was assumed that the Ar–"Y" resonance effect would be essentially constant for such a reaction series. The reaction series of this class for which data were available are listed in Table A2.II. The linear free energy relationships for these reactivities

TABLE A2.II

p-X—C_6H_4—CH_2—Y Reactivities Used to Define σ^0 Values
for *para*-Substituted Phenyl Groups

Reaction[a]		ρ
B.1	Ionization, $ArCH_2CO_2H$, H_2O, 25°	+0.46
B.2	Rate of saponification, $ArCH_2CO_2Et$, 88% aq. EtOH, 30°	1.00
B.3	Ionization $ArCH_2CH_2CO_2H$, H_2O, 25°	+0.24
B.14	Rate of saponification, $ArCH_2OCOCH_3$, 60% aqueous acetone, 25°	+0.73

[a] Reaction designations refer to those given in ref. 10.

appear to be as precise for *para*- as for *meta*-substituted phenyl substituents (with all sigma values precise to a standard error of approximately ± 0.03). The σ^0 values so obtained for the *para*-substituted phenyl groups are listed in Table A2.I.

In Table A2.I the substituted phenyl groups listed above the standard phenyl group contain $-R$ *meta* or *para* substituents by Wheland's classification (16,17), i.e., the substituents, X, release charge by resonance interaction with the ring. The substituted phenyl groups appearing below the phenyl group contain substituents, X, which are $+R$ in character. A negative sign of σ^0 indicates a net charge releasing effect relative to the standard, and a positive sign a net charge accepting effect. It is significant to note for the $-R$ substituted groups that the value of σ^0 is in each case more negative for the *para* than the *meta* substitution. In turn, for the $+R$ substituted groups the value of σ^0 is more positive for the *para* than the *meta* substitution.

The σ^0-values for groups containing $+R$ *para* substituents (but not those with $-R$ *para* substituents, cf. Section C-6) apparently are not distinguishable from those obtained by Hammett within a standard error of ± 0.03. Consequently, the σ^0-values listed for these substituents in Table A2.I (and A2.III) are the average values obtained from all reactivities except those with nucleophilic character.[10] In obtaining these averages it was apparent that the σ^0 values for the $+R$ substituents show a dependence on solvent. The values given in the main column of Table A2.I pertain to aqueous solution and those listed in parentheses pertain to reactivities in the designated media. If no value is given in parentheses, the σ^0 value apparently holds generally independent of solvent.

If the σ^0 values for *para*-substituted phenyl groups applied only to precise correlations of reactivities of specifically the ArCH$_2$Y type, their utility would be so limited as to be of little value. However, a substantial number of other reactivities are correlated by σ^0 values to a precision of ± 0.03 sigma units (15). Two typical examples are given in Figures A2.4

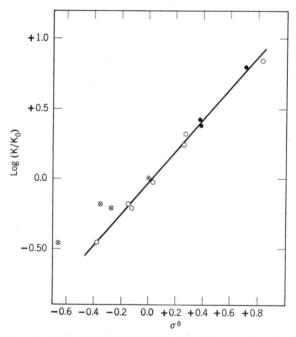

Fig. A2.4. Ionization of ArPO$_2$(OH$^-$), H$_2$O, 25°, correlated by σ^0 values. Cf. reaction B.6, ref. 10. The crossed circles designate points based upon original Hammett σ-values where these differ in a significant way (cf. Table III) from the σ^0 values. $\rho = +1.19$. (●) *Meta* substituent, (○) *para* substituent, (⊙) origin.

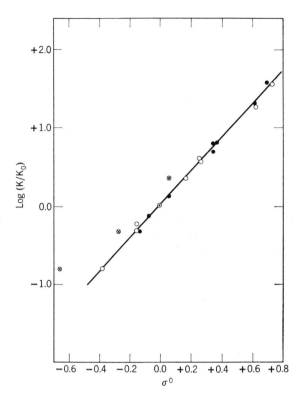

Fig. A2.5. Ion-pair equilibria, $ArCO_2H$ with 1,3-diphenylguanidine, benzene, 25°, correlated by σ^0 values. Cf. reaction A.13, ref. 10. The crossed circles designate points based upon original Hammett σ-values where these differ in a significant way from the σ^0 values. $\rho = +2.24$. ($\bigcirc$) *Para* substituent, ($\bullet$) *meta* substituent, ($\odot$) origin.

and A2.5. The crossed circles in these figures designate points for sigma values based upon the ionization of benzoic acids in water (original Hammett σ values).

6. Evaluation of Specific Ar–Y Resonance and Polarization Effects

A number of authors have expressed the belief that the sigma value which Hammett derived from the ionization of *para*-methoxybenzoic acid in water, for example, has a measurable contribution from an Ar–Y resonance effect (18). Such a contribution will result if the isovalent conjugation in the acid

is more important than that in the anion (as is reasonable on several grounds).

In accord with this expectation the σ values obtained from the ionization of benzoic acids in water are distinctly exalted compared to the corresponding σ^0 values for the *para*-substituted phenyl groups containing strong $-R$ substituents. The two sets of sigma values (for aqueous solutions) are compared in Table A2.III. The difference, $\sigma-\sigma^0$, is the Ar–Y resonance effect for ionization of benzoic acids in water (19).

TABLE A2.III

Comparison of σ^0 and Hammett σ Values for *para*-Substituted Phenyl Groups

Substituent	σ^0	σ
$-C_6H_4N(CH_3)$	-0.44	-0.82
$-C_6H_4NH_2$	-0.38	-0.76
$-C_6H_4OCH_3$	-0.12	-0.27
$-C_6H_4F$	$+0.17$	$+0.06$
$-C_6H_4Cl$	$+0.27$	$+0.23$
$-C_6H_4Br$	$+0.26$	$+0.23$
$-C_6H_4I$	$+0.27$	$+0.28$
$-C_6H_4CH_3$	-0.15	-0.17
$-C_6H_5$	0.00	0.00
$-C_6H_4CN$	$+0.69$	$+0.66$
$-C_6H_4CO_2R$	$+0.46$	$+0.45$
$-C_6H_4COCH_3$	$+0.46$	$+0.50$
$-C_6H_4NO_2$	$+0.82$	$+0.78$

It is apparent from Table A2.III that the σ^0 values for the groups containing the weak $-R$ *para* substituents such as, Cl, Br, CH_3, are not really distinguishable from the Hammett σ-values.

The precise correlation (Fig. A2.5) of the data of Davis and Hetzer (20) on ion-pair formation of benzoic acids in benzene by σ^0 values is especially noteworthy. The correlation implies that there is essentially no Ar–Y resonance effect (for strong $-$R para-substituted phenyl groups) as in the ionization of benzoic acids in water. The absence of Ar–Y resonance effects for this reaction does not require that isovalent conjugation in para-methoxybenzoic acid, for example, be frozen out in benzene solution. Dipole moment analysis of Rogers (21) indicates that such an interaction exists under these conditions. The correlation implies only that such interaction takes place to the same extent in the ion-pair product as in the reactant benzoic acid.

It is highly probable, however, that such a condition arises as a consequence of the fact that the fractional contribution of the dipolar isovalent resonance form (above) to the resonance hybrid of the substituted benzoic acid does make a substantially smaller contribution in benzene solution than in the more ionizing aqueous solution. Solvent effects of this kind have been proposed earlier by Gutbezahl and Grunwald (22) and have been considered by Davis and Hetzer in interpretation of their results.

It is of interest in this connection that unequivocal evidence has been reported based upon the shielding parameters from the ^{19}F nuclear magnetic resonance spectra of meta- and para-substituted fluorobenzenes which shows that dipolar resonance structures, e.g.,

make increasingly greater contributions to the resonance hybrid in more ionizing media (23).

The σ^0 values have general utility in the identification and study of specific polarization and Ar–Y resonance effects which are dependent upon both solvent conditions and reaction type. Deviations from the relationship $\log (k/k_0) = \sigma^0 \rho$ provide a measure of such effects. This use of σ^0 values has been illustrated (10,11). The Ar–Y resonance effects obtained by this procedure have been represented (3) by the symbol ψ.

D. Concerning ρ Values

The reaction constants, ρ, obtained from the precise linear free energy relationship for the select group of meta-substituted phenyl substituents in acid ionization equilibria in aqueous solution are instructive. All of the available values conform to the relationship $\rho = (2.8 \pm 0.5)^{1-i}$, where $i =$ the number of "saturated" links intervening between the benzene ring and the atom at which there is a unit decrease in formal charge on ionization of the proton (24). This relationship is followed in a manner roughly independent of the charge type of the acid or the kind of atoms involved. It represents a special case of the Branch and Calvin scheme for treating inductive effects in acid ionization equilibria (25). This relationship, besides having value as a useful empirical tool, serves to show that ρ is a measure of the change (for the reaction process) of the electron density at the reaction center.

Hine has proposed (26) that the value of ρ can be obtained from the σ values of substituent groups involved at the functional centers, Y and Y'. Evidently, the change in formal charge of an atom can be represented approximately by a substituent parameter (about 0.75 sigma unit for each unit increase in formal charge of the first atom of the functional center).

E. The Separation of Log (k/k_0) Values to Contributing X–Ar Inductive and Resonance Effects

1. Linear Inductive Effect Relationships in the Aliphatic Series

It is useful now to consider the alternate viewpoint of the effect of structure on the reactivity of *meta-* and *para-*substituted derivations of benzene. That is, the substituent (R) of eq. (2) is now to be regarded as the *meta* or *para* substituent, X, and the functional group (Y or Y') is a side-chain substituted phenyl group. By adopting this point of view we turn attention to the effects on reactivity which result from interactions taking place (formally) through the X–Ar bond (instead of the Ar–Y bond which we have been considering to this point). Such an empirical analysis is permitted by results obtained from the consideration of reactivity effects in the aliphatic series.

The inductive effects of substituent groups in reactions in the aliphatic series (i.e., R in R–Y) are correlated by an equation of the same form as Hammett's $\sigma\rho$ relationship (i.e., $\sigma^*\rho^*$) (27). Although reactivities in the aliphatic series are frequently complicated by specific steric, resonance and other effects, an appreciable variety of reactivity types has been found in which these effects are apparently constant and the relationship, $\log (k/k_0) = \sigma^* \rho^*$, holds. These inductive effect correlations have been reviewed recently (28).

2. Parameters Applicable to X–Ar Inductive Effects

The generality of the correlations of inductive effects in the aliphatic series leads to the hypothesis that the same inductive order holds for substituents bonded to an aromatic as to an aliphatic carbon atom (27,28).

The aliphatic series inductive sigma values (for the substituent, X) on a scale for direct comparison with aromatic sigma values, can be obtained by the defining equation (27).

$$\sigma_I \equiv (1/6.23)[\log (k/k_0)_{OH^-} - \log (k/k_0)_{H^+}]$$

where k's refer to rate constants for the ester hydrolysis reactions

$$XCH_2CO_2Et + H_2O \underset{k_{H^+}}{\overset{k_{OH^-}}{\longrightarrow}} XCH_2CO_2H + EtOH$$

The logarithmic difference on the right-hand side of this equation is a measure of the substituent effect on the free energy difference between two transitions states of closely identical steric requirements but of opposite charge types. Ingold had suggested previously that such a quantity would provide a good experimental measure of the inductive effect (29). The basis for the 1/6.23 factor has been discussed (28). For present purposes, suffice it to say that it is essentially equal to (coincidentally) an empirical fall-off factor for the inductive effect of the substituent X, acting through an interposed benzene ring.

Table A2.IV shows the periodic relationship of the σ_I parameters. Comparison is made between the values of σ_I and the relative electronegativities of Pauling (30) and the inductive constants, I_a, of Branch and Calvin (25). The latter were obtained from acidities of X–OH compounds in water. Although there is generally a satisfactory qualitative correlation of σ_I and electronegativity, the relationship is by no means quantitative. The reasons for the lack of a better quantitative correlation are not fully understood at present (31). On the other hand, the correlation with Branch and Calvin constants is satisfactorily quantitative, as shown in Figure A2.6.

The order of σ_I values is in general distinctly different from the order of observed effects of *meta* and *para* substituents in any known reactivity. This unique character has been previously illustrated (32),

TABLE A2.IV

Periodic Relationship of the Inductive Constants, σ_I and I_a,
and Pauling's Electronegativities, X

	CH_3	NH_2	OH	F
σ_I	-0.05	$+0.10$	$+0.25$	$+0.52$
I_a	-0.4	$+1.3$	$+4$	$+9$
$X - X_H$	$+0.4$	$+0.9$	$+1.4$	$+1.9$
	SiH_3	PH_2	SH	Cl
σ_I	-0.10	$+0.06$	$+0.25$	$+0.47$
I_a	—	$+1.1$	$+3.4$	$+8.5$
$X - X_H$	-0.3	0.0	$+0.4$	$+0.9$
		AsH_2	SeH	Br
σ_I		$+0.06$	$+0.16$	$+0.45$
I_a		$+1.0$	$+2.7$	$+7.5$
$X - X_H$		-0.1	$+0.3$	$+0.7$
			TeH	I
σ_I			$+0.14$	$+0.38$
I_a			$+2.4$	$+6$
$X - X_H$			—	$+0.4$

for example, by the wide scattering of points in a plot of $\log (K/K_0)$ values for the ionization of benzoic acids in water versus corresponding σ_I-values. Values of σ_I are tabulated in reference 24.

Two independent direct lines of evidence have been obtained which support the hypothesis that σ_I values are acceptable empirical measures of the inductive order of substituent groups bonded to aromatic carbon, i.e., of X–Ar inductive effects. In both cases physical property–reactivity correlations are involved.

Wepster (33) has reported a linear relationship (Fig. A2.7) between the effects on the ultraviolet extinction coefficient and on the base strength produced by the introduction of groups which sterically inhibit the resonance interaction of the nitro group in substituted *para*-nitroanilines. If it is assumed that complete steric inhibition of resonance of the nitro group will reduce the extinction coefficient to approximately zero, then by a very short extrapolation of this relationship it may be estimated that the effect on base strength (which must now be due to the inductive effect only) is 1.8 log units. This value divided by the reaction constant, ρ, gives the "aromatic" σ_I-value for the nitro group, i.e., a value of $+0.62$. The σ_I value for aliphatic series reactivities is $+0.63$, an altogether satisfactory agreement. By a similar means Taft

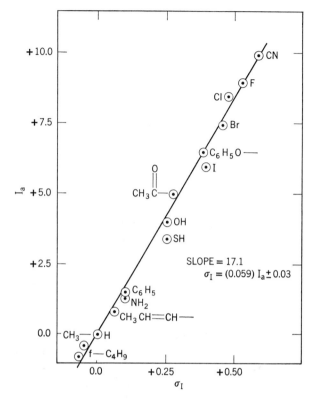

Fig. A2.6. Relationship between Branch and Calvin inductive constant, I_a, and σ_I-values.

and Evans (34) have shown that complete steric inhibition of resonance of the OCH_3 and $N(CH_3)_2$ groups give sigma values equal to σ_I.

The second line of direct evidence is the correlation of the shielding parameters for *meta* substituents in the ^{19}F nuclear magnetic resonance spectra of fluorobenzenes by σ_I-values. This correlation is shown in Figure A2.8. The shielding parameters (in ppm) shown in Figure A2.8 have been obtained in very dilute carbon tetrachloride solutions (15,35). Nearly equivalent results have been obtained for very dilute methanol and aqueous methanol solutions (35). The correlation is remarkable in view of the wide range of resonance capacities of the substituents included in the relationship, a point to be considered further in Section G. The relationship of Figure A2.8 is particularly significant since the

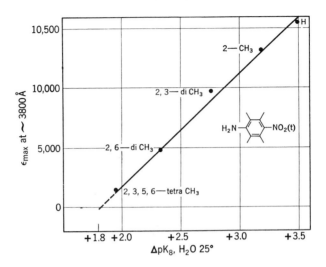

Fig. A2.7. Wepster's correlation of the effects of steric inhibition of resonance of the nitro group in methyl substituted *para*-nitroanilines on ultraviolet absorption and base strength.

NMR parameters are apparently measures of the inductive perturbations of the *meta* substituents on the charge density in the immediate vicinity of the fluorine atom. These parameters therefore are measures of the inductive order in a single state (the fluorobenzene) rather than in the difference between two states as in always involved in reactivities.

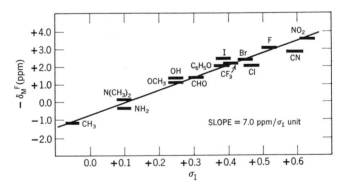

Fig. A2.8. Correlation of NMR shielding parameter, δ_M^F, for *meta*-substituted fluorobenzenes with inductive parameter, σ_I.

For these and reasons which follow, it is considered a satisfactory and useful approximation to treat generally the effects of *meta* and *para* substituents on the reactivities of benzene derivatives according to the additive relationships (10,24).

$$\log (k^m/k_0) = I + R^m = \sigma_I \rho + R^m$$

and

$$\log (k^p/k_0) = I + R^p = \sigma_I \rho + R^p$$

The term $\sigma_I \rho$ is the estimated effect (hereafter called the X–Ar inductive effect) on $\log (k/k_0)$ values resulting formally from inductive interaction through the X–Ar bond for the hypothetical molecule which involves no resonance interactions through this bond. The R value is the total effect on the $\log (k/k_0)$ value resulting from resonance (or additional) interactions which take place through the X–Ar bond. It is the sense of this proposed separation that any inductive effects which arise as a consequence of the resonance interaction are included in the R values (10).

In further support of the hypothesis that σ_I-values are applicable to aromatic reactivities in general, Taft and Lewis have shown that the assumption of a roughly linear relationship between corresponding R^m

TABLE A2.V

"Aromatic" σ_I Values Derived from Values of Log (k^m/k_0) and Log (k^p/k_0) by Method of Roberts and Jaffé[a]

Subst.	A.1 Ioniz. $ArCO_2H$ H_2O, 25°	C.9 Ioniz. $ArNH_3{}^+$ H_2O, 25°	E.4 Decomp. $ArN_2{}^+$ H_2O, 29°	C.4 Ioniz. ArSH 48% aq. EtOH, 35°	F.2 $ArCH_3$ $+Cl_2$, 70°	"Ali- phatic" σ_I
CH_3	−0.03	−0.02	−0.01	−0.04	−0.06	−0.05
NH_2	+0.04	+0.11		+0.10		+0.10
C_6H_5	+0.09	+0.14	+0.07		+0.09	+0.10
OCH_3	+0.28	+0.27	+0.26	+0.32		+0.29
CH_3CO	+0.32	+0.27		+0.27		+0.28
CF_3	+0.38	+0.32				+0.41
Br	+0.45	+0.42		+0.44		+0.45
Cl	+0.42	+0.43	+0.45	+0.46	+0.48	+0.47
F	+0.45	+0.55				+0.52
CN	+0.52	+0.56			+0.50	+0.58
NO_2	+0.68	+0.60	+0.66	+0.66		+0.63

[a] Reaction designation refer to those as listed in Table II of ref. 10.

and R^p values for a *given reaction series* permits an approximate evaluation of "aromatic" σ_I values (10,24). The method has been somewhat modified by Roberts and Jaffé (36), who show statistically its wide applicability. In Table A2.V are shown some typical "aromatic" σ_I values obtained in the manner of Roberts and Jaffé from the log (k/k_0) values for the indicated reaction series. The agreement between the σ_I values is of a precision closely on the order of a standard error of ± 0.03.

It follows consequently that the dispersion (especially of *para* substituents) in linear free energy plots of aromatic series reactivities (e.g., Figure A2.2 and A2.3) is attributable to good approximation to the dependence of R values on reaction type and conditions (10).

3. The Evaluation of X–Ar Resonance Effects (R Values)

The use of the $\sigma_I\rho$ term to evaluate resonance contributions (R values), i.e., $R = \log(k/k_0) - \sigma_I\rho$, offers two generally promising avenues of attack on reactivity problems.

First, precise empirical relationships between R values for similar reactivities provide a means of making accurate predictions and correlations of reactivities. These empirical relationships *and their limitations* provide a means of studying the factors which determine R values. The specific nature of R values offers a valuable tool in establishing the nature of transition states in mechanism studies. Secondly, the empirical evaluation of R values provides an "observable" which may be given theoretical consideration, for example, by the π-electron model. An example of such a treatment is given brief comment in later discussion related to Figure A2.10.

4. X–Ar Resonance Effect Parameters

At the considerable risk of vulnerability in the matter of discussing the perplexities of nature in terms of a profusion of symbols, it is frequently convenient to express R values in sigma units (10). This may be done by dividing through the above relationship by the ρ-value. Thus, in general, the "effective" $\bar{\sigma}_R$-value is obtained from the "effective" $\bar{\sigma}$-value by subtracting the value of σ_I: $\bar{\sigma}_R = 1/\rho[\log(k/k_0) - \sigma_I\rho] = \bar{\sigma} - \sigma_I$. If the value belongs to one of the various classes having limited generality, a resonance parameter characteristic of that class of reactivity is obtained. If used within its appropriate reactivity class, this resonance parameter is capable of giving precise correlations and predictions.

For example, an X–Ar resonance effect parameter is obtained for the limited group of "electrophilic" reaction series which are correlated by the σ^+ parameter (8), i.e., $\sigma_R^+ \equiv \sigma^+ - \sigma_I$. Similarly, for the limited group of "nucleophilic" reaction series which are correlated by σ^- values (7,28), one obtains the X–Ar resonance effect parameter $\sigma_R^- \equiv \sigma^- - \sigma_I$. Sigma values corresponding to what has been called the resonance polar effect (3) are obtained from the Ar–Y inductive constants, σ^0, i.e., $\sigma_R^0 \equiv \sigma^0 - \sigma_I$.

Since the σ_R^0 parameters for the select group of *meta* substituents have very great general applicability, it is of especial interest to examine these X–Ar resonance effect parameters. This may be done by reference to Figure A2.9 in which σ_R^0 values for *meta* substituents, X, are plotted versus σ_R^0 values for corresponding *para* substituents. Values of σ_R^0 are tabulated in ref. 10. The order of substituents along the correlation line in Figure A2.9 is Ingold's well-known mesomeric order (37). The $-R$ substituents (*meta* or *para*) have negative resonance parameters and those for the $+R$ substituents are positive. The relatively highly precise correlation of these two resonance parameters (10) (which is exact within the precision of the parameters or their solvent dependence) is of interest. This result is reasonable on the basis of conclusions reached earlier (10,11) concerning the parent sigma values, namely, that neither of the parameters is affected by direct specific interactions between the substituent, X, and sidechain reaction centers, Y and Y'. A similar plot of corresponding σ^0 values leads to a wide scattering of points indicating

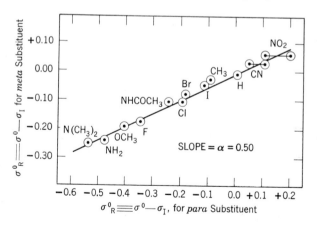

Fig. A2.9. Resonance effect fall-off factor, α, between *meta* and *para* positions for the σ^0 scale.

that the separation to σ_R^0 and σ_I values is essential to achieving the simple relationship of Figure A2.9.

Both the nonzero values of σ_R^0 for *meta* substituents and the slope (α) of ~ 0.5 obtained in Figure A2.9 may be accounted for by classical valence bond structures and an empirical inductive effect fall-off per bond. This is illustrated below by the structures in which equal charges are shown localized in the positions *ortho* and *para* to the substituent, X. If the usual fall-off factor per bond of approximately $\frac{1}{3}$ is employed (25) it follows that the net charge accumulated on Y is six-tenths as great when the substituent, X, is in the *meta* as in the *para* position. While this relationship undoubtedly is oversimplified, it appears to deal adequately with the approximate empirical relationship of Figure A2.9

$$\text{``}\alpha\text{''} = \frac{\text{``effective charge'' from \textit{meta} X}}{\text{``effective charge'' from \textit{para} X}} = \frac{2q(1/3) + 2q(1/3)^3}{q + 2q(1/3)^2} \cong 0.6$$

5. Further Remarks on Specific Character of X–Ar Resonance Effects

In general, it is found that precise correlations of "effective" $\bar{\sigma}_R$ parameters are limited to closely related reactivities and reaction conditions (10). An example of the dispersion which may appear in plots of X–Ar resonance effects is illustrated by a plot of σ_R^+ versus σ_R^0 for $-R$ substituents, as shown in Figure A2.10. In this plot only a limited precise linear relationship is found.

In unpublished work Ehrenson, Goodman, and Taft have derived models for consideration by the naive LCAO–MO method of the exalted resonance effects of $-R$ *para* substituents in σ^+ as compared to σ^0 reactivities. The method makes the basic assumption that the σ_R^+ and σ_R^0 parameters are proportional to the change in the total π-electronic energy accompanying the reaction process. The exaltation factor for the former parameters predicted by the treatment is that shown as the theoretical slope in Figure A2.10. The theory anticipates that the linear relationship of Figure A2.10 should not pass through the origin and that it should include the five substituents shown. Only these five substituents involve simple $\pi(p\text{-}p)$-conjugation of the first

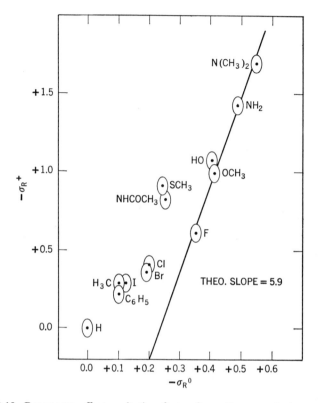

Fig. A2.10. Resonance effect exaltation factor for $-R$ *para* substituents in σ^+ and σ^0 reactivities.

atom of the substituent with the benzene ring, for which conjugative electron-release decreases regularly with increasing effective electronegativity (or σ_I value). The deviating substituents all have available open orbitals which undoubtedly are involved in additional conjugation e.g.,

$$\underset{\cdot\cdot}{\overset{\text{H}}{-\text{N}}}-\overset{\overset{\displaystyle\text{O}}{\parallel}}{\text{C}}-\text{CH}_3 \longleftrightarrow \underset{\oplus}{\overset{\text{H}}{-\text{N}}}=\overset{\overset{\displaystyle\text{O}^-}{\diagup}}{\underset{\diagdown}{\text{C}}}_{\displaystyle\text{CH}_3}$$

The halogen atoms, Cl, Br, and I, have d orbitals of their valence shell which may exert acceptor action in $\pi(p\text{-}d)$ conjugation. This possibility was first pointed out to the author several years ago by Professor J. G.

Aston, and is strongly supported by the theoretical calculations reported in his symposium paper by Goodman (38).

F. Summary of Schematics for Classification and Treatment of Reactivities

In Table A2.VI the schematics for the treatment of the log (k/k_0) values for the reactivities of *meta-* and *para*-substituted derivatives of benzene as discussed herein have been summarized.

G. Correlation of Electronic Distribution Parameters from Reactivities with Those Obtained by Physical Means

In conclusion, brief consideration will be made of how empirical physical property–reactivity relationships greatly add to our confidence that both measured properties do provide useful information on electronic distributions. The shielding parameters for *meta* and *para* substituents in the ^{19}F nuclear magnetic resonance spectra of fluorobenzene have found especial utility and may be regarded as illustrative in this connection. These shielding parameters were first investigated by Gutowsky and his students (39). Their work in the present connection has been so valuable (40) that additional investigations have been initiated and are being further pursued in our laboratory.

Table A2.VII lists values of shielding parameters (in ppm) for some typical *meta* and *para* substituents obtained in "infinitely dilute" carbon tetrachloride solution (15,36). Positive values of $\int^F$ indicate that resonance occurs at a higher field than for unsubstituted fluorobenzene and, accordingly, that the ^{19}F nucleus "sees" a greater density of electronic charge (41). It is immediately apparent that all of the substituents of Table A2.VII exert very different effects from the *para* than the *meta* position. Figure A2.8 shows that the $\int_M^F$ values are quite precisely correlated by the inductive constants, σ_I (*meta* σ^0 values give a considerably poorer correlation).

The quantity $-(\int_P^F - \int_M^F)$ may be considered an accurate measure of the perturbations in electron density detected by the ^{19}F nucleus as the result of resonance interaction of the *para* substituent with the fluorobenzene system (15,39,40).

It is expected according to classical valence bond theory that the $-R$ *para* substituents cannot conjugate with the fluorine atom of the

TABLE A2.VI

Schematics for Treatment of Log (k/k_0) Values for Reactivities of *meta-* and *para-* Substituted Derivatives of Benzene

Point of view: $X\!-\!Ar\overset{\rightarrow}{}Y$ general substituent: $X\!-\!C_6H_5\!-\!$ standard substituent: C_6H_5 $\log(k/k_0)$		Point of view: $X\overset{\rightarrow}{}Ar\!-\!Y$ general substituent: $X\!-\!$ standard substituent: $H\!-\!$ $\log(k/k_0)$	
Resonance effects $(Ar\!-\!Y)$	Inductive effects $(Ar\!-\!Y)$	Resonance effects $(X\!-\!Ar)$	Inductive effects $(X\!-\!Ar)$
Frequently estimated by $\psi \equiv \log(k/k_0) - \sigma^0\rho$. In general ψ values are specific to the nature of Y and reaction conditions. However, for a correlation of ψ values for electrophilic reactivity cf. Yukawa and Eaborn[b]	Frequently represented by σ^0; relationship followed: $\log(k/k_0) = \sigma^0\rho$	Represented by $\bar\sigma_R \equiv \bar\sigma - \sigma_I$. In general, $\bar\sigma_R$ values show a considerable degree of specificity to reaction type and conditions. Represented in appropriate special cases by $\sigma_R{}^0$, $\sigma_R{}^-$, $\sigma_R{}^+$. Limited relationships followed: $R = \rho'\sigma_R{}^{(0,+,-)}$	Represented by σ_I; relationship followed: $I = \sigma_I\rho$

[a] Y. Yukawa and Y. Tsuno, *Bull. Chem. Soc. Japan*, 32, 971 (1959).
[b] J. D. Dickinson and C. Eaborn, *J. Chem. Soc.*, *1959*, 3036.

TABLE A2.VII
^{19}F Shielding Parameters for *meta*- and *para*-Substituted Fluorobenzenes
in Infinitely Dilute Carbon Tetrachloride Solution (Exp. Error $= \pm 0.1$
ppm)

para		meta	
Subst.	$\int_P^F$, ppm	Subst.	$\int_M^F$, ppm
NMe_2	$+15.6$	CH_3	$+1.2$
NH_2	$+14.2$	NH_2	$+0.4$
OCH_3	$+11.5$	NMe_2	-0.1
OH	$+10.8$	CO_2H	-0.9
OC_6H_5	$+7.4$	OCH_3	-1.1
F	$+6.8$	OH	-1.3
CH_3	$+5.4$	CHO	-1.3
C_6H_5	$+2.9$	CH_3OCO	-1.5
Cl	$+3.1$	C_6H_5O	-2.0
Br	$+2.5$	Cl	-2.0
I	$+1.5$	CF_3	-2.1
H	0.0	COCl	-2.1
CF_3	-5.1	Br	-2.3
CH_3CO	-6.6	I	-2.4
CN	-9.2	CN	-2.8
NO_2	-9.3	F	-3.0
CHO	-9.4	NO_2	-3.5

fluorobenzene so that the charge released by resonance interaction of these substituents must be inductively transmitted to the ^{19}F nucleus through the Ar–F bond. Support for this theory is provided by the precise correlation shown in Figure A2.11 between values of

$$-\left(\int_P^F - \int_M^F\right)$$

and the appropriate reactivity resonance parameters, σ_R^0 values (any of the other classes of resonance parameters of Section E-4 give poorer correlations).

Conjugation between $+R$ *para* substituents and the fluorine atom of fluorobenzene is anticipated by valence bond theory, and in accord with this theory these substituents shield the ^{19}F nucleus substantially less than expected by correlation with the σ_R^0 parameter (15,40). This relationship is indicated in Figure A2.11 by the deviations for the NO_2 and CN substituents [the extent of deviation is substantially greater in hydroxylic solvents (23)].

The correlations of Figure A2.8 and A2.11 cannot be reasonably explained unless the classical notion (12) that the resonating substituent

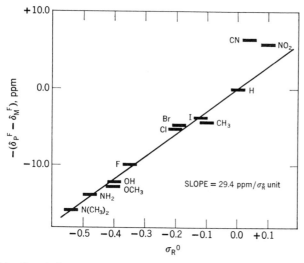

Fig. A2.11. Correlation of resonance contribution to the NMR shielding parameter for *para*-substituted fluorobenzene, $-(\delta_P{}^F - \delta_M{}^F)$, with the resonance parameter, $\sigma_R{}^0$.

localizes charge alternately on the *ortho* and *para* positions of the ring is a remarkably exact relationship. These correlations demonstrate that the shielding parameters and the reactivities which follow the σ^0-scale are both determined to good approximation by the same generalized quantitative orders of charge-release through the X–Ar bond by the inductive and resonance mechanisms. Further, the separation of the X–Ar inductive and resonance contributions shows that the susceptibilities of the two observables to these factors are very appreciably different.

This is illustrated by the relationships which summarize the several correlations:

For *para* substituents:

$$\sigma^0 \equiv \sigma_R{}^0 + \sigma_I$$

$$-\int_P^F = (29.7)\sigma_R{}^0 + (7.0)(\sigma_I) + c$$

For *meta* substituents:

$$\sigma^0 = \sigma_R{}^0 + \sigma_I = (0.5)\sigma_{R^0\text{(para)}} + \sigma_I$$

$$-\int_M^F = (7.0)\sigma_I + c$$

By definition, the relative susceptibility of σ^0-values for *para* substituents to σ_R^0 and σ_I is unity. However, the shielding parameters for *para* substituents are *five* times more susceptible to σ_R^0 than σ_I. For *meta* substituents, σ^0 values are, in contrast, 0.5 as susceptible to σ_R^0 (for *para* substituents) as σ_I, whereas the shielding parameters show no measurable dependence upon σ_R^0.

It is puzzling that the shielding parameters for *meta* substituents show a dependence upon σ_I but no detectable dependence upon σ_R^0. Nevertheless it is clear that the grossly different behavior must arise as a consequence of different susceptibilities of the physical observable to the total electronic distribution. The shielding parameter depends upon electron density in the very close vicinity of the ^{19}F nucleus (perhaps the electron density localized on the F atom). The reactivities apparently depend upon charges localized more widely throughout the substituted benzene ring (as discussed in connection with Figure A2.9).

These relationships point up the urgent need for knowledge of accurate detailed electronic distributions in molecules. Only when this information becomes available will it be possible to develop rigorous theories of structural effects on the electronic distribution in molecules and of the effects of these electronic distributions on various observables.

References and Footnotes to Appendix 2

1. This work was supported in part by the Office of Naval Research.
2. L. P. Hammett, *Physical Organic Chemistry*, McGraw-Hill, New York, 1940, p. 76.
3. R. W. Taft, Jr., *Steric Effects in Organic Chemistry*, M. S. Newman, Ed., Wiley, New York, 1956, (a) pp. 563–570; (b) pp. 660–665.
4. (a) K. J. Laidler, *Trans. Faraday Soc.*, 55, 1725 (1959); (b) R. P. Bell, *The Proton in Chemistry*, Cornell University Press, Ithaca, New York, 1959, pp. 69–73.
5. L. P. Hammett, *Chem. Rev.*, 17, 125 (1935).
6. Ref. 2, pp. 184–187.
7. (a) H. H. Jaffé, *Chem. Rev.*, 53, 191 (1953); cf. also *J. Am. Chem. Soc.*, 81, 3020 (1959); (b) cf. reference 26.
8. H. C. Brown and Y. Okamoto, *Chem. Rev.*, 79, 1913 (1957); 80, 4979 (1958); cf. also N. C. Deno and A. Schriesheim, *ibid.*, 77, 3051 (1955); and N. C. Deno and W. L. Evans, *ibid.*, 79, 5804 (1957).
9. G. E. K. Branch and M. Calvin, *The Theory of Organic Chemistry*, Prentice-Hall, New York, 1941, pp. 246–257, 416–419.
10. R. W. Taft, Jr., and I. C. Lewis, *J. Am. Chem. Soc.*, 81, 5343 (1959).
11. H. Van Bekkum, P. E. Verkade, and B. M. Wepster, *Rec. Trav. Chim.*, 78, 815 (1959). These authors use the symbol σ^n instead of ρ^0 used in the present

paper. In general, corresponding values of σ^0 and σ^n do not differ significantly. However, for the m- and p-$N(CH_3)_2$ and NH_2 substituents and the p-F substituent there are, for example, significant differences. In these instances, the author prefers σ^0 values for the reasons set forth in reference 10 and in C-5, C-6, and E-4 of this paper.

12. C. K. Ingold, *Chem. Rev.*, *15*, 225 (1934); G. W. Wheland and L. Pauling, *J. Am. Chem. Soc.*, *57*, 2086 (1935).
13. Evidence of effects of this kind has been obtained from the polarographic potentials for oxidation of *meta*- and *para*-substituted anilines to corresponding radical ions, I. Fox, R. W. Taft, Jr., and J. M. Schemf, abstracts of papers, American Chemical Society Meeting, Atlantic City, New Jersey, Sept. 1959, p. 72-P.
14. H. K. Hall, Jr., *J. Am. Chem. Soc.*, *79*, 5441 (1957).
15. R. W. Taft, Jr., S. Ehrenson, I. C. Lewis, and R. E. Glick, *J. Am. Chem. Soc.*, *81*, 5352 (1959).
16. G. W. Wheland, *Resonance in Organic Chemistry*, Wiley, New York, 1955, p. 429.
17. The Wheland classification of a $-R$ effect is equivalent to the $+T$ effect, in Ingold's classification (C. K. Ingold, *Structure and Mechanism in Organic Chemistry*, Cornell University Press, Ithaca, New York, 1953, p. 64). The conformity of both rate and equilibrium processes to the linear free energy relationships provides no evidence for the necessity to distinguish the T effects for the former processes as E effects and those for the latter as M effects.
18. Cf., for example (a) ref. 9; (b) F. G. Bordwell and P. J. Boutan, *J. Am. Chem. Soc.*, *78*, 854 (1956); *79*, 719 (1957); (c) ref. 3, p. 576–578; (d) ref. 11, p. 842.
19. Several attempts have been made to estimate quantitatively the Ar—Y resonance effects in the ionization of benzoic acids from calculations of the Ar—Y inductive effects by field electrostatic theory, cf., for example, J. N. Sarmousakis, *J. Chem. Phys.*, *12*, 277 (1944). In general agreement between this method and the empirical σ^0 method is poor. However, in the case of the four *para* halogen substituted phenyl groups the agreement between the values given by Sarmousakis and the σ–σ^0 values is reasonably quantitative.
20. M. M. Davis and H. B. Hetzer, *J. Res. Natl. Bur. Std.*, *60*, 569 (1958).
21. M. T. Rogers, *J. Am. Chem. Soc.*, *77*, 3681 (1955).
22. B. Gutbezahl and E. Grunwald, *J. Am. Chem. Soc.*, *75*, 559 (1953).
23. R. W. Taft, Jr., R. E. Glick, I. C. Lewis, I. Fox, and S. Ehrenson, *J. Am. Chem. Soc.*, *82*, 756 (1960).
24. R. W. Taft, Jr., and I. C. Lewis, *J. Am. Chem. Soc.*, *80*, 2436 (1958); cf. also ref. 10 and additional examples provided in Table II of this reference.
25. Reference 9, pp. 201–225.
26. J. Hine, *J. Am. Chem. Soc.*, *81*, 1126 (1959).
27. (a) R. W. Taft, Jr., *J. Am. Chem. Soc.*, *75*, 4321 (1953); (b) ref. 3, Chapter 13; (c) R. W. Taft, Jr., *J. Chem. Phys.*, *26*, 93 (1957).
28. R. W. Taft, Jr., N. C. Deno, and P. S. Skell, *Ann. Rev. Phys. Chem.*, *9*, 292 (1958).
29. C. K. Ingold, *J. Chem. Soc.*, *1930*, 1032.
30. L. Pauling, *The Nature of the Chemical Bond*, Cornell University Press, Ithaca, New York, 1944, p. 60.

31. Cf. discussion of ref. 27.
32. R. W. Taft, Jr., and I. C. Lewis, *Tetrahedron*, 5, 213 (1959).
33. B. M. Wepster, *Rec. Trav. Chim.*, 76, 335, 357 (1957).
34. R. W. Taft, Jr., and H. D. Evans, *J. Chem. Phys.*, 27, 1427 (1957).
35. Unpublished results.
36. J. L. Roberts and H. H. Jaffé, *J. Am. Chem. Soc.*, 81, 1635 (1959).
37. C. K. Ingold, *Structure and Mechanism in Organic Chemistry*, Cornell University Press, Ithaca, New York, 1953, Chapter VI.
38. (a) L. Goodman, *J. Phys. Chem.*, 64, 1816 (1960); (b) the order of net charge *release* by conjugation through the X—Ar bond is F > Cl > Br > I > H = 0 (cf. Fig. 11). The order expected for $-(p-p)$ conjugation (based upon ionization potentials) is F < Cl < Br < I. The $\pi(p-d)$ conjugation is expected to give the following order of charge *withdrawal*: Cl < Br < I. Consequently, a combination of $\pi(p-p)$ and $\pi(p-d)$ conjugation in which the former predominates can explain both the net charge releasing effect and the observed halogen order. Apparently, the relative contributions of $\pi(p-p)$ and $\pi(p-d)$ conjugation for a given halogen are not strongly dependent upon the side-chain functional group (Y, of X—Ar—Y). This conclusion is dictated by the σ^0 and σ_R^0 correlations, e.g., Fig. 11. The interpretation given here receives additional support for nuclear quadrupole coupling studies of BX_3 compounds. In these compounds $\pi(p-d)$ acceptor action is not possible and the % double bond character of the B—X bond increases in the order expected for increasing $\pi(p-p)$ conjugation, i.e., Cl < Br < I, cf. W. G. Laurita and W. S. Koskai, *J. Am. Chem. Soc.*, 81, 3179 (1959).
39. H. S. Gutowsky, D. W. McCall, B. R. McGarvey, and L. H. Meyer, *J. Am. Chem. Soc.*, 74, 4809 (1952).
40. R. W. Taft, Jr., *J. Am. Chem. Soc.*, 79, 1045 (1957).
41. J. A. Pople, W. G. Schneider, and H. J. Bernstein, *High-Resolution Nuclear Magnetic Resonance*, McGraw-Hill, New York, 1959.

VIII. APPENDIX 3. ON THE SEPARABILITY OF INDUCTIVE AND RESONANCE EFFECTS WITHIN THE FRAMEWORK OF LINEAR FREE ENERGY RELATIONSHIPS (1)*†

A. Introduction

A program for *a priori* statistical refitting of aromatic reactivity data to the Taft linear free energy relationship has recently been outlined (2). The principal purpose for reexamination of this relationship, generally expressible as $\log(k/k_0) = I + R$, was to determine what

* By S. Ehrenson, Chemistry Department, Brookhaven National Laboratory, Upton, L.I., N.Y.
† References and footnotes for this appendix will be found on pp. 241-243.

physical significance attaches to imposition of the constraints which transform it to its operational forms (3)

$$\log (k^p/k_0) = \rho_I \sigma_I + R^p \equiv \rho_I(\sigma_I + \sigma_R)$$
$$\log (k^m/k_0) = \rho_I \sigma_I + R^m \equiv \rho_I(\sigma_I + \alpha \sigma_R) \tag{1}$$

The corollary question of what, if anything, is accomplished by relaxation or modification of these constraints was also considered. Accomplishment, in the latter sense, has implications both to the understanding of divisions to total observable substituent effects, if possible in terms of physically justifiable reaction system models, as well as to improvement of statistical fitting. In the present discussion, concern with correlational improvement will be limited to questions of divisibility of effects and of applicability of such models, and not directly to how the correlations may be improved for their own sakes.

The completely general dual substituent equation,

$$\log (k/k_0) = \rho_I \sigma_I + \rho_R \sigma_R \tag{2}$$

may be mathematically constrained in a variety of ways to conform to models of physical reality. In Sections C–E it is shown that at least four independent constraints are required to yield unique solutions in a mathematical sense if ρ_I, ρ_R, σ_I, and σ_R are independent variables. Taft's equations (1) arise from more stringent constraints than minimally required, i.e., he takes $\rho_I^p = \rho_I^m$, $\alpha \rho_R^p = \rho_R^m$ and a standard reaction (benzoic acid ionizations in H_2O at 25°) for which $\rho_I \equiv 1.000$ are three plus the additional specification that $\rho_I = \rho_R$. The latter condition removes one independent set of variables.

Several important physical demands attend imposition of these constraints. Independent of reaction,

1. there is a constant ratio of sensitivity to inductive and resonance effects, and independent of reaction and substituent,

2. inductive effects are the same from the *meta* and *para* positions,

3. there is a constant ratio of resonance effects from the *meta* and *para* positions.

Of these, *3* is known to be unrealistic, although within general groupings where valence-bond-represented conjugation between the substituent and reaction center is or is not possible, some correlation is noted. For the former, α's from 1/5 to 1/10 pertain, for the latter, 1/2 to 1/3. As for *1* and *2*, these are suspect; *2* has had some examination, but only in a limited manner and under conditions which imply *1* (3). If *1* and *2* are physically realistic, relaxation of the constraints which imply them should not substantially improve the statistics for fitting the reactivity data to which they pertain. If they are not, their interactions with each other and with *3* are crucial. The bearing of all three upon such questions as the mechanism of transmission of the *I*-effect in aromatic versus aliphatic systems, the related proportionality of σ_I and σ^*, and upon the overriding question of whether *I* and *R* can be satisfactorily separated at all, are the paramount physical issues of the investigation.

Here we constrain by requiring $\rho_I{}^p = K_I\rho_I{}^m$, $\rho_R{}^p = K_R\rho_R{}^m$, by defining a standard reaction and by holding at least one σ value. The standard reaction is the same as Taft's if it is in the group of reactions considered; with σ_R for one substituent held, the minimum constraint conditions are met (see Sec. C). The correlation equations employed are then,

$$
\begin{aligned}
\log\left(k^p/k_0\right) &= K_I\rho_I\sigma_I + K_R\rho_R\sigma_R \\
\log\left(k^m/k_0\right) &= \rho_I\sigma_I + \rho_R\sigma_R
\end{aligned}
\tag{3}
$$

with all parameters assumed independent and with the σ's solely functions of substituent and not of position. Of the former physical conditions, *3* remains, *2* is weakened to read, . . . proportional . . . rather than the same, and *1* is removed.

These appear physically more consistent under the presumptions that substituent and reaction center effects are truly separable, and, also, that a given substituent may contribute different proportions of its total effect to *I* and *R* for different reactions. The notion of separability conforms to the assignment of a single σ_I and σ_R dependent only upon the nature of the substituent, and a single ρ_I and ρ_R for a given reaction group, which determines ρ_I/ρ_R, with modification due to geometry (position) reflected in K_I and K_R. It should, of course, be noted that the greater flexibility of the present equations is accompanied by computational complexity; this is, however, surmountable by use of nonlinear statistical analysis procedures programmed for digital computer handling.

The statistical procedures employed have been outlined in ref. 2 and discussed in more detail in Section V of ref. 4. A modification of the program was made part way through the present work to more efficiently utilize computer storage. Rather than simultaneous optimization of all parameters by the Gauss-Newton procedure (5), the so-called block diagonal modification was employed which varies groups of parameters in consecutive fashion (6). The general order of variation was, upon input of guesses for the σ's and K's, to determine best ρ's (by least-squares). Using the latter and the K's, best σ's were generated followed by determination of the K's from the first order ρ's and σ's. This process was continued cyclically until either all ρ and σ have converged within prespecified limits (0.005) or the precision of fit to the data no longer changes upon cycling (threshold of 0.05% of the standard deviation). In all cases tested, where initial guesses were sufficiently good to provide monotonic convergence by the original method, the converged results of the block-diagonal method were found to agree with the latter.

Again, as in ref. 2, the groups, OCH_3, CH_3, F, Cl, and NO_2 (hydroxylic medium) were chosen as test substituents. The choice of these groups was manifoldly motivated: they appear from other correlations to be the least sensitive to specific effects such as solvation, and, at the same time, they have been sufficiently widely employed so as to contribute significantly to, if not to essentially determine the correlations of all aromatic reaction series studied to date. As well, a wide range of electron donor–acceptor capacity is covered by these substituents, both in the inductive and resonance effect sense.

A total of $17\sigma^0$- and 24 "benzoic-type," hereafter called σ-type or simply σ-, reactivities have been examined, both within these groupings and together. For the former and latter sets, 120 and 211 individual $\log k/k_0$ data respectively were available. These comprise $\sim 70\%$ and $\sim 90\%$ of all possible data for the reactions and substituents considered and yield average data-per-parameter ratios of ~ 3 and 3.8 for the σ^0- and σ-type reactivities (7).

B. Results and Discussion

1. Statistics for σ^0- and σ-Reactivities

As indicated in Section A, interpretation of the separated I- and R-effects by the Taft equation must be accompanied by justification of the physical constraints by which the separation is effected, and hopefully also by statistical evidence to assure that the separation is unique. In extending previously described work (2), several results suggesting the Taft equation is generally valid were obtained but some disquieting and incompletely understood features were also uncovered.

Incorporation of 5 new σ^0-reactivities (reactions 13–17 in Sec. D) into the previously examined set indicate the best K_I and K_R to be 0.95 ± 0.01 and 1.97 ± 0.03 independent of initial input K's with σ_I

held at Taft's values (8). Here and elsewhere, the initial choices for σ_R are Taft's σ_R^0 values. These results agree with those obtained on the smaller σ^0-set: the standard deviation of fitting (SD) is improved by inclusion of these reactions from 0.031 to 0.029. Fitting of the σ-type and σ^0-type sets together under similar conditions yielded essentially the same K_I results but $K_R \sim 2.2$. The SD's for both sets were 0.038–0.039. These results are taken to suggest that distinctions drawn between reactivities wherein the reaction group can and cannot conjugate with the aromatic framework is useful in the sense of precision of fit but less so in terms of K_R (or α). Intuitively-directed inclusion of the 5 new reactivities into the σ^0-set significantly changed none of the results previously obtained for this set, e.g., K_I and K_R values were 0.97 and 1.9 (2). At the same time, the σ^0 reactivities are found to correlate with eq. (3) significantly better than do the σ-type although both fit within precisions expected if the experimental data contain on the average 5% uncertainty per measured point. With reference to distinctions between the σ^0- and σ-reactivities on the basis of K_R, the optimum value for this parameter is found to agree with Taft's assignment for the σ^0-set ($\alpha = \frac{1}{2}$), but for the σ-set, the value of 2.2 for K_R is somewhat better, if anything, than 3.0 (Taft's $\frac{1}{3}$) for which SD = 0.043. It is likewise pertinent to note that although K_I is found to be somewhat smaller than unity, this is probably not significant, contrary to a suggestion in ref. 2. Maintaining $K_I = 1.00$, the SD's of fit for the σ^0-, σ- and σ^0-plus σ-sets are 0.030, 0.039, and 0.039.

When fitting is attempted holding only to the minimum constraint set, with $\sigma_R(CH_3) = -0.10$, a broad range of equally good fitting K_I and K_R values are obtained. Over the range of initial K_I employed above and the extended K_R range down to 1.8, a variety of parameter sets were evolved all of which fit the σ^0-set to SD's of 0.028–0.030. Table A3.I presents some of these results in terms of initial and optimized K values. The final K_I's are seen to span the range 0.85 to 1.15 and are generally coupled so that small K_I values accompany large K_R's. The various optimized ρ and σ sets are quite different between sets, cf., Table A3.II where this is illustrated in terms of the σ's. Similar multiple optima results are obtained for the σ- and mixed sets. Here the SD's are 0.034 ± 0.002 as long as the initial K_I is unity or less: for larger K_I, convergence is to SD $\simeq 0.040$. In contrast, a test case where $\log k/k_0$ values were constructed from hypothetical ρ and σ values, this statistical method was able to effectively reproduce the parameters, even to the extent where the data are invested with up to 3% random errors.

TABLE A3.I
Initial and Optimized K_I and K_R Values for σ^0 Reactivities
under Minimum Constraint
Optimized K_I, K_R
[10^2 SD]

	Initial K_R			
Initial K_I	1.8	2.0	2.3	3.0
0.75	—	0.88, 2.25	0.88, 2.53	0.85, 3.19
		[2.85]	[2.83]	[2.95]
1.00	—	0.99, 2.04	—	0.99, 2.92
		[2.84]		[2.75]
1.15	1.06, 1.78	1.06, 1.94	1.06, 2.19	—
	[2.93]	[2.88]	[2.81]	
1.30	1.14, 1.70	1.14, 1.86	1.14, 2.09	1.13, 2.66
	[3.09]	[3.04]	[2.97]	[2.81]

It is apparent from these results that eq. (3), when minimally constrained (in a mathematical sense) cannot produce a unique parameter set for the reactivity series considered. This may be due to one, or both, of two fundamentally different reasons. The first is concerned with the freedom allowed in eq. (3), i.e., that all the parameter independence assumed does not exist. The second involves the interplay of data error and degree of correlation between various parameters, and is purely a statistical problem where the first is physical-statistical. If I and R are in some way dependent, there clearly will be more than one way eq. (3) can fit the data: if data errors are significant and trends in log (k/k_0) associated with change in one parameter are producible

TABLE A3.II
σ^0 Values for Converges Sets with Different K_I and K_R Values

		Substituents				
K_R	K_I	OCH$_3$	CH$_3$[a]	F	Cl	NO$_2$
2.53	0.88	0.20, −0.34	−0.05, −0.10	0.42, −0.23	0.43, −0.13	0.64, 0.28
2.04	0.99	0.28, −0.44	−0.05, −0.10	0.51, −0.36	0.51, −0.26	0.68, 0.17
2.09	1.14	0.32, −0.47	−0.05, −0.10	0.59, −0.46	0.61, −0.39	0.86, −0.10
Taft's σ values		0.26, −0.42	−0.05, −0.10	0.52, −0.33	0.46, −0.19	0.63, 0.18

[a] Both σ_I and σ_R held at Taft's values, the former only for convenience. When allowed to go free, $0.00 > \sigma_I > -0.08$ with insignificant effects on precision of fit.

in part by another, or others in concert, multiple solutions are likewise possible.

Partial resolution of this question is forthcoming from the following observations. When the last statistical analysis is repeated holding K_I constant at unity and also at 0.87, optimized parameter sets which vary with initial choice of K_R but which yield essentially identical fitting are again obtained. The choice of 0.87 as one fixed value for K_I is connected with analysis on the I effect discussed below; the results obtained with this value are identical for present purposes with those obtained using K_I of unity. Upon initial choice of 2.0, 2.5, and 3.0 for K_R, with $K_I = 0.87$, converged values for the former of 2.02, 2.47, and 2.92 were obtained with corresponding SD values of 0.0295, 0.0288, and 0.0285 for the σ^0-set. The same results are obtained for the σ- and mixed reactivity sets. While the fit is independent of K_R, the converged σ's and ρ_R^m's are not. The former quite precisely follow the relationship, $\sigma_I + \sigma_R = \text{constant } \{f(\text{substituent})\}$ among the various converged sets, while the latter change proportionally among these sets. Minor variations are noted among the other ρ's: the overall effect appears to be one where parts of I and R are shifted into each other, both for different K_R solutions at constant K_I and among the different K_I solutions.

Examination of the constraint equations in Section C2 shows the two observations, i.e., on $\sigma_I + \sigma_R$ and on ρ_R^m, to be consistent as long as ρ_R^p/ρ_I^p is essentially constant over all reactions and near unity if there is no change in any ρ's but ρ_R^m. Table III shows the constancy of $\sigma_I + \sigma_R$ among the various solutions for the σ^0-set. The ratio of ρ_R^p/ρ_I^p is unity with apparently unimportant deviations, e.g., for $K_I = 0.87$, $K_R = 2.02$, this ratio is 1.041 ± 0.095, with $K_R = 2.92$ the ratio is 1.050 ± 0.116. In Table A3.IV some reactions chosen randomly from the σ^0-set illustrate the shifting under change of K_I between I and R (9). This shifting, it should be pointed out, leaves the total predicted effect, $\log (k/k_0)$, unchanged not only in the statistical sense (SD for all reactions) but, because of the extra freedom introduced by the ρ_R^p, ρ_I^p dependency, in the individual points as well.

We are led therefore to conclude that Taft's assumption of proportional sensitivity to inductive and resonance effects is correct, at least for the reactions thus far considered. Whether this formalism may be carried over to the reactivities where direct conjugation is possible between substituent and reaction center is not presently known. It is not unreasonable to speculate, however, that if it does hold, at least a

TABLE A3.III
σ^0 Values for Sets with K_I held at 0.87 and 1.00 with Various K_R

Substituent	$K_I = 1.00$			$K_I = 0.87$		
	$K_R = 2.02$	2.94	3.88	$K_R = 2.02$	2.47	2.92
OCH_3						
σ_I	0.279	0.178	0.142	0.246	0.197	0.168
σ_R	−0.422	−0.313	−0.278	−0.382	−0.330	−0.302
CH_3						
$\sigma_I{}^a$	−0.050					
σ_R	−0.076	−0.082	−0.086	−0.083	−0.086	−0.090
F						
σ_I	0.514	0.428	0.397	0.456	0.421	0.402
σ_R	−0.362	−0.269	−0.238	−0.265	−0.227	−0.207
Cl						
σ_I	0.506	0.444	0.422	0.442	0.422	0.410
σ_R	−0.253	−0.187	−0.165	−0.149	−0.127	−0.116
NO_2						
σ_I	0.668	0.704	0.717	0.578	0.620	0.644
σ_R	0.147	0.108	0.095	0.314	0.269	0.245

[a] G_1 held at -0.050 in all runs.

TABLE A3.IV
Shifts between I and R upon Change of K_I[a]

Reaction,[b]	Substs.	$K_I = 1.00$		$K_I = 0.87$	
		I^m	R^m	I^m	R^m
	OCH_3	0.134,	−0.094	0.122,	−0.081
2	F	0.247,	−0.080	0.227,	−0.056
	NO_2	0.321,	0.033	0.288,	0.066
	OCH_3	0.300,	−0.231	0.271,	−0.198
4	F	0.553,	−0.198	0.502,	−0.137
	NO_2	0.719,	0.080	0.637,	0.163
	OCH_3	−0.185,	0.171	−0.160,	0.147
16	F	−0.341,	0.147	−0.296,	0.102
	NO_2	−0.444,	−0.060	−0.376,	−0.121

[a] K_R (SD) for both sets of 2.02 (0.0290 $\pm$ 0.0006); I^p obtainable from $K_I I^m$, R^p from $K_R R^m$.

[b] Reactions numbered as in Section F.

different (larger) constant for ρ_R^p/ρ_I^p will pertain. Some of the present difficulties in defining a single σ_R for a substituent may be overcome by this expedient. Work is presently underway to see if modification of the constraint is sufficient to improve correlation or whether this constraint must be entirely removed for σ^+ and σ^- reactivities.

2. Interaction Modes and Models for the I Effect

From the results of Table A3.III it is apparent that σ_I sets may be derived which fit the data satisfactorily, but which are not proportional to Taft's σ_I set and thereby are not proportional to σ^*. Again, from Section C2, where K_I is held and σ_I is a set equal or proportional to Taft's, there are primed sets where,

$$\sigma_I' = \sigma_I - \frac{K_I(1 - K_R/K_R')}{K_R' - K_I}\sigma_R \qquad (4)$$

Therefore, unless σ_I and σ_R are proportional, σ_I' is not proportional to σ^*. This result raises several interesting questions; i.e., why should σ_I for aromatic reactivities and σ^* for aliphatic and alicyclic reactivities be proportional, what contributes to σ_I above and beyond the factors determining σ^*, and are these effects separable. Further, are the latter clearly associated with the π electrons of the framework and, if so, how are they related to the conjugative effects presumably responsible for σ_R.

Two major effects are generally supposed to characterize induction in aromatic systems. The first, also present in saturated molecules, is the classical inductive effect which may be due either to electrostatic interaction through space of the substituent X and the reaction center Y, or to transmission via the framework σ bonds with damping, or more likely by both, if, in fact, the distinction is meaningful (4). If simple models pertain, the effect in this mode of a substituent in the *meta* position should be greater than in the *para* position (10). The inductive capacity of a substituent will also polarize the π-framework in aromatic molecules. This is the π-inductive effect, variously called the inductoelectromeric (11) and nonclassical inductive effect (12), which is expected by qualitative valence-bond and intuitive notions to be larger from the *para* than the *meta* position. *The two effects, then, operationally inseparable, in contrast to separations of I and R by such expedients as twisting of the substituent relative to the ring framework (13), may yield total inductive effects of equal magnitude from the meta and para position, but can only in total be proportional to the effects in aliphatic systems if they are proportional to each other!* This proportionality has been assumed by various workers without much justification (14); it is possible, however, to semiquantitatively indicate where it may be expected to hold.

Taking the monopole or dipole moment charge induced by X at the ring site of attachment equal to e_{RX}, the electrostatic energy is given as,

$$E_{El} = e_Y e_{RX}/Dr \quad \text{or} \quad e_Y \mu_{RX} \cos \theta / Dr^2 \tag{5}$$

where D is the averaged dielectric for the medium between sites RX and Y, r is the distance between these sites, μ_{RX} is the dipole moment of the ϕ_X dipole considered to be located at the ring site of attachment, and θ is the angle between the dipole direction of ϕ_X and the line joining sites RX and Y. If through-the-bonds transmission also enters, e_{RX} and e_Y are damped upon progression along the framework by a factor ϵ per bond, as is usually assumed, and if the interactions from these perturbed ring sites is also electrostatic, the total classical induction energy is,

$$E_{Cl} = \frac{e_Y e_{RX}}{D} \left[\frac{1}{r} + \epsilon \sum \frac{1}{r'} + \epsilon^2 \sum \frac{1}{r''} + \cdots \right] \tag{6}$$

Here the sums are over all interaction configurations arising from the bond transmission models, where the number of primes indicates the number of bond dampings for Y and RX in total. D is assumed the same for all paths and ϵ the same independent of the nature of the bond (15).

The π-inductive effect, on the other hand, arises from the perturbation at RX of the π orbital at this site followed by resonance distribution to the rest of the π framework. This effect at RX is reasonably Ce_{RX}, where C is the coupling factor relating the perturbation of all AO's at this site to the π-AO of proper symmetry.

We may therefore picture the entire inductive effect as,

$E_I =$ electrostatic + through-the-bonds + π-inductive
   $= f(r, e_{RX}, e_Y; p, \epsilon) + f(\pi \text{ framework}, e_{RX}, e_Y \text{ or } e_{RY})$

where r, e_{RX}, e_Y, and ϵ are as previously defined, p expresses the number of paths and bonds for the framework induction mode, and e_Y or e_{RY} allows for the cases of conjugating and nonconjugating reaction centers. Specifically, as derived in footnote 15 and in ref. 4, Section IV, respectively, for *meta*- and *para*-disubstituted benzenes,

$$E_{CI} = (e_Y e_{RX}/Dr)(1 + 3.85\epsilon) \qquad (7)$$

$$E_{\pi I} = A_I \overline{\nabla}^X \overline{\Delta}^Y \qquad (8)$$

Here, $\overline{\nabla}^X = 2\delta_{RX}$, $\overline{\Delta}^Y = \delta_{RY'} - \delta_{RY}$ (for nonconjugating Y) and A_I is a function of the framework and positions of substitution (19). For example, for the σ^0-reactivities (nonconjugating Y),

where the standard substituent, X_0, is taken as sp^2CH_2 as in ref. 4, our interest is in the following models for the π-inductive effect.

For these models, $A_{I\,meta} = -6.9 \times 10^{-3}$ and $A_{I\,para} = -14.4 \times 10^{-3}$, as obtained by first- and second-order perturbation theory with self-consistent charge redistribution effects considered (20).

Upon reexamination of the assumptions leading to these results, both the classical and π-inductive effects may be shown to be similarly dependent upon coulomb integral properties of X, X_0, Y, and Y', much in the manner previously demonstrated for resonance effects (4), and thereby to be simple combinable.

$$\Delta\Delta E(\text{total induction}) = \left[\frac{1 + 3.85\epsilon}{2L_Y L_R Dr} + A_I\right] \overline{\nabla}^X \overline{\Delta}^Y \qquad (9)$$

under the following assumptions.

a. The charge at the ring site of attachment of Y may be related to the effective charge of Y by $e_{RY} = \epsilon(e_Y - e_R)$, where e_R represents the unperturbed charge at this ring site and ϵ is invariant to the nature of Y. This often-made assumption has been previously employed in the combination of electrostatic and through-the-bonds models (15); some recent, as well as long standing observations relating electronegativities of atoms and the bonds formed between them (21), especially for grossly different X's and Y's, suggests, however, that this may be less secure an approximation than desired.

b. The effective coulomb integral for the π-AO at the substituted ring site can be coupled proportionately to the perturbation experienced in the σ orbital at this site due to substitution. If $e^\pi_{RY} = Ce^{\sigma+\pi}_{RY}$ and $e^\pi_{RX} = Ce^{\sigma+\pi}_{RX}$, then by the familiar linear relationship between coulomb integral and charge density, i.e., $\delta = \omega\beta(q_0 - q)$,

$$\delta_{RY} = \omega\beta(1 - C\epsilon(e_Y - e_R)) \tag{10}$$

Therefore, $L_Y = -\omega\beta C\epsilon$ and $L_X = -\omega\beta C$, as used in eq. (9), and

$$\overline{\Delta}^Y = 2L_Y(e_{Y'} - e_Y)$$
$$\overline{\nabla}^X = L_X(e_{RX} - e_{RX_0}) = L_X\epsilon(e_X - e_{X_0}) \tag{11}$$

In the event these assumptions hold, the inductive effects in aromatic systems will understandably be proportional to those noted in aliphatic systems where π induction is not possible. All else being equal, enhancement due to the latter effect would be $(A_I + B)/B$, where B is the classical effect: other changes, e.g., in geometry may, of course, introduce further proportionality factors.

Order of magnitude estimations of the contributions of the classical and π-inductive effects may be made as follows. With $\epsilon = 1/10$ (22) and C taken as unity (equal sensitivity of σ- and π-AO's to inductive perturbation) leaves only ω for determination of the L parameters. The proportionality constant relating the coulomb integral and charge density changes should be somewhere between the values used for the framework considered as a whole, where distant electron repulsion effects are averaged in ($\omega = 1.2$–1.4) (23), and that for the carbon atom in isolation ($\omega = 4.6$) (24). This argument is based on the understanding that we are here considering C–X (or C–Y) fragments, where some effect of X (or Y) upon carbon is expected: the value, $\omega = 3$, roughly the average of extremes, is employed.

Therefore, taking the effective electronic charge as 1×10^{-10} esu, following Ketelaar (25) and $D = 2.0$ (4), the classical term coefficient of eq. (9) is roughly -0.10 for *meta* position cases and -0.087 for *para* position cases. These values are for the energy in units of β ($\sim -2_+$ eV); the negative signs of the coefficients reflect the increase in energy for $(\delta_{RY'} - \delta_{RY})$ and $(\delta_{RX} - \delta_{RX_0})$ of the same sign. Addition of the π-inductive effects as expressed in A_I raises the ratio of the total effect to near unity, i.e., $p/m = -0.10_3/-0.10_7$.

What seems especially significant, then, is that the direct or classical inductive effects are formally and physically inseparable from the π-inductive effects which they, in a sense, produce, not only for a given substituent but over all substituents. The principal physical manifestation of the mixing is to make the total effects about the same from the *meta* and *para* positions (26).

3. Theory of Differences between I and R

Equation 9 which relates the total inductive effect to coulomb integral differences at the ring sites of attachment of X and Y is similar in form to equations developed in ref. 4 for resonance effects. There is however a strong operational difference between these equations which resides in the origin of the coulomb integrals considered!

In the resonance term, δ_X and δ_Y refer to the coulomb integrals for the π electrons in the $p\pi$ orbitals of X and Y which can conjugate with the framework. The π-inductive effect on the other hand, and hence the total inductive effect by the relationships here developed, depends upon the change in coulomb integrals induced at the ring site by inductive feeding or withdrawal by X and Y, i.e., δ_{RX} and δ_{RY} are inductively caused perturbations due to X and Y. These deltas, and as a result ∇^x (resonance) and $\overline{\nabla}^x$ (induction), are not necessarily the same, nor even necessarily simply proportional. They may in fact be of opposite sign.

It is well established in heteronuclear diatomic molecule theory that σ bonds may be polarized differently than π bonds in the same molecule, i.e., in the molecule AB, as an extreme case, A may appear electronegative with respect to B as regards charge density distributions in the σ MO's while appearing electropositive in the π-MO's (27). In the systems of present concern this conceptual difference is reflected in qualitative thinking about various substituents, e.g., methoxy while being a good electron withdrawer inductively is a donor by resonance; the nitro group, in contrast, withdraws by both modes. Valence bond structures such as $F \leftarrow C\big\langle$ and $F^+ = C\big\langle \ominus$ have similar implications as of course do σ_I and σ_R for a given substituent when these parameters are of different sign.

This is of course not to imply that interactions in the different MO (symmetry) dimensions are not related. In truth they must be, albeit not simply: here, whatever relationship there is may be additionally clouded by the models chosen. Substituent groups in many cases are unrealistically viewed as quasi-atoms, e.g., NO_2. When comparisons are drawn between substituents where the quasi-atom is a more sensible model, some of the apparent incoherence disappears, e.g., among the isoelectronic groups, F, OH, NH_2, $\Delta\sigma_I = \Delta\sigma_R$, and presumably ∇^X and $\overline{\nabla}^X$ are simply related, even though we continue to see donor–acceptor properties in the resonance–inductive mode. In general, however, as pictured, the inductive and resonance capabilities of substituents appear unrelated, and thereby ∇^X and $\overline{\nabla}^X$ (and σ_I and σ_R) are assumed independent variables.

As a final point in this regard, it is of interest to note that the resonance model coefficients, -12.5 and -15.1×10^{-3} for *meta-* and *para-*position substitution (19) are considerably smaller than those for the inductive model. If these estimates are of the correct order, it might be argued that in making a change from sp^2 carbon to a group, X, the sensitivity of the π orbital on X to the change is several times ($\sim 8 \times$) that experienced at the neighboring ring atom, or about equal, with $\epsilon = 1/10$, to that felt in the σ orbitals of X. This number is obtained by adjustment of the resonance coefficient to equal the inductive coefficient, as in Taft's *para-*substituent cases, thereby equally scaling the deltas; the result seems intuitively quite reasonable and consistent with respect to the value of unity previously taken for C, the π-, δ-coupling factor.

4. Statistical Test of the I-Effect Theory

Having rationalized the proportional carryover of σ^* to σ_I for model systems, implications of this result to real reactivities were statistically examined. Table A3.V contains the least-squares results for the σ^0 reactivities with K_I held at 0.87 and 1.00, and with σ_I held to Taft's values. In both cases, K_R was initially specified as 2.0, although upon further testing with values as high as 4.0, the fit upon convergence was found to be essentially the same. In both cases shown, the optimum K_R was found to be 2.10 ± 0.05 and the SD of fit was 0.030. Somewhat higher converged values of K_R were found with higher starting values for this parameter, and were generally accompanied by σ_R values of greater magnitude.

The comparisons of most interest are the changed proportion of the total effect found in I and R in going from a K_I of unity to 0.87. For the *meta-*position cases, there is an average of a 12% decrease in I— increase in R; for the *para-*position cases the shift is $\sim 17\%$. These numbers are derived by averaging the shifts in I with those in R between the two K_I-set results, recognizing that decreased I and increased R (or vice versa, depending on the signs of the parameter couples, $\rho\sigma$)

TABLE A3.V

Comparative Statistics for the σ^0 Reactivity Series with
$K_I = 1.00$ and $0.87^{a,b}$

	$\rho_I{}^m$		$\rho_R{}^m$	
Reaction	For $K_I = 1.00$	0.87	1.00	0.87
1	2.05^c	1.97	1.01	0.53
2	0.48	0.46	0.22	0.11
3	0.22	0.20	0.13	0.07
4	1.07	1.02	0.55	0.28
5	0.70	0.67	0.37	0.19
6	1.03	0.97	0.58	0.30
7	-2.68	-2.53	-1.40	-0.71
8	0.38	0.38	0.15	0.08
9	0.95	0.94	0.37	0.19
10	0.64	0.61	0.29	0.15
11	0.52	0.50	0.26	0.14
12	0.82	0.79	0.41	0.21
13	1.01	0.96	0.54	0.28
14	1.26	1.20	0.60	0.30
15	-0.29	-0.26	-0.20	-0.10
16	-0.66	-0.61	-0.41	-0.21
17	-0.20	-0.19	-0.10	-0.05

[a] Substituent constants

	OCH_3	CH_3	F	Cl	NO_2
σ_I (held)	0.26	-0.05	0.52	0.46	0.63
σ_R (initial)	-0.42	-0.10	-0.33	-0.19	0.18
For $K_I = 1.00$	-0.40	-0.08	-0.37	-0.19	0.20
$= 0.87$	-0.65	-0.15	-0.54	-0.23	0.54

[b] $\rho_I{}^p = K_I \rho_I{}^m$; $\rho_R{}^p = K_R \rho_R{}^m$. K_R (initial) $= 2.0$ for both runs; K_R (SD) converged $= 2.04$ (0.0299) and 2.15 (0.0298) for $K_I = 1.00$ and 0.87, respectively.

[c] Standard reaction, this $\rho_I{}^m$ held.

shifts one of these effects into the other. The comparable shifts from theory where the π-inductive effect is removed from I and placed in R are, for *meta* cases $\sim 7\%$ (7 parts in 107) and for the *para* cases $\sim 16\%$ (16 parts in 103).

The agreement between the statistical and theoretical results are satisfactory; interpretation of the meanings of these results is, however, not comparably satisfying. As previously noted for the runs where σ_I's were allowed to go free, $\rho_R{}^p/\rho_I{}^p$ is again essentially constant for a given

K_I, but varies upon change of K_I, i.e., $\rho_R{}^p/\rho_I{}^p \simeq 1$ for $K_I = 1.00$ and 2/3 for $K_I = 0.87$. As well, the sets of σ_R generated appear essentially proportional. When the transformation equations in Section C.3 are examined, the mechanics of the shift are understood. Examination of

TABLE A3.VI
Comparative Statistics for the Benzoic-Type Reactivity Series
with $K_I = 1.00$ and 0.87[a,b]

Reaction	$\rho_I{}^m$		$\rho_R{}^m$	
	For $K_I = 1.00$	0.87	1.00	0.87
18	1.00	0.91	0.60	0.17
19	1.58	1.49	0.82	0.24
20	1.72	1.59	1.05	0.31
21	1.22	1.14	0.71	0.21
22	1.36	1.27	0.74	0.22
23	1.48	1.39	0.80	0.23
24	1.26	1.19	0.69	0.20
25	1.39	1.31	0.76	0.22
26	1.56	1.47	0.80	0.23
27	1.52	1.43	0.82	0.24
28	1.47	1.37	0.81	0.24
29	0.92	0.86	0.50	0.14
30	2.34	2.16	1.37	0.40
31	2.48	2.30	1.44	0.42
32	2.57	2.40	1.42	0.42
33	0.44	0.40	0.27	0.08
34	1.26	1.17	0.73	0.21
35	0.75	0.69	0.46	0.13
36	0.65	0.60	0.40	0.12
37	3.08	2.84	1.89	0.55
38	1.79	1.62	1.25	0.36
39	1.69	1.59	0.88	0.26
40	0.40	0.37	0.27	0.08
41	0.44	0.41	0.27	0.08

[a] Substituent constants

	OCH_3	CH_3	F	Cl	NO_2
σ_I (held)	0.26	-0.05	0.52	0.46	0.63
σ_R (initial)	-0.42	-0.10	-0.33	-0.19	0.18
For $K_I = 1.00$	-0.39	-0.06	-0.32	-0.16	0.14
$= 0.87$	-1.18	-0.23	-0.82	-0.32	0.76

[b] Footnotes of Table A3.V apply with K_R (SD) converged $= 2.28$ (0.0389) and 2.35 (0.0388) for $K_I = 1.00$ and 0.87, respectively.

Table A3.VI indicates substantially identical results are obtained for the σ-type reactivities.

In the absence of the transformation relationships shown in the Section C, it would have been difficult to escape the conclusion that σ_I and σ_R were dependent. They would, in fact, have had to be linearly related, so that shifting of I into R could be accomplished without change in precision of fit. This would have provided anomalies as regards the theoretically rationalized differences between I and R and, more directly, a plot of σ_I versus σ_R for the substituents of present concern in particular, and all substituents in general. These anomalies are happily avoided.

At the same time, we are presented with the challenge of interpreting the variation in $\rho_R{}^p/\rho_I{}^p$ with change in division of I and R. Upon hypothesizing a separation where all effects either propagated by or transmitted by the π framework are included in R, the relative sensitivity to resonance (as now defined, compared to classical inductive) effects is found to decrease, compared to the more usual separation. If the statistics are meaningful, and they appear to be so as evidenced by the essentially identical fittings obtained in the different (K_I value) cases, this may be an entirely reasonable result. Since all effects of a substituent must be fundamentally related, and since the linear free energy equation can only recognize the simplest of relations between effects which are empirically separated, e.g., linearity in σ_I and σ_R, variations in sensitivity factors to the separated effects may be expected upon change in the divisions of such effects.

The implications of this analysis to the ultimate question of whether I and R are truly separable is obvious. Extension of these methods to reactivity systems where enhanced and levelled effects are known to occur is desirable, both to test concepts and to better define the practical limits of the statistical methods employed. Such extensions are presently under consideration.

C. Conditions for Obtaining Unique Parameter Sets

1.   General Relations

For

$$\log (k/k_0) = \rho_I\sigma_I + \rho_R\sigma_R$$

with $\rho = f$ (reaction, position of substitution) and $\sigma = f$ (substituent) and where the ρ's and σ's are assumed to be totally independent of each other and exactly fitted to the data or best fitted by least-squares procedures, one may construct alternative sets of parameters, ρ', by linear combinations,

$$\rho_I' = a\rho_I + b\rho_R$$
$$\rho_R' = c\rho_I + d\rho_R$$

which are also functions only of reaction and substitution position. The corresponding σ parameters are then,

$$\sigma_I' = (d\sigma_I - c\sigma_R)/(ad - bc)$$
$$\sigma_R' = (a\sigma_R - b\sigma_I)/(ad - bc)$$

Unique determination of parameter sets thereby requires conditions which fix a, b, c, and d. In contrast, the Hammett equation, $\log (k/k_0) = \rho\sigma$, requires only one constraint, i.e., if $\rho' = \bar{a}\rho$, then $\sigma' = \sigma/\bar{a}$ and only $\bar{a}$ must be fixed. The latter is accomplished by definition of a standard reaction (ρ for benzoic acid ionizations in H_2O at $25°$ taken as unity).

Under the Taft constraints, the "mixing parameters" $a–d$ are fixed as follows:

$$\left.\begin{aligned} \rho_I{}^m = \rho_I{}^p &= \rho_I \\ \rho_R{}^m &= \alpha\rho_R{}^p \\ \rho_R{}^p &= \rho_I \\ (\rho_I)_{SR} &\equiv 1.000 \end{aligned}\right\} \rightarrow \left\{\begin{aligned} a + b\alpha &= (\rho_I^{m'})_{SR} \\ c + d\alpha &= (\rho_R^{m'})_{SR} \\ a + b &= (\rho_I^{p'})_{SR} \\ c + d &= (\rho_R^{p'})_{SR} \\ \frac{\alpha\rho_I^{p'} - \rho_I^{m'}}{\alpha\rho_R^{p'} - \rho_R^{m'}} &= \text{constant} = \frac{a}{c} \end{aligned}\right.$$

Here SR denotes the standard reaction; the last relationship on the right implies constancy over all reactions and corresponds to $\rho_R{}^m = \alpha\rho_R{}^p$. In the present studies

$$\left.\begin{aligned} \rho_I{}^p &= K_I\rho_I{}^m \\ \rho_R{}^p &= K_R\rho_R{}^m \\ (\rho_I^m)_{SR} &\equiv 1.000 \\ (\sigma_R)_{SS} &= \sigma_R(\text{Taft})_{SS} \end{aligned}\right\} \rightarrow \left\{\begin{aligned} d(\rho_I^{m'})_{SR} - b(\rho_R^{m'})_{SR} &= ad - bc \\ \frac{K_I\rho_I^{m'} - \rho_I^{p'}}{K_I\rho_R^{m'} - \rho_R^{p'}} &= \text{constant} = \frac{b}{d} \\ \frac{K_R\rho_I^{m'} - \rho_I^{p'}}{K_R\rho_R^{m'} - \rho_R^{p'}} &= \text{constant} = \frac{a}{c} \\ b(\sigma_I')_{SS} + d(\sigma_R')_{SS} &= \sigma_R(\text{Taft})_{SS} \end{aligned}\right.$$

SS signifies a substituent chosen as standard with σ value (here σ_R) held invariant.

2. Where Different Parameter Sets are Generated with Different K_R at Constant K_I

Assume $\sigma_I + \sigma_R = $ constant $\{f(\text{substituent})\} = \sigma_I' + \sigma_R'$ where $\rho_R^p/\rho_I^p = \rho_R^{p'}/\rho_I^{p'} = 1$ and $\rho_I^p = \rho_I^{p'}$ then

$$
\left.
\begin{aligned}
\rho_I^{p'} &= (a + b)\rho_I^p \\
\rho_R^{p'} &= (c + d)\rho_I^p \\
\rho_I^{m'} &= \left(\frac{a}{K_I} + \frac{b}{K_R}\right)\rho_I^p \\
&= \rho_I^{p'}/K_I \\
\rho_R^{m'} &= \left(\frac{c}{K_I} + \frac{d}{K_R}\right)\rho_I^p \\
&= \rho_R^{p'}/K_R'
\end{aligned}
\right\}
\rightarrow
\left\{
\begin{aligned}
&a + b = 1 \\
&c + d = 1 \\
&a + \frac{b}{K_R} = 1 \\
&\frac{c}{K_I} + \frac{d}{K_R} = \frac{1}{K_R'}
\end{aligned}
\right.
$$

or, $a = 1$, $b = 0$, $c = (y - x)/(1 - x)$, $d = (1 - y)/(1 - x)$, where $y = K_I/K_R'$ and $x = K_I/K_R$.

$$\sigma_I + \sigma_R = \sigma_I' + \sigma_R' = [(d - b)\sigma_I + (a - c)\sigma_R]/(ad - bc)$$

$$\sigma_I' + \sigma_R' = \frac{\{(1 - y)/(1 - x)\sigma_I + [1 - (y - x)/(1 - x)]\sigma_R\}}{(1 - y)/(1 - x)}$$

$$= \sigma_I + \sigma_R$$

The transformation conditions check; σ_I and σ_R are analytically transformable knowing K_R'.

3. Where Different Parameter Sets are Generated, Holding σ_I Constant, for Different K_I.

Here, $\rho_R^{p'}/\rho_I^{p'} = l'$ for K_I' set and $\rho_I^{p'} = m\rho_I^p$. In a similar fashion to C2.

$$a + b = K_I'm \qquad c + d = l'K_I'm$$
$$a/K_I + b/K_R = m \qquad c/K_I + d/K_R = l'K_I'm/K_R$$

With $K_I = 1.00$

$$a = m(K_R - K_I')/(K_R - 1) \qquad b = m(K_I' - 1)K_R/(K_R - 1)$$
$$c = 0, \qquad d = ml'K_I'$$
$$\sigma_I' = \sigma_I/a \qquad \sigma_R' = (1/d)[\sigma_R - (b/a)\sigma_I]$$

For the case discussed, $l' \simeq \frac{2}{3}$, $m = 1/1.04$ with $K_I' = 0.87$ such that $K_R = K_R' = 2.1$. Therefore, $a \simeq 1.0$ or $\sigma_I' = \sigma_I$, as required, and

$$\sigma_R' = 1.8(\sigma_R + 0.26\sigma_I)$$

When σ_R' versus $(\sigma_R + 0.26\sigma_I)$ is plotted, fitting by a straight line is somewhat better than σ_R' versus σ_R.

F. σ^0- and σ-Reactivity Sets

Reactions 1–12 are as listed in ref. 2. Reactions 13–17 are respectively: ionization of $ArCH_2NH_2^+$ in H_2O; hydrogen bond equilibrium in CCl_4 of substituted phenols with pyridine; $\cdot CCl_3$ addition and abstraction reactions with $X\phi CH_2CH=CH_2$ and addition to $X\phi(CH_2)_2CH=CH_2$. These comprise the σ^0 set.

Reactions 18–41 are σ- or benzoic-reactivities and are in order: ionization (ion) of benzoic acid in H_2O and EtOH and ion of 2-Me benzoic acid in EtOH at 50°; ion of benzoic acid in 26.5, 43.5, 73.5% dioxane, in Et glycol, in MeOH, EtOH, i-PrOH and t-BuOH at 25°; $ArCO_2H$ + diphenyldiazomethane; saponification of $ArCO_2Et$ in aqueous acetone and EtOH; *trans* esterification of $ArCO_2C_{10}H_{19}$ with OMe^-; ion of cinnamic acids in H_2O, saponification of its ethyl esters in EtOH; ion phenyl propiolic acids in aqueous dioxane and EtOH; hydrolysis of acetic anhydrides; saponification of phthalides; $ArCO_2H$ dissociation in Me cellosolve; cinnamic acid + diphenyldiazomethane in EtOH at 25° and 35°.

References and Footnotes to Appendix 3

1. Work conducted under the auspices of the U.S. Atomic Energy Commission.
2. S. Ehrenson, *Tetrahedron Letters*, 7, 351 (1964).
3. See R. W. Taft, Jr., and I. C. Lewis, *J. Am. Chem. Soc.*, 81, 5343 (1959) as a prime reference relating to the work under discussion here.
4. S. Ehrenson, "Theoretical Interpretations of the Hammett and Derivative Structure-Reactivity Relationships, in *Progress in Physical Organic Chemistry*, Vol. 2, S. G. Cohen, A. Streitwieser, Jr., and R. Taft, Eds., Interscience, New York, 1964.
5. H. O. Hartley, *Technometrics*, 3, 269 (1959).
6. Cf. L. I. Hodson and J. S. Rollett, *Acta Cryst.*, 16, 329 (1963) where the block-diagonal procedure is discussed and additional references mainly concerned with crystal structure determinations, are presented.
7. With m reactions and n substituents, with one ρ_I^m set and held for a standard reaction and with one σ_R (or σ_I) set and held for a standard substituent, there are $(2m - 1)$ ρ's and $(2n - 1)$ σ's plus 2 K's to determine. Further constraints of course increase the data per parameter ratio.
8. Input K_I and K_R, respectively, 0.75–1.3 and 2.0–3.0.

9. For changing K_R at constant K_I, from Appendix C2, *para* $\rho_I{}^p\sigma_I \rightarrow \rho_I{}^p\sigma_I'$, $\rho_R{}^p\sigma_R \simeq \rho_I{}^p\sigma_R \rightarrow \rho_I\sigma_R'$; *meta*, $\rho_I{}^m\sigma_I = \rho_I{}^p\sigma_I/K_I \rightarrow \rho_I{}^p\sigma_I'/K_I = \rho_I{}^m\sigma_I'$, $\rho_R{}^m\sigma_R = \rho_R{}^p\sigma_R/K_R \simeq \rho_I{}^p\sigma_R/K_R \rightarrow \rho_I{}^p\sigma_R'(K_R/K_R')/K_R = \rho_I{}^p\sigma_R'/K_R' \simeq \rho_R{}^p\sigma_R'/K_R' = \rho_R{}^m\sigma_R'$.

10. In the case of electrostatic transmission assuming r^{-1} dependence (in the monopole–monopole modification of the Kirkwood-Westheimer model), with poles located at the ring sites, the *para-meta*-effect ratio corresponds to $K_I = 0.87$; for the K.W. model with the moment of the dipole on the ring, the result is again 0.87; for the center of the dipole one-half a benzene bond length from the ring, 0.94. The same average dielectric is assumed for paths from both positions. Similar results are obtained in the bond transmission model with damping factors per bond of $\sim 1/2$.

11. Cf. C. K. Ingold, *Structure and Mechanism in Organic Chemistry*, Cornell University Press, Ithaca, New York, 1953, p. 65.

12. K. B. Everard and L. E. Sutton, *J. Chem. Soc.*, *1951*, 2821.

13. R. W. Taft, Jr., and H. D. Evans, *J. Chem. Phys.*, 27, 1427 (1957).

14. (a) D. Peters, *J. Chem. Soc.*, *1957*, 2654, (b) O. Exner, *Tetrahedron Letters*, 6, 815 (1963).

15. The monopole–monopole model is used for all electrostatic interactions; it certainly is preferable as the distance between the interacting sites decreases (16). An extremely important point concerning the size of ϵ is also raised. If ϵ is large (weak damping), the effects arising at X and Y are propagated for substantial distances. This being the case, the electrostatic model would be expected to break down and, more importantly from an observational point of view, the effects of X and Y would necessarily be inseparable, especially as regards linear free energy substituent–reaction center effects, because of X,Y cross-terms. If, on the other hand, ϵ is small, equation 6 may be truncated after the first-power term in ϵ, and in neither *meta-* or *para*-substituted benzene cases would cross-terms be of importance. (The implications to small molecules where X and Y are proximate are clear.) It is further of interest that this model predicts a *para/meta*-classical *I*-effect ratio essentially the same as that from the simple electrostatic models previously discussed, and, is as a result, operationally indistinguishable from pure through-the-bonds models (with weak damping), on the basis of this ratio alone (4). The present model is, however, capable of rationalizing the need for strong damping to explain such phenomena as quadrupole-coupling effects of substituents (17), which neither the simple electrostatic models nor single-ϵ through-the-bonds models reasonably can explain.

For *meta*-disubstituted benzene as pictured, with Y as site 1, RX as site 4, all bonds of benzene C—C length, r_0, and ϵ independent of the nature of the bond (*vide infra*),

$$\frac{DE_{CI}}{e_{RX}e_Y} = \frac{1}{r_{14}} + \epsilon\left(\frac{1}{r_{13}} + \frac{1}{r_{15}} + \frac{1}{r_{24}}\right) = \frac{1}{r_0}(0.358 \pm 1.488\epsilon)$$

and for the *para*-disubstituted molecules with the same substituents,

$$\frac{DE_{CI}}{e_{RX}e_Y} = \frac{1}{r_{15}} + \epsilon\left(\frac{2}{r_{14}} + \frac{1}{r_{25}}\right) = \frac{1}{r_0}(0.333 + 1.256\epsilon)$$

The factor $1 + (3.85 \pm 0.09)\epsilon$ may be factored out of each quantity in parentheses, leaving a *para/meta*-ratio of ~ 0.87, essentially independent of ϵ. While indistinguishable, because of this ratio, from the pure framework transmission model with $\epsilon \sim 1/2$ (see discussion on this point for the simple electrostatic model in ref. 4), the cross-term problem is here avoided for small ϵ ($\sim 1/10$). Further, it is entirely reasonable that for reactivities and F NMR shifts (18), both inductive effects in combination contribute, requiring large ϵ if the framework model is to be substituted to account for the whole effect. On the other hand, quadrupole coupling constants may importantly depend only upon the polarization of the bond to the atom whose quadrupole moment is observed, i.e., to interactions having a given symmetry(σ-, as here postulated). In this case, the framework model is alone correct, and the small ϵ value pertains.

16. M. J. S. Dewar and P. J. Grisdale, *J. Am. Chem. Soc.*, *84*, 3539 (1962).

17. H. O. Hooper and P. J. Bray, *J. Chem. Phys.*, *33*, 334 (1960).

18. R. W. Taft, Jr., *J. Am. Chem. Soc.*, *79*, 1045 (1957); however, also see H. Spieseke and W. G. Schneider, *J. Chem. Phys.*, *35*, 722, 731 (1961) where strong damping is required to rationalize ^{13}C NMR effects.

19. A_l is proportional to the atom–atom polarizabilities for sites RX and RY (or Y) in the zeroth-order (non-self-consistent) approximation.

20. The A_l coefficients obtained as described in ref. 4 were checked by complete Wheland-Mulliken secular equation solution with self-consistent charge redistribution for benzyl radical-type structures perturbed at the terminal site and at its neighbor. The comparable A_l results of -6.6 and -15.3×10^{-3} were obtained. Of further interest, it was found that the perturbation theory presumption of additivity of terminal atom and neighbor atom effects is essentially correct: the complete solution for both positions perturbed yields -16.1 and -28.3×10^{-3} for the *meta*- and *para*-case coefficients, respectively and for the terminal perturbation alone, -9.5 and -13.0×10^{-3}. The perturbation theory results for the latter are -12.5 and -15.1×10^{-3}.

21. (a) R. P. Iczkowski and J. L. Margrave, *J. Am. Chem. Soc.*, *83*, 3547 (1961); (b) R. S. Mulliken, *J. Chem. Phys.*, *2*, 782 (1934).

22. Cf. H. C. Longuet-Higgins and C. A. Coulson, *J. Chem. Soc.*, *1949*, 971. Values on this order are required to correlate bond dipole moments and the previously mentioned quadrupole coupling effects.

23. S. Ehrenson, *J. Phys., Chem.*, *66*, 706, 712 (1962).

24. N. Muller, L. W. Pickett, and R. S. Mulliken, *J. Am. Chem. Soc.*, *76*, 4770 (1954).

25. J. A. A. Ketelaar and G. W. van Oosterhout, *Rec. Trav. Chim.*, *65*, 448 (1946).

26. The specific numerical illustration for the σ^0 reactivities may be generalized. The relative contributions of the two terms is not expected to change much for other reactivities. The A_l values for *meta*- and *para*-substituted benzyl-type structures, which reflect the π-inductive effects for xylylene-type molecules (models for conjugating X and Y groups) are not much different from those for the benzene-type structures considered above for π induction in the σ^0-reactivities.

27. Cf. E. Clementi, *J. Chem. Phys.*, *36*, 33 (1962), also see further discussion of this question with regard to alkyl group effects in S. Ehrenson, *J. Am. Chem. Soc.*, *86*, 847 (1964).

IX. APPENDIX 4. SUBSTITUENT EFFECTS IN NAPTHALENE.
II. THE STRENGTHS OF THE 4-, 5-, 6-, 7-, AND
8-SUBSTITUTED 2-NAPHTHOIC ACIDS*†‡§

A. Introduction

The syntheses of a series of substituted naphthoic acids were described in Part I of this series (1). Apparent dissociation constants for 50% v/v aqueous ethanol at 25° are now reported for these acids, the unsubstituted naphthoic acids, and six substituted 1-naphthoic acids. The latter were examined to extend the data for the α-series and to check our results against other studies (2).

B. Experimental (omitted)

C. Results

Tables A4.I and A.4II list the apparent pK_a values obtained as the mean of three determinations unless otherwise indicated.

TABLE A4.I

pK_a in 50% v/v Aqueous Ethanol at 25°: 2-Naphthoic Acids

Substituent	4β	5β	6β	7β	8β
NO_2	4.78 ± 0.00	5.05 ± 0.00	4.98 ± 0.01	5.15 ± 0.01	5.28 ± 0.01[a]
CN	4.80 ± 0.01	5.10 ± 0.00	5.12 ± 0.01	5.12 ± 0.00	5.33 ± 0.01[a]
F			5.55 ± 0.00	5.44 ± 0.01	5.58 ± 0.01
Cl	5.30 ± 0.00		5.42 ± 0.01	5.39 ± 0.02	5.61 ± 0.01
Br	5.32 ± 0.00	5.39 ± 0.01	5.40 ± 0.01	5.37 ± 0.01	5.61 ± 0.00
I	5.36 ± 0.00		5.43 ± 0.01	5.39 ± 0.01	5.63 ± 0.01
CH_3	5.83 ± 0.00		5.74 ± 0.01	5.73 ± 0.00	5.81 ± 0.01
OH	5.71 ± 0.00	5.70 ± 0.01		5.84 ± 0.01	5.92 ± 0.01[b]
OCH_3	5.72 ± 0.01	5.67 ± 0.00	5.82 ± 0.01	5.67 ± 0.01	5.93 ± 0.00[b]
NH_2	5.83 ± 0.00			5.93 ± 0.01	5.71 ± 0.01[b]
$N(CH_3)_2$			6.10 ± 0.01	5.86 ± 0.01	

[a] Mean of six determinations.

[b] Mean of four determinations.

* Part I, *Aust. J. Chem.*, **18**, 1351, (1965).

† By P. R. Wells and W. Acdock, Department of Chemistry, University of Queensland, Brisbane.

‡ Reprinted from the *Australian Journal of Chemistry*, **18**, 1365 (1965).

§ References for this appendix will be found on p. 255.

TABLE A4.II
pK_a in 50% v/v Aqueous Ethanol at 25° of
1-Naphthoic Acids

Substituent	3α	4α	5α
NO_2	4.54 ± 0.01	4.22 ± 0.01[a]	4.68 ± 0.01
CH_3	5.55 ± 0.01		
Cl	5.01 ± 0.01		
F		5.39 ± 0.01	

[a] Mean of six determinations.

Three different glass electrodes were employed.

(i) For the 6 and 7 series, with the exception of 7-NO_2, 7-OH, and 7-NH_2:
  pK_a of 2-naphthoic acid = 5.66 ± 0.01 (8 determinations; maximum deviation 0.03).
  pK_a of benzoic acid = 5.67 ± 0.01 (8; 0.01)

(ii) For the 8β, the 4β (except 4-CN), and three of the 7β series:
  pK_a of 2-naphthoic acid = 5.70 ± 0.02 (8; 0.03)
  pK_a of benzoic acid = 5.73 ± 0.02 (8; 0.03)

(iii) For the 5β-series, 4-cyano-2-naphthoic acid, and the α-series:
  pK_a of 2-naphthoic acid = 5.66 ± 0.01 (6; 0.02)
  pK_a of benzoic acid = 5.67 ± 0.01 (8; 0.02)
  pK_a of 1-naphthoic acid = 5.74 ± 0.01 (5; 0.01)

ΔpK_x values, recorded in Tables A4.III and A4.IV, are obtained as $\Delta pK_x = pK_0 - pK_x$, where the pK_0 employed is the pK_a of the unsubstituted naphthoic acid obtained employing the appropriate electrode system.

TABLE A4.III
ΔpK Values (β-Naphthoic acids)

Subst.	4β	5β	6β	7β	8β
NO_2	0.92	0.61	0.68	0.56	0.42
CN	0.86	0.56	0.54	0.54	0.37
F			0.11	0.22	0.12
Cl	0.40		0.24	0.27	0.09
Br	0.38	0.27	0.26	0.29	0.09
I	0.34		0.23	0.27	0.07
CH_3	-0.13		-0.08	-0.07	-0.11
OH	-0.01	-0.04		-0.14	-0.22
OCH_3	-0.02	-0.01	-0.16	-0.01	-0.23
NH_2	-0.13			-0.23	-0.01
$N(CH_3)_2$			-0.44	-0.20	

TABLE A4.III

ΔpK Values (α-Naphthoic Acids)

Present results in *italics*, others from Dewar and Grisdale (2)

Subst.	3α	4α	5α	6α	7α
NO_2	*0.93*[a]	*1.25*[b]	*0.79*[c]	0.62	0.55
CN	0.90	1.20	0.70	0.50	0.47
F		*0.08*			
Cl	*0.46*	0.39	0.44	0.25	
Br	0.52	0.45	0.45	0.28	0.11
CH_3	*−0.08*	−0.21	0.02	−0.07	−0.11
OH	0.09	−0.79	−0.10	−0.12	−0.15
OCH_3		−0.55	−0.02	−0.09	−0.12
NH_2		−1.09	−0.20		

[a] 0.93.

[b] 1.31.

[c] 0.82 (2).

In Table A4.IV the present results are given in italics. The remaining results for the α-series are from Dewar and Grisdale (2). Their ΔpK values for the 3-, 4-, and 5-nitro-1-naphthoic acids are given in the footnote to Table A4.IV.

D. Discussion

It is the ΔpK values (abbreviated to Δ) and not the pK values themselves that are the important results of these studies. They can be used as measures of the relative polar effects of the substituents.

There is a considerable variation in the reported pK of benzoic acid in 50% aqueous ethanol solution. McDaniel and Brown (9) have discussed this observation and it appears to be associated with the behavior of the glass electrode in alcohol/water mixtures. We have observed that for a particular glass electrode a definite pK value is consistently obtained with satisfactory precision. However, another glass electrode can yield a slightly different value with equal consistency. These variations do not appear to affect the Δ values, provided that the differences are taken within a set of pK measurements made using the same electrode assembly (cf. refs. 10,11). Variability of the asymmetry potential of the electrode has been suggested as the origin of these small changes (12).

One intuitively expects that the Hammett relationship will apply to the effect of substituents on reactivity in the naphthalene system. A limited test is possible using the data of Price, Mertz, and Wilson (13), and this suggests that the relationship holds. In most previous studies (2,14–16) of the naphthalene system the reactive reactivities have been converted to σ values through equation (1)

$$\sigma = \log (k/k_0)/\rho \tag{1}$$

in which the required ρ value has been taken to be that for the corresponding reaction of benzene derivatives. This assumption is unjustifiable and there is evidence that the true ρ values are significantly different. We shall report in Part III of this series on the application of the Hammett relationship to naphthalene reactivities and the choice of appropriate ρ values. For the present we shall not therefore convert the Δ values to σ values.

Dewar and Grisdale (2) have suggested two relatively simple expressions for σ values in aromatic systems. The more successful of these can be written in the form of eq. (2)

$$\Delta_{ij} = F/r_{ij} - M\pi_{ij} \tag{2}$$

The first term accounts for direct electrostatic interactions and the second term for π-electronic effects (induction and resonance). F and M are characteristic of the substituent and independent of the aromatic system to which it is attached.

The factor, $1/r_{ij}$, where r_{ij} is the distance between the points of attachment of the substituent and the side-chain containing the reaction site, accounts for the transmission of electrostatic interactions. A rationale can be presented for the use of this factor instead of the customary $\cos (\theta_{ij})/R_{ij}^2$ and for the neglect of differences in intramolecular dielectric constant (θ_{ij} is the angle between the substituent dipole vector and the line of length R_{ij} joining the reaction site and the midpoint of the dipole). Certainly both factors are likely to be rough approximations, although the latter takes some account of orientation. A comparison of the two factors is made in Table A4.V.

It is evident that, with the possible exception of 7α, the two factors would be expected to perform equally well in the correlation of the benzoic and the α-naphthoic series. Indeed Dewar and Grisdale find their equations are least satisfactory in the 7α case. They attribute this failure to differences in intramolecular dielectric constant. This has been assumed the same for all other cases where the intervening medium is the aromatic system itself. In the 7α case a substantial part of the intervening medium is outside the aromatic system (see Fig. A4.1). The

TABLE A4.V
Relative Electrostatic Transmission
($meta = 1$)

Position	$\cos{(\theta_{ij})}/R_{ij}^2$	$1/r_{ij}$	Position	$\cos{(\theta_{ij})}/R_{ij}^2$	$1/r_{ij}$
meta	1	1	para	0.90	0.87
3α	1	1	4β	1	1
4α	0.90	0.87	5α	0.35	0.58
5α	0.64	0.65	6β	0.44	0.48
6α	0.65	0.58	7β	0.41	0.50
7α	0.52	0.65	8β	0.13	0.65

appropriate dielectric constant may well be higher, yielding a lower

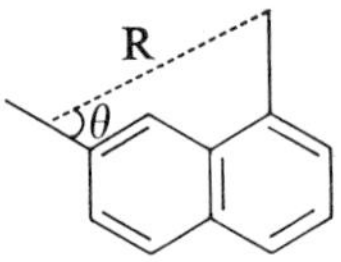

Fig. A4.1. 7α interaction. 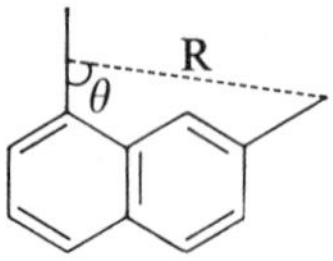   Fig. A4.2. 8β interaction.

electrostatic effect and hence a lower Δ value for acidic strengthening dipolar substituents. (All except the methyl group are of this type.)

The β-naphthoic series clearly provides a better test of the $1/r_{ij}$ factor since it contains the unique 8β position. In this case the dipole vector makes a large angle with $R_{8\beta}$, hence $\cos{\theta_{8\beta}}$ is small and the electrostatic effect is small (see Fig. A4.2). Added to this is the dielectric constant effect discussed for the 7α case.

In Table A4.VI the observed Δ values and those calculated (see Section E) from eq. (2) (the same for both positions) are compared for the 7α and 8β positions.

The fact that the failure of eq. (2) is considerably more serious for the 8β than the 7α series suggests that it is largely the orientational factor which is responsible. Figures A4.1 and A4.2 suggest that the dielectric constants will be very similar for the two positions.

Table A4.V also indicates that the orientational effect, but presumably not the dielectric effect, should lead to failures in the 5β case. Our study of this series is rather limited (Table A4.VII).

As in the 7α and 8β series, the acid strengthening effect of the C–O

TABLE A4.VI
7α and 8β Substituent Effects

Subst.	$\Delta_{7\alpha}$ (obs.)	$\Delta_{8\beta}$ (obs.)	Δ (calc.)
NO_2	0.55	0.42	0.81
CN	0.47	0.37	0.70
F	—	0.12	0.22
Cl	—	0.09	0.33
Br	0.11	0.09	0.36
I	—	0.07	0.37
CH_3	−0.11	−0.11	−0.11
OH	−0.15	−0.22	−0.13
OCH_3	−0.12	−0.23	−0.08

TABLE A4.VII
5β Substituent Effects

	NO_2	CN	Br	OH	OCH_3
$\Delta_{5\beta}$: Obs.	0.61	0.56	0.27	−0.04	−0.01
Calc.	0.62	0.54	0.31	0.12	0.12

TABLE A4.VIII
3α, 4β, and 4α Substituent Effects

Subst.	$\Delta_{3\alpha}$	$\Delta_{4\beta}$	Δ Calc.	$\Delta_{4\alpha}$	Δ Calc.
NO_2	0.93	0.92	1.06	1.25	1.27
CN	0.90	0.86	0.93	1.20	1.13
F				0.08	−0.03
Cl	0.46	0.40	0.58	0.39	0.34
Br	0.56	0.38	0.61	0.45	0.38
I		0.34	0.55		
CH_3	−0.08	−0.13	−0.10	−0.21	−0.23
OH	0.09	−0.01	0.26	−0.79	−0.80
OCH_3		−0.02	0.24	−0.55	−0.62

dipole, and perhaps the C–Br dipole, has been reduced. Equation (2) appears to predict the Δ values for the 5-nitro- and 5-cyano-2-naphthoic acids very well; however, if the ρ value for this series were larger than that for the benzoic series then the calculated Δ values would all be larger

TABLE A4.IX
5α and 6α Substituent Effects

	$\Delta_{5\alpha}$		$\Delta_{6\alpha}$	
Subst.	Obs.	Calc.	Obs.	Calc.
NO$_2$	0.79	0.78	0.62	0.62
CN	0.70	0.67	0.50	0.54
Cl	0.44	0.35	0.25	0.32
Br	0.45	0.37	0.28	0.31
CH$_3$	−0.07	−0.06	−0.11	−0.11
OH	−0.12	0.12	−0.15	−0.13
OCH$_3$	−0.09	0.12	−0.12	−0.08

and there would be an overall failure. For the present these limited data must be regarded as somewhat anomalous.

From the point of view of electrostatic interactions one may reasonably expect eq. (2) to perform quite well for all other positions. The π-electronic term though undoubtedly an oversimplification proves generally satisfactory as indicated by the comparisons in Tables A4.VIII–A4.X.

TABLE A4.X
6β and 7β Substituent Effects

	$\Delta_{6\beta}$		$\Delta_{7\beta}$	
Subst.	Obs.	Calc.	Obs.	Calc.
NO$_2$	0.68	0.62	0.56	0.56
CN	0.54	0.52	0.54	0.48
F	0.11	0.13	0.22	0.25
Cl	0.24	0.24	0.27	0.28
Br	0.26	0.26	0.29	0.30
I	0.27	0.27	0.27	0.28
CH$_3$	−0.08	−0.09	−0.07	−0.06
OH	—	—	−0.14	0.06
OCH$_3$	−0.16	−0.11	−0.01	0.07

There are, however, systematic deviations from eq. (2) that merit some comment.

For the pseudo-*meta* positions, 3α and 4β, the calculated Δ values appear to be too large. A reduction of ca. 10% would bring the calculated Δ values of nitro- and cyano-groups more or less in line with the observed results. This could arise if the ρ values for then aphthoic series were smaller by ca. 10% than that for the benzoic series. It would not, however, correct for the finding that the observed Δ values for the remaining groups are smaller than expected. All these groups have electron-releasing resonance effects. The π-electronic term based upon atom–atom polarizabilities takes no direct account of *secondary resonance effects*, i.e., π-electron density changes at atoms adjacent to the attachment of the side-chain. Presumably such effects are present and have been included in some way into F and M. But these are derived from benzoic acid data. If resonance interactions in naphthalene are greater than in benzene, and the π_{ij} values indicate this is so, then proper account has not been taken of them in eq. (2). In support of this conclusion is the fact that the discrepancies are larger for the 4β than for the 3α position. Other evidence suggests that π-electronic interactions are greater from α than from β positions.

Equation (2) gives a good account of the 4α series. In comparison with the *para*-substituted benzoic acids the Δ values are larger in the sense of increased π-electronic effects.

Tables A4.IX and A4.X demonstrate that the performance of eq. (2) is at its best for the 5α, 6α, 6β, and 7β series and particularly so in the latter two cases for which the most extensive data have been obtained. In these cases the constancy of the intramolecular dielectric constant is a good approximation, dipole orientations are very similar, and secondary resonance interactions are at a minimum. The best correlated series are 6β and 7β. In these cases both the substituents and the carboxyl group are in β positions. They will suffer no interference due to the *peri*-hydrogen. This is a factor not present in the *meta* and *para* benzene series but present in all other naphthalene series. Its effect cannot be accurately judged at this stage although one may expect to find modified ρ values owing to interference with the carboxyl group, and modified π-electronic and electrostatic effects owing to interference with substituents (for example, twisting and bending out of the aromatic plane).

The status of eq. (2) has not been changed by this further test using the β-naphthoic acid data. Dewar and Grisdale (2) recognized several possible shortcomings. Some of these have been clarified, particularly the influence of dielectric constant, dipole orientation, and secondary resonance effects. Certain refinements are obvious but one must guard against the introduction of further empirical terms into eq. (2). Improvements in the correlation of substituent effects will inevitably follow without necessarily improving our understanding.

Dewar and Grisdale (2) have not explicitly considered π-inductive effects, i.e., interactions arising from the distortion of the π-electronic system due to the electronegativity of the substituent. The second term in eq. (2) is taken to correspond to indirect resonance interactions. They omit consideration of π-inductive effects on the basis of the fact that charged substituents, e.g., Me_3N^+ and Me_2S^+, have a greater acid strengthening effect in the *meta* than in the *para* position of benzoic acid. However, this is not necessarily relevant in the case of dipolar substituents. The distance and orientational dependences of the two types of substituent are different. What evidence there is suggests that substituents of the type XCH_2, which cannot exert a resonance effect, are more effective from the *para* than the *meta* position in benzene derivatives (17). On the basis of properly scaled aliphatic σ values it can be shown that these purely inductive substituent effects are larger than those in the geometrically similar, but saturated, 4-substituted bicyclo-[2,2,2]octane-1-carboxylic acid system. The inevitable conclusion is that inductive effects are also transmitted by way of the π-electron system. It is likely that the π_{ij} factor will give a good account of the transmission of π-inductive effects so that the empirically determined M values presumably are measures of both resonance and π-inductive effects.

Ritchie and Sager (18) have pointed out that the F values should be proportioned to the inductive σ values (σ_I). The correlation is not entirely satisfactory. Part of the trouble may of course arise from inadequacies of the σ_I values; part may arise from the inaccuracy of the $1/r_{ij}$ and π_{ij} factors. In the empirical determination of the F and M values mutual correction of the two terms in eq. (2) undoubtedly occurs with the result that neither term is "purely" electrostatic nor π-electronic.

A correlation of the available benzene and α-naphthalene data with σ_I values is possible. Substituents are divided into two classes, "normal" and lone-pair types for which the following are reported (18)

Normal substituents:

$$\sigma_m = \sigma_{3_a} = \sigma_I$$
$$\sigma_p = \sigma_{4_a} = 1.44\sigma_I - 0.08$$
$$\sigma_{5_a} = 0.70\sigma_I + 0.06$$
$$\sigma_{6_a} = 0.62\sigma_I - 0.02$$
$$\sigma_{8_a} = 0.53\sigma_I$$

Lone-pair substituents:

$$\sigma_m = \sigma_{3_a} = 1.14\sigma_I - 0.17$$
$$\sigma_p = \sigma_{4_a} = 2.08\sigma_I - 0.73$$
$$\sigma_{5_a} = 1.05\sigma_I - 0.22$$

No indication of the precision of these correlations is given but the results given in Table A4.VIII suggest they cannot be very precise. We feel this approach is simply an exercise in data fitting and is not particularly constructive. The σ_I values derived from aliphatic reactivities depend predominantly on the relative magnitude of direct electrostatic interactions, and will be more or less proportional to bond dipole moments. It can be readily shown that bond dipole moments are dependent upon relative electronegativities as are π-inductive effects and, to some extent, resonance effects. It is therefore not surprising that a rough proportionality between the σ_I values and a large part of the total aromatic substituent effect can be found.

Although a number of amino- and dimethylamino-substituted acids were examined in this study the correlation of their Δ values by eq. (2) has not been described. F and M values can be obtained but the performance of eq. (2) for these substituents is variable. One probable cause of this has been discussed by Bryson (19) in terms of the zwitterion equilibrium:

$$H_2NArCO_2H \xrightleftharpoons{K_1} H_2NArCO_2^-$$

with K_3 and K_2:

$$H_3N^+ArCO_2^- \xrightleftharpoons{K_2}$$

While the acidity constant required as a measure of the substituent effect of the amino group is K_1 the observed acidity constant is

$$K = K_1(1 + K_3) = K_1K_2/(K_1 + K_2) \tag{3}$$

Thus until K_3, which will vary with substituent position, has been determined a correlation by eq. (2) cannot be attempted. Alternatively substituent parameters may be derived from a reaction series whose zwitterion equilibria are not involved, e.g., ester hydrolysis (to be dealt with in Part III). K_1 may then be calculated and some indication of the extent of zwitterion formation can be obtained indirectly.

Acknowledgments

The authors wish to thank Professor H. H. Jaffé, University of Cincinnati, U.S.A., for supplying atom–atom polarizabilities of the naphthalene system and Professor J. Vaughan, University of Canterbury, Christchurch, N.Z., for samples of 3-nitro-, 4-fluoro-, 3-chloro-, and 3-methyl-1-naphthoic acids. One of us (W.A.) is indebted to the Rothman's University Endowment Fund for the award of a Fellowship.

E. Calculated Δ Values

F and M values were calculated from the Δ values for *meta-* and *para*-substituted benzoic acids using

$$\Delta_m = 0.578F - 0.009M$$
$$\Delta_p = 0.500F + 0.102M \tag{4b}$$

and are listed in Table A4.XI. (These are all ca. 1.55 times as large as the listing of Dewar and Grisdale (2) who calculated σ values.)

TABLE A4.XI
Calculation of F and M Values

Subst.	Δ_m	Δ_p	F	M
NO$_2$	1.10[a]	1.21[b]	1.93	2.30
CN	0.95	1.05	1.67	2.07
F	0.53	0.09	0.86	−3.30
Cl	0.57	0.37	0.97	−1.07
Br	0.60[c]	0.41[d]	1.02	−0.92
I	0.54	0.47	0.95	0.15
CH$_3$	−0.11	−0.20[e]	−0.21	−0.92
OH	0.19[f]	−0.57[f]	0.22	−6.52
OCH$_3$	0.19[f]	−0.42[f,g]	0.24	−5.29

[a] Ref. 10, 1.09.
[b] Ref. 10, 1.22.
[c] Ref. 10, 0.53.
[d] Ref. 10, 0.40.
[e] Ref. 10, −0.20; ref. 11, −0.19.
[f] From ref. 1.
[g] Ref. 11, −0.32.

Δ_m and Δ_p were calculated from the σ values of Wells (20) or McDaniel and Brown (9) by multiplying by 1.55, the ρ value correlating the data of Roberts and co-workers for benzoic acid dissociation in 50% v/v aqueous ethanol at 25°.

Calculated Δ values were obtained using the factors listed in Table A4.XII.

TABLE A4.XII
Calculated Δ Values

Position	r_{ij}	π_{ij}	Position	r_{ij}	π_{ij}
3α	$\sqrt{3}$	0.018	4β	$\sqrt{3}$	0.018
4α	2	−0.139	5β	3	0.007
5α	$\sqrt{7}$	−0.023	6β	$\sqrt{13}$	−0.033
6α	3	0.007	7β	$2\sqrt{3}$	0.000
7α	$\sqrt{7}$	−0.033	8β	$\sqrt{7}$	−0.033

References to Appendix 4

1. W. Adcock and P. R. Wells, *Australian J. Chem.*, *18*, 1351 (1965).
2. M. J. S. Dewar and P. J. Grisdale, *J. Am. Chem. Soc.*, *74*, 3539 (1962).
9. D. H. McDaniel and H. C. Brown, *J. Org. Chem.*, *23*, 420 (1958).
10. J. D. Roberts, E. A. McElhill, and R. J. Armstrong, *J. Am. Chem. Soc.*, *71* 2923 (1949).
11. J. D. Roberts and C. M. Regan, *J. Am. Chem. Soc.*, *75*, 4102 (1953).
12. R. G. Bates, M. Paabo, and R. A. Robinson, *J. Phys. Chem.*, *67*, 1833 (1963).
13. C. C. Price, E. C. Mertz, and J. Wilson, *J. Am. Chem. Soc.*, *76*, 5131 (1954).
14. P. R. Wells and E. R. Ward, *Chem. Ind. (London)*, *1958*, 528.
15. E. Berliner and E. H. Winicov, *J. Am. Chem. Soc.*, *81*, 1630 (1959).
16. A. Bryson, *J. Am. Chem. Soc.*, *82*, 4862 (1960).
17. O. Exner, *Tetrahedron Letters*, *13*, 815 (1963).
18. C. D. Ritchie and W. F. Sager, *Progr. Phys. Org. Chem.*, *2*, 323 (1964).
19. A. Bryson and R. W. Matthews, *Australian J. Chem.*, *14*, 237 (1961).
20. P. R. Wells, *Chem. Rev.*, *63*, 171 (1963).

X. APPENDIX 5. FLUORINE NUCLEAR MAGNETIC RESONANCE SHIELDING IN *meta*-SUBSTITUTED FLUOROBENZENES. THE EFFECT OF SOLVENT ON THE INDUCTIVE ORDER (1)*†‡

Substantial effects of solvent on ^{19}F nuclear magnetic resonance shielding have been observed (4,5). These solvent effects are a potential source of information on the nature of solute–solvent interactions. In

* By Robert W. Taft, Elton Price, Irwin R. Fox (2), Irwin C. Lewis, K. K. Andersen (3) and George T. Davis, College of Chemistry and Physics, The Pennsylvania State University, University Park, Pennsylvania.

† Reprinted from the *Journal of the American Chemical Society*, **85**, 709 (1963). Copyright 1963 by the American Chemical Society and reprinted by permission of the copyright owner.

‡ References and footnotes for this appendix will be found on pp. 291-293.

general, however, there are substantial contributions to the solvent effects from diamagnetic susceptibility and other effects of magnetic origin, which greatly complicate evaluation of the effects of normal solute–solvent interactions (6).

In the present work on *meta*-substituted fluorobenzenes we have made use of the relatively effective transmission of electronic interactions through the benzene ring to provide information, both qualitative and quantitative, on the interaction between a large variety of organic and inorganic functional groups and solvents. The fluorine atom may be regarded as a distant but sensitive observer removed from the confusion of the "battlefield" (i.e., field of interaction of solvent and substituent) by the rigid benzene ring.

The shielding of the ^{19}F nucleus may be treated as the sum of intramolecular and intermolecular contributions (7). The use of unsubstituted fluorobenzene as an internal standard of reference provides a fluorine atom which is very nearly identical with that of the *meta*-substituted fluorobenzene. Shielding relative to this standard consequently may be expected (and has been shown by the present work) to measure the intramolecular shielding effect of the substituent and to contain little if any contribution from certain complicating intramolecular terms such as neighboring anisotropy effects (7). The relatively high sensitivity of the fluorine NMR shielding to intramolecular changes, however, provides information on the solute–solvent interaction which leads to a modification in the electronic character of the *meta*-substituent group.

An analysis of the π-electronic shielding in *para*-substituted fluorobenzenes indicates that the π-charge density on the fluorine atom in the carbon–fluorine bond is primarily responsible for intramolecular ^{19}F shielding within a series of such molecules (8,9). Although a satisfactory theory of ^{19}F shielding in *meta*-substituted fluorobenzenes has not been developed, a leading term appears to involve the ionic character of the C–F σ-bond (8b,9). This conclusion is apparently supported by the fact that the shielding parameters are in the inductive order and a relatively precise correlation with the inductive substituent constant, σ_1, has been observed (10).

The present investigation has had as an important objective the determination of the dependence of σ_I values on solvent as implied by the ^{19}F shielding solvent effects. There is very little previous information regarding the effect of solvent on the inductive order since most reactivity investigations have been carried out in hydroxylic solvents.

However, the wide variation in solvent permitted by the highly precise modified Hammett equation, $\log (k/k_0) = \sigma^0 \rho$, suggests that for several well-behaved substituents there is very slight (if any) modification in the Ar–Y inductive order due to solute–solvent interactions (10). On the other hand, *para*-substituent effects involving direct interaction dipolar quinoid forms show substantial dependence on solvent (10). A recent investigation of solvent effects on the ionization constants of a series of 4-substituted [2,2,2]bicyclooctanecarboxylic acids has indicated that solute–solvent interaction may cause appreciable variations in σ_I values for some substituent groups (11). It is of interest to determine whether the σ_I values obtained from ^{19}F shielding reproduce similar trends. Such a correspondence would be expected of characteristic specific group–solvent interactions (that is, σ_I values which are characteristic of both the substituent group and the solvent). In this endeavor, ^{19}F NMR shielding offers the practical advantages of a ready and systematic investigation, including substituents too reactive to be included in usual structure–reactivity investigations.

In the present work ^{19}F NMR shielding parameters for *meta*-substituted fluorobenzenes relative to internal fluorobenzene and to a fixed external standard have been determined in 5% (vol.) solutions of 20 widely varying pure solvents at room temperature. It is demonstrated (with few exceptions) that this dilution corresponds within the experimental error to the infinitely dilute solution. Twenty-one widely varying substituent groups have been investigated under these conditions. In addition, shielding parameters for some 30 additional substituent groups have been determined largely in hydroxylic solvents, both pure and mixed. The solvent variation studied included hydroxylic solvents of acidities varying between that of t-C_4H_9OH and CF_3CO_2H. Pyridine is included as an example of a moderately basic solvent.

A. Experimental (omitted)

B. Results

1. Shielding Parameters Relative to Fixed External Standard

Table A5.I lists the observed shielding parameters for tetrachlorotetrafluorocyclobutane and for fluorobenzene relative to a fixed external standard [20% (vol.) *para*-difluorobenzene in carbon tetrachloride] in twenty-five solvents at high dilution. Also listed are the corresponding values for the susceptibility corrected shielding parameters. (55).

Some of the types of solvent–fluorine atom interactions which could reasonably contribute to the observed shielding are (56): (*1*) van der Waals dispersion forces (57,58), (*2*) hydrogen bonding (59), and (*3*) dipole interactions (57,58). In addition, solvent magnetic anisotropy effects may contribute (60). In an attempt to determine if one of these

TABLE A5.I

Shielding Parameters of Tetrachlorotetrafluorocyclobutane and Fluorobenzene in Infinitely Dilute Solutions Relative to a Fixed External Standard

[20% (vol) *para*-Difluorobenzene in CCl_4]

Solvent	$X_v \times 10^6$	$-\int_{\text{ext. ref.}}^{\text{TCTFCB}}$ ppm		$-\int_{\text{ext. ref.}}^{C_6H_5F}$ ppm	
		Obs.	Cor.	Obs.	Cor.
1 Cyclohexane	0.612[b]	4.6	4.8	6.0	6.2
2 Benzene	0.612[b]	5.1	5.3	5.7	5.9
3 Carbon tetrachloride	0.691[b]	5.9	5.9	7.1	7.1
4 Methylene iodide	1.156[a]	9.8	8.8	11.3	10.3
5 Diethyl ether	0.526[a]	3.8	4.2	4.3	4.7
6 Tetrahydrofuran	0.641[d]	4.4	4.5	4.6	4.7
7 Dioxane	0.658[d]	4.7	4.8	5.1	5.2
8 Diethyl maleate	0.573	4.8	5.1	4.9	5.2
9 Acetic anhydride	0.439[d]	4.6	5.1	4.5	5.0
10 Acetone	0.461[a]	4.0	4.5	3.9	4.4
11 Pyridine	0.609	5.0	5.2	5.4	5.6
12 Dimethylformamide	0.607[d]	4.7	4.9	4.4	4.6
13 Monomethylformamide	0.856[d]	4.5	4.2	4.6	4.3
14 Nitrobenzene	0.602[a,b]	4.8	5.0	5.3	5.5
15 Nitromethane	0.391[b]	3.6	4.2	3.0	3.6
16 Acetonitrile	0.522	4.5	4.9	4.0	4.4
17 Dimethyl sulfoxide	—	6.0	—	6.1	—
18 Methanol	0.530[a]	3.7	4.1	3.5	3.9
19 Formic acid	0.527[a]	4.4	4.8	4.0	4.4
20 Trifluoroacetic acid	0.773[d]	2.6	2.4	2.1	1.9
21 3-Methylpentane	0.592[a]	4.3	4.5	5.7	5.9
22 Ethyl acetate	0.549[a]	4.2	4.5	4.2	4.5
23 Benzonitrile	0.635[c]	5.0	5.2	5.4	5.6
24 *o*-Dichlorobenzene	0.791[d]	5.9	5.7	6.7	6.5
25 75% (vol.) aq. methanol	0.578[e]	4.1	4.4	3.9	4.2

[a] S. Broersma, *J. Chem. Phys.*, *17*, 873 (1949).

[b] F. A. Badder and J. Sugden, *J. Chem. Soc.*, *1950*, 308.

[c] C. M. French, *Trans. Faraday Soc.*, *50*, 1320 (1954).

[d] Estimated from Pascal's constants.

[e] Estimated from Wiedemann's additivity law (cf. ref. 7. p. 18.)

factors clearly predominates, the corrected shielding parameters have been plotted vs. the following parameters: (1) the refractive index function (57,58) $(n_D^2 - 1)/(2n_D^2 + 1)$, (2) the molar polarizability of the solvent (5), (3) the Onsager reaction field parameter (56,61), and (4) the Kirkwood solvation energy (62). All of these plots show such substantial scatter than none of the parameters individually is capable of dealing with the experimental results. The most successful parameter is the refractive index function which reproduces the largest overall trends of the data. For example, trifluoroacetic acid has both the lowest value of $\int_{ext.ref.}^{C_6H_5F}$, while methylene iodide has the largest values of these two quantities. Thus it appears that dispersion forces figure predominantly but not exclusively in the intermolecular shieldings.

The intermolecular shielding due to solvent variation is apparently rather specific to the fluorine atom involved. Thus, although corresponding values of the corrected shielding parameters for tetrachlorotetrafluorocyclobutane and fluorobenzene show similar trends, one obtains substantial scatter in a plot of the one shielding parameter vs. the other. The order of solvent effects on the shielding of either compound bears no recognizable relationship to the order of solvent effects on the shielding parameters for any *meta*- or *para*-substituted fluorobenzene relative to internal fluorobenzene. This fact makes critical the evidence presented in the following section for cancellation of the fluorine atom intermolecular shielding in the latter parameters.

2. Intramolecular Shielding Due to meta-Substituents

Table A5.II lists the results of an essentially complete survey of the shielding effects of 21 widely varying *meta* substituents in twenty pure solvents. Among the various substituents, shielding varies over a range of 7.0 ppm. In general, however, closely similar shielding parameters are observed for a given substituent in any solvent. This result demonstrates that the shielding relative to fluorobenzene as an internal standard must be determined almost solely by intramolecular terms. The variations of the shielding parameters with solvent which are observed are systematic and bear definite relationships to the general chemical properties of the solvent and the substituent. Most notable of these are the shifts to lower field strength in trifluoroacetic acid for substituents having measurable base strengths.

In Table A5.III are listed the results for seven *meta* substituents which will be generally accepted as chemically inert groups. The

TABLE A5.IIA

Shielding Parameters, $\int_{H}^{m-x}$, for *meta*-Substituted Fluorobenzenes[a]

Solvent	*meta* Substituent									
	—CH₃	CH₂=CH—	C₆H₅	—SCH₃	—CO₂C₂H₅	—COCH₃	—OCH₃	—CHO	—NO	—OC₆H₅
1 Cyclohexane	+1.23	+0.63	+0.15	−0.23	−0.15	−0.60	−0.98	−1.30	−1.73	−1.88
2 Benzene	+1.15	+0.60	+0.08	−0.35	−0.43	−0.68	−1.18	−1.10	−1.73	−2.20
3 Carbon tetrachloride	+1.18	+0.65	+0.15	−0.38	−0.13	−0.73	−1.05	−1.35	−1.78	−1.95
4 Methylene iodide	+1.18	+0.63	+0.23	−0.38	−0.20	−0.63	−1.10	−1.28	−1.75	−1.83
5 Diethyl ether	+1.10	+0.68	+0.13	−0.33	−0.33	−0.73	−1.05	−1.23	−1.88	−2.00
6 Tetrahydrofuran	+1.10	—	+0.05	−0.35	−0.40	−0.58	−1.15	−1.08	−1.83	−2.05
7 Dioxane	+1.20	+0.65	+0.15	−0.30	−0.30	−0.35	−1.13	−0.85	−1.43	−1.93
8 Diethyl maleate	+1.08	+0.50	−0.13	−0.53	−0.65	−0.80	−1.20	−1.30	−1.98	−2.20
9 Acetic anhydride	+1.13	+0.43	−0.08	−0.55	−0.60	−0.75	−1.23	−1.18	−1.85	−2.10
10 Acetone	+1.13	+0.55	−0.13	−0.45	−0.58	−0.68	−1.28	−1.13	−1.83	−2.13
11 Pyridine	+1.18	+0.60	+0.03	−0.38	−0.43	−0.60	−1.15	−1.13	−1.68	−2.10
12 Dimethylformamide	+1.10	+0.45	−0.15	−0.55	−0.70	−0.68	−1.25	−1.13	−1.88	−2.15
13 Monomethylformamide	+1.20	—	+0.05	−0.38	−0.50	−0.70	−1.13	−1.10	−1.78	−2.00
14 Nitrobenzene	+1.20	+0.55	0.00	−0.35	−0.38	−0.73	−1.15	−1.13	−1.70	−2.08
15 Nitromethane	+1.13	+0.43	−0.10	−0.53	−0.58	−0.73	−1.30	−1.20	−1.70	−2.18
16 Acetonitrile	+1.13	+0.30	−0.23	−0.53	−0.65	−0.73	−1.28	−1.23	−1.78	−2.20
17 Dimethyl sulfoxide	+1.05	+0.38	−0.23	−0.55	−0.70	−0.60	−1.30	−1.25	−1.88	−2.10
18 Methanol	+1.15	+0.58	0.00	−0.40	−0.73	−1.00	−1.38	−1.50[c]	−2.03	−2.05
19 Formic acid	+1.35	R[b]	+0.15	−0.53	−0.90	−1.30	−1.53	−1.78	−1.78	I[b]
20 Trifluoroacetic acid	+1.05	R	−0.10	−1.80	−1.88	−2.65	−3.05	−3.18	R	−2.93

[a] In ppm relative to fluorobenzene; exptl. error = ±0.08.

[b] I designates insufficient solubility; R designates that a fast reaction occurs.

[c] Most intense signal. A second weaker signal at higher field was also observed; cf. Discussion.

TABLE A5.IIB

Shielding Parameters, $\int_{H}^{m-x}$, for meta-Substituted Fluorobenzenes[a]

Solvent		—COF	CF₃	—Br	—COCF₃	—CN	—COCN	—F	—SO₃C₂H₅	—OCF₃	—NO₂	—SF₅
						meta Substituent						
1	Cyclohexane	−2.20	−2.10	−2.43	−2.48	−2.73	−3.25	−3.03	−2.83	−3.28	−3.43	−3.10
2	Benzene	−2.03	−2.28	−2.60	−2.60	−2.75	−2.78	−3.08		−3.25	−3.25	−3.45
3	Carbon tetrachloride	−2.15	−2.13	−2.30	−2.63	−2.75	−3.33	−3.03	−2.98	−3.33	−3.45	−3.13
4	Methylene iodide	−1.98	I	−2.40		−2.73		−2.88	−3.10		−3.35	I
5	Diethyl ether	−2.30	−2.38	−2.48	−2.90	−2.85	−3.15	−3.05	−3.15	−3.43	−3.55	−3.38
6	Tetrahydrofuran		−2.35	−2.60		−2.75		−3.23			−3.25	−3.43
7	Dioxane	−1.70	−1.95	−2.45	−2.15	−2.30	−2.23	−2.98	−2.85	−3.28	−2.78	−3.75
8	Diethyl maleate	−2.35	−2.53	−2.63	−2.88	−3.00	−2.98	−3.25	−3.60	−3.53	−3.60	−3.63
9	Acetic anhydride	−2.20	−2.40	−2.63	−2.63	−2.78		−3.15	−3.48	−3.43	−3.40	−3.50
10	Acetone	−2.20	−2.38	−2.65	−2.63	−2.80	−2.70	−3.20	−3.45	−3.43	−3.40	−3.50
11	Pyridine	−1.98	−2.25	−2.58	+0.65	−2.78	−2.55	−3.08		−3.33	−3.23	−3.43
12	Dimethylformamide	−2.08	−2.48	−2.63	−2.65[c]	−2.78	−2.45	−3.15	−3.53	−3.45	−3.35	−3.55
13	Monomethylformamide		−2.30	−2.60		−2.80		−3.10	−3.45		−3.30	−3.40
14	Nitrobenzene	−2.10	−2.20	−2.55	−2.65	−2.85	−2.90	−3.05	−3.48	−3.33	−3.40	−3.53
15	Nitromethane	−2.10	−2.25	−2.63	−2.55	−2.73	−2.80	−3.15	−3.53	−3.33	−3.38	−3.40
16	Acetonitrile	−2.08	−2.30	−2.65	−2.60	−2.68	−2.78	−3.20		−3.33	−3.33	−3.45
17	Dimethyl sulfoxide	−1.90	−2.33	−2.63	−2.35[c]	−2.58	−2.13	−3.13		−3.25	−3.15	−3.45
18	Methanol	−2.35[b]	−2.50	−2.63	+0.03	−3.10	−3.85	−3.38	−3.75	−3.50	−3.60	−3.58
19	Formic acid	−2.08	−1.98	−2.48	−2.50	−3.13		−3.03	−3.68	−3.05	−3.50	−3.10
20	Trifluoroacetic acid	−3.40	−2.28	−2.55	−3.75	−4.63	−4.55	−3.23	−5.68	−3.25	−5.10	−3.43

[a] In ppm relative to fluorobenzene; exptl. error = ±0.08.
[b] Initial signal; on standing, signal due to ester appears.
[c] A second weak signal at +0.64 ppm was also observed; cf. Discussion.

shielding parameters for these substituents cover a range of 4.5 ppm, but the value for each substituent is the same in all twenty solvents to a precision closely on the order of the experimental error. This point is illustrated by the mean value of the shielding parameter listed at the bottom of Table A5.III. The average deviations from the mean values for all of the substituents (± 0.05 to ± 0.09 ppm) are of essentially the same magnitude as the individual experimental errors (± 0.08 ppm). Although a few of the largest deviations from the mean value (note especially $-CF_3$ and $-SF_5$ in dioxane and cf. subsequent discussion) are apparently outside of experimental error and bear a relationship to solvent effects observed for substituents of a similar class, most deviations are not reliably outside of experimental error and clearly are of second-order character.

Since the shielding for fluorobenzene is subject to the same experimental error (or small nearly random second order solvent effects) as for any *meta*-substituted fluorobenzene, the results listed in Table A5.III are based upon fluorobenzene values adjusted slightly (as indicated) to minimize the deviations from the mean for all substituents. The high precision of the mean values (essentially the experimental error) as well as the consistency of the magnitude of average deviations for all of the substituents testifies to the fact that only intramolecular shielding contributes to the shielding parameters.

Eleven more substituents from Table A5.II define mean values of the same precision as those of Table A5.III if the results in certain acidic and certain basic solvents are excluded. This fact is illustrated in Table A5.IV. All of the substituents of Table A5.IV are measurably basic and for each (in contrast to the results in Table A5.III) the trifluoroacetic acid result does not conform to the mean. Similar but smaller deviations occur in other acidic hydroxylic solvents. The solvents excluded in obtaining the mean values listed at the bottom of Table A5.IV are indicated by the designation S.E. (or I, if insoluble; R, if a fast reaction occurs). It is significant that measurable deviations from the precise means of Tables A5.III and A5.IV occur for the basic solvents, e.g., dioxane, dimethyl sulfoxide, and pyridine, only in the case of $+R$ substituent groups. No such deviation occurs for any of the $-R$ substituent groups (e.g., CH_3, F, SCH_3, OCH_3, etc.).

C. Discussion

The character of these results offers convincing evidence that such solvent effects as are observed in the substituent shielding parameters

TABLE A5.III

Solvent Insensitive Shielding Parameters, $\int_{J_H}^{m-x}$, for Fluorobenzene with Chemically Inert *meta* Substituents[a,b]

				meta Substituent				
Solvent	CH$_3$	C$_6$H$_5$	H	Br	CF$_3$	F	OCF$_3$	SF$_5$
1 Cyclohexane	+1.15	+0.08	-0.08	-2.43	-2.18	-3.10	-3.35	-3.18
2 Benzene	+1.10	+0.03	-0.05	-2.60	-2.33	-3.13	-3.30	-3.50
3 Carbon tetrachloride	+1.18	+0.15	0.00	-2.30	-2.13	-3.03	-3.33	-3.13
4 Methylene iodide	+1.10	+0.15	-0.08	-2.40	I[b]	-2.95	—	I
5 Diethyl ether	+1.10	+0.13	0.00	-2.48	-2.38	-3.05	-3.43	-3.38
6 Tetrahydrofuran	+1.10	+0.05	0.00	-2.60	-2.35	-3.23	—	-3.43
7 Dioxane	+1.08	+0.03	-0.13	-2.45	-2.08	-3.10	-3.28	-3.20
8 Diethyl maleate	+1.23	+0.03	+0.15	-2.63	-2.38	-3.10	-3.38	-3.48
9 Acetic anhydride	+1.20	0.00	+0.08	-2.63	-2.33	-3.08	-3.35	-3.43
10 Acetone	+1.18	-0.08	+0.05	-2.65	-2.33	-3.15	-3.38	-3.45
11 Pyridine	+1.18	+0.03	0.00	-2.58	-2.25	-3.08	-3.33	-3.43
12 Dimethylformamide	+1.18	-0.08	+0.08	-2.63	-2.40	-3.08	-3.38	-3.48
13 Monomethylformamide	+1.20	+0.05	0.00	-2.60	-2.30	-3.10	—	-3.40
14 Nitrobenzene	+1.20	0.00	0.00	-2.55	-2.20	-3.05	-3.33	-3.53
15 Nitromethane	+1.13	-0.10	0.00	-2.63	-2.25	-3.15	-3.33	-3.40
16 Acetonitrile	+1.13	-0.23	0.00	-2.65	-2.30	-3.20	-3.33	-3.45
17 Dimethyl sulfoxide	+1.05	-0.23	0.00	-2.63	-2.33	-3.13	-3.25	-3.45
18 Methanol	+1.23	+0.08	+0.08	-2.63	-2.43	-3.20	-3.43	-3.50
19 Formic acid	+1.18	0.00	-0.15	-2.48	-2.13	-3.18	-3.20	-3.25
20 Trifluoroacetic acid	+1.05	-0.10	0.00	-2.55	-2.28	-3.23	-3.25	-3.43
Mean value	+1.15	0.00	0.00	-2.55	-2.30	-3.10	-3.33	-3.40
Av. dev.	±0.05	±0.08	±0.05	±0.08	±0.08	±0.05	±0.05	±0.09

[a] In ppm relative to fluorobenzene; exptl. error = ±0.08.

[b] I designates insufficient solubility.

TABLE A5.IV

Normal Shielding Parameters, $\int_{H}^{m-x}$, for Fluorobenzene with Chemically Active *meta* Substituents[a,b]

Solvent					*meta* Substituent						
	—CH=CH₂	—SCH₃	—OCH₃	—OC₆H₅	—COCH₃	—CHO	—NO	—COF	—COCF₃	—CN	—NO₂
1 Cyclohexane	+0.55	−0.30	−1.05	−1.95	−0.68	−1.38	−1.80	−2.28	−2.55	−2.80	−3.50
2 Benzene	+0.55	−0.40	−1.23	−2.25	−0.73	−1.15	−1.78	−2.08	−2.65	−2.80	−3.40
3 Carbon tetra-chloride	+0.65	−0.38	−1.05	−1.95	−0.73	−1.35	−1.78	−2.15	−2.63	−2.75	−3.45
4 Methylene iodide	+0.55	−0.45	−1.18	−1.90	−0.70	−1.35	−1.83	−2.05	—	−2.80	−3.43
5 Diethyl ether	+0.68	−0.33	−1.05	−2.00	−0.73	−1.23	−1.88	−2.30	−2.65	−2.85	−3.55
6 Tetrahydro-furan	—	−0.35	−1.15	−2.05	−0.58	−1.08	−1.83	—	—	−2.75	−3.25
7 Dioxane	+0.53	−0.43	−1.25	−2.05	S.E.	S.E.	S.E.	S.E.	S.E.	S.E.	S.E.
8 Diethyl maleate	+0.65	−0.38	−1.05	−2.05	−0.65	−1.15	−1.83	−2.20	−2.73	−2.85	−3.45
9 Acetic anhydride	+0.50	−0.48	−1.15	−2.03	−0.68	−1.10	−1.78	−2.13	−2.55	−2.70	−3.33
10 Acetone	+0.60	−0.40	−1.23	−2.08	−0.63	−1.08	−1.78	−2.15	−2.58	−2.75	−3.35
11 Pyridine	+0.60	−0.38	−1.15	−2.10	−0.60	−1.13	−1.68	S.E.	S.E.	−2.78	S.E.

12 Dimethyl-formamide	+0.53	−0.48	−1.18	−2.08	−0.60	−1.05	−1.80	−2.00	−2.65	−2.70	−3.28
13 Monomethyl-formamide	—	—	−1.13	−2.00	−0.70	−1.10	−1.78	—	—	−2.80	−3.30
14 Nitrobenzene	+0.55	−0.38	−1.15	−2.08	−0.73	−1.13	−1.70	−2.10	−2.65	−2.85	−3.40
15 Nitromethane	+0.43	−0.53	−1.30	−2.18	−0.73	−1.20	−1.70	−2.10	−2.55	−2.73	−3.38
16 Acetonitrile	+0.30	−0.53	−1.28	−2.20	−0.73	−1.23	−1.78	−2.08	−2.60	−2.68	−3.33
17 Dimethyl sulfoxide	+0.38	−0.55	−1.30	−2.10	−0.60	−1.15	−1.88	S.E.	S.E.	S.E.	S.E.
18 Methanol	+0.65	−0.33	−1.30	−1.98	S.E.	S.E.	S.E.	S.E.	S.E.	S.E.	S.E.
19 Formic acid	R	S.E.	I	S.E.	S.E.	S.E.	S.E.	−2.23	−2.65	S.E.	S.E.
Mean value	+0.47	−0.47	−1.23	−2.09	−0.68	−1.14	−1.77	−2.10	−2.62	−2.75	−3.34
Av. dev.	±0.08	±0.06	±0.07	±0.07	±0.05	±0.08	±0.04	±0.08	±0.05	±0.05	±0.07

[a] In ppm relative to fluorobenzene; exptl. error = ±0.08.

[b] I designates insufficient solubility; R designates that a fast reaction occurs; S.E. effect designates that a deviation of at least 0.20 ppm from the mean value occurs.

of Table A5.II result from the alteration in intramolecular shielding produced by certain substituent–solvent interactions. That is, to a precision on the order of the experimental error, these solvent effects result from solvent modification in the electronic character of the *meta* substituents.

Our extensive investigation has disclosed four distinct categories of solvent effects on the substituent shielding parameters. The first involves hydrogen bonding and proton transfer interactions between acidic and basic groups (either $+R$ or $-R$). The second limited category involves polarity effects on amide and ester functions. The third limited category involves the formation of methanol–carbonyl addition compounds. The fourth category involves interactions of Lewis-acidic ($+R$) substituents with solvents of basic character. Each of these types of behavior, which generally may be regarded as chemical rather than physical in nature, are considered in detail in following sections.

The precision of the solvent independence of the intramolecular shieldings of Tables A5.III and A5.IV is of importance in connection with mode of transmission of the electron effects of the *meta* substituents to the region of detection by the fluorine nucleus. The twenty solvents investigated include non-associated, associated, polar and nonpolar liquids. Dielectric constant variation of nearly two orders of magnitude has been carried out, ranging from that of the non-polar liquids, e.g., cyclohexane, CCl_4 and benzene (D^{25} 2.0–2.3) (63), through highly polar liquids, e.g., CH_3NO_2, $(CH_3)_2SO$ (D^{25} 36.7 and 45, resp. (64), to that of the associated liquid, $HCONHCH_3$ (D^{25} 182) (65). In the reactivities of side-chain derivatives of benzene and similar systems, such solvent variation produces very substantial changes in the magnitude of substituent effects (11,66).

In the latter situation, the solvent effects are attributed (at least in part) to the fact that the lines of force between the substituent dipole and the pole or dipole of the reaction center tranverse the region of the bulk solvent (67). This results in an effective dielectric constant which will generally lie between that of the internal hydrocarbon cavity and the bulk solvent (67). The present results show that the intramolecular shielding is (within experimental error) completely independent of the dielectric constant of the bulk solvent. Consequently field effects of *meta* substituents either make no practical contribution to the intramolecular shielding or the effective dielectric constant is independent of solvent. The Westheimer-Kirkwood model (67) does not anticipate the latter condition.

Our results appear to be understandable on the basis that the polarizing force (68) exerted on the C–F bond moment by the distant substituent dipole moment is inversely proportional to r^4 and thus is too weak to be effective in intramolecular ^{19}F shielding. The observed substituent effects on shielding can therefore arise only by an intra-molecular electronic transmission involving either or both the σ- and π-electrons of the benzene ring.

The results for unsaturated substituents, e.g., –COCH₃, NO, COF, COCF₃, CN and NO₂ (Table A4.IV), imply that in most aprotic media the contribution of the ionic resonance form to the resonance hybrid (and therefore the group dipole moment) is essentially invariant with the polarity or the internal pressure of the medium at room tempera-tures. A similar conclusion may be reached concerning resonance

$$(-)\!\!\left\langle\!\!\begin{array}{c}=\\=\end{array}\!\!\right\rangle\!\!=\overset{+}{O}-$$

forms, e.g., for the –OC₆H₅ substituent. In both instances at least one of the charge centers of the dipolar resonance form is buried within the molecular cavity (69).

1. Hydrogen Bonding Effects

Table A5.V presents the shielding parameters of three weakly basic substituent groups in a variety of protonic solvents. It is apparent that for all such solvents the resonance signal is shifted to lower applied field strength (implying greater inductive electron-withdrawal by the substituent) compared to the precise means established for these substituent groups in Table A5.IV. These shifts are to be attributed to either partial or complete hydrogen bond formation between substituent and solvent. The effects are not attributable to measurable "complete" proton transfer equilibria since the strongest base *meta*-fluoroaceto-phenone has a pK_A of ~ -7 (70), and the estimated H_0 of trifluoro-acetic acid (71), the most acidic solvent, is -4. Further, the downfield shifts on proton transfer to substrates such as these have been observed (72) to be much larger (at least 5.0 ppm) than any of those given in Table A5.V.

The results presented in Table A5.V give several interesting qualitative results regarding the effects of the competitive hydrogen bonding which is involved. The magnitudes of the shifts to lower field

TABLE A5.V

Effects of Hydrogen Bonding on $\int_{H}^{m-x}$

Solvent	meta Substituent		
	CH_3CO	$-CN$	$-NO_2$
Normal[a]	-0.68	-2.78	-3.38
CH_3OH	-0.93	-3.03	-3.53
C_2H_5OH	—	—	-3.73
$t\text{-}C_4H_9OH$	-1.08	—	-3.68
$HCF_2CF_2CH_2OH$	-1.40	-3.40	-3.95
CH_2Cl_2	—	—	-3.83
$HCCl_3$	—	—	-3.98
CH_3CO_2H	—	—	-3.40
HCO_2H	-1.45	-3.28	-3.65
95% (vol.) aq. acetone	-0.80	-2.95	-3.48
80%	-0.83	-3.00	-3.53
75%	-0.90	-3.08	-3.63
70%	-0.90	-3.08	-3.60
95% (vol.) aq. CH_3OH	-0.95	-3.13	-3.60
90%	-1.00	-3.18	-3.63
80%	-1.10	-3.23	-3.70
75%	-1.38	-3.30	-3.70
83% (vol.) aq. $t\text{-}C_4H_9OH$	-1.08	—	-3.73
83% (vol.) aq. $HCF_2CF_2CH_2OH$	-1.33	—	-3.90
CF_3CO_2H	-2.65	-4.63	-5.10

[a] Mean value from Table A5.IV.

do not correlate well with a perhaps naïvely expected trend for increasing effects of hydrogen bonding with increased acidity of the solvent. In fact, there is a notable tendency for the reverse to occur frequently. It is quite apparent from the examples provided in Table A5.V that the effect of increased strength of hydrogen bonding to a given substrate with increased acid strength of the solvent molecule (73,74) is frequently compensated by self-association of the solvent. The very weakly acidic solvents, CH_2Cl_2 and $HCCl_3$, are unquestionably less self-associated than the more acidic solvents, e.g., CH_3OH or HCO_2H. Consequently, even though a weaker hydrogen bond is formed by these less acidic solvents with a given substrate [producing potentially less downfield shift] (75), their potential for hydrogen bond formation is actually greater. Thus larger downfield shifts are frequently observed for the (intrinsically) weaker acid solvents.

The relatively large apparent effect of hydrogen bonding by methylene chloride is perhaps surprising in view of its expected weakly acidic nature. In contrast, methylene iodide gives nearly normal shifts (cf. Tables A5.II–A5.IV). The effect on acidity of the greater electro-negativity of chlorine than iodine and the relative insensitivity of the standard free energy of hydrogen bond formation to acid strength (74) are factors involved, but their relative role is not made definite by the present results.

In the mixed aqueous solvents, the methanol–water system consistently produces larger solvent shifts than acetone–water of the same water content. This effect is no doubt associated with the fact that the equilibrium (76) $CH_3OH + H_2O = CH_3O^- + H_3O^+$ effec-tively enhances hydrogen bonding to the substrate.

Results in protonic solvents for 15 additional $+R$ substituent groups are given in Tables A5.VI and A5.VII. Similar trends are observed. Although quantitative information on base strength is not available for most of these substituents, it is apparent that there is a qualitative trend for the effects of a given acidic solvent to increase with increasing base strength. Notable in this connection are the relatively large solvent shifts in formic acid and aqueous methanol for substituents such as $-CONH_2$, $-SOCH_3$, $-CO_2C_2H_5$ and $-SO_3C_2H_5$ and the relative small shifts (from normal) observed in these solvents for the $-NO$, $-COF$ and $-COCF_3$ groups.

The results given in Table A5.II for the SCH_3, OCH_3, OC_6H_5 substituents in formic and trifluoroacetic acids provide similar examples of shifts to lower fields due to hydrogen bonding to weakly basic $-R$ substituents. It is apparent from the results given in Table A5.III that the $-R$ substituents Br, F and OCF_3 are too weakly basic to give rise to measurable H-bonding effects. Also no measurable effects of hydrogen bonding are found for the SCH_3, OCH_3 and OC_6H_5 substituents in methanol (Table A5.IV) and in 75% (vol.) aqueous methanol (un-published results). In the more donor alcohol $HCF_2CF_2CH_2OH$, however, the $-OCH_3$ shielding is -0.57 ppm to lower field than normal and the shielding of the $-NH_2$ group is -1.13 ppm to lower field than that observed in hydrocarbon solvents (unpublished results). Especially interesting, however, are the results given in Tables A5.VIII and A5.IX which provide additional examples of effects produced in competitive hydrogen bonding.

Except for the non-protonic substituents $N(CH_3)_2$, $C_6H_5N=N-$, $-OCOCH_3$, of Table A5.VIII, the substituents listed could potentially

TABLE A5.VI

Effects of Hydrogen Bonding on $\int_H^{m-x}$

Solvent		meta Substituent							
	—CHO	—NO	—SOCH$_3$	—CO$_2$C$_2$H$_5$	—SO$_3$C$_2$H$_5$	—COF	—COCF$_3$	—COCN	—CH=CHNO$_2$
Normal[a]	−1.18	−1.80	−2.90	−0.23[b]	−2.90	−2.13	−2.62	−3.33[b]	−1.13[c]
HCCl$_3$	—	−2.20	—	—	—	−2.53	−2.90	−3.65	−1.35
CH$_3$OH	−1.45	−1.95	−3.48	−0.65	−3.68	−2.28	R	R	—
75% (vol.) aq. CH$_3$OH	−1.53	−2.08	−3.65	−1.00	−4.00	−2.40	—	—	—
HCF$_2$CF$_2$CH$_2$OH	−1.80	−2.13	−4.23	—	—	−2.43	−2.88	−3.43	—
HCO$_2$H	−1.63	−1.93	−3.93	−1.05	−3.83	−2.23	−2.65	—	−1.35
CF$_3$CO$_2$H	−3.18	R	−6.48	−1.88	−4.68	−3.40	−3.75	−4.55	—

[a] Mean value from Table A5.IV unless otherwise designated.

[b] Mean of shielding parameters observed in carbon tetrachloride and in cyclohexane; precision of mean, 0.10 ppm or less.

[c] Mean of shielding parameters observed in benzene, ethyl acetate, diethyl maleate, acetone, pyridine, dimethylformamide, nitrobenzene, nitromethane, and acetonitrile; average deviation from mean ±0.08 ppm.

TABLE A5.VII

Effects of Hydrogen Bonding on $\int_H^{m-x}$

Solvent	meta Substituent					
	—CONH₂	—CO₂C₆H₅	—SO₂NH₂	—SO₂CH₃	—SO₂Cl	—SO₂F
H.C.[a]	—0.03[b]	—0.73	I	—3.30	—5.10	—4.73
CH₃OH	—0.73	—1.25	—2.50	—3.60	—5.43	—5.10
75% (vol.) aq. CH₃OH	—0.98	I	—2.90	—3.95	—5.53	—5.23

[a] Mean of shielding parameters observed in carbon tetrachloride and in cyclohexane; precision of the mean, 0.10 ppm or less.
[b] Shielding parameter for tetrahydrofuran.

hydrogen bond by serving either as the proton donor or acceptor. The results in Table A5.VIII fall into two characteristic patterns. The nonprotonic substituents experience downfield shifts (indicative of loss of electronic charge to the solvent) in the two hydroxylic relative to hydrocarbon solvents, the effect being greater for aqueous methanol than pure methanol (consistent with the results for the $+R$ substituents of Tables A5.VI–A5.VII as discussed above). In contrast, the acidic substituents $-CO_2H$ and $-OH$, for which hydrogen bonding by proton donation unquestionably predominates, experience shifts to higher field strengths (indicative of loss of positive charge to the solvent) under the same conditions. Furthermore, the shift is larger for methanol than aqueous methanol, indicating a greater effect of proton acceptor action toward uncharged substituents by methanol. These two distinct categories of behavior quite apparently can be used to distinguish the predominant mode of hydrogen bonding by amphoteric substituents.

The $-CH_2CO_2H$, $-CH_2OH$ and $-CH(CH_3)OH$ substituents show the same trends as for the functional groups (CO_2H and OH, resp.) directly attached to the aromatic ring; i.e., results apparently characteristic of predominant proton donor effects are observed. These results are not completely unambiguous, however. At the concentration used ($\sim 0.5M$), these compounds are appreciably associated in the hydrocarbon solvents and the association probably produces measurable shielding effects.

In contrast to the above behavior, the interposition of a methylene group between the ring and the $-NH_2$ group produces a rather dramatic change from one category of behavior to the other. The much more basic benzylamine shows behavior characteristic of predominant proton–acceptor action (the effect is larger than that observed for the $N(CH_3)_2$ substituent). The more-acidic and less-basic aniline, on the other hand, clearly shows in methanol the results characteristic of *predominant* proton-donor action. Additional examples of this behavior are provided by the $-NHNH_2$, $-SH$ and $-B(OH)_2$ substituents.

In Table A5.IX are given the effects of hydrogen bonding for a series of charged substituents on change from methanol to 75% (vol.) aqueous methanol. The results for all of the anions correspond to the previously discussed greater proton acceptor action of the aqueous methanol. The dramatically larger downfield shifts observed for the $N(CH_3)_3{}^+$ than the $NH_3{}^+$ substituent attests to the importance of delocalization of the positive charge of the latter by hydrogen bonding to the hydroxylic solvents (77–79). It is further of interest that the

TABLE A5.VIII

Effects of Hydrogen Bonding on $\int_H^{m-x}$

Solvent				*meta* Substituent			
	—N(CH₃)₂	—CH₂NH₂	C₆H₅N=N—	—OCOCH₃	—NHCOCH₃	—NH₂	—NHNH₂
H.C.[a]	−0.08	+0.58	−0.78	−1.33	−0.80[b]	+0.50	−0.38
CH₃OH	−0.10	+0.33	−1.18	−1.60	−1.35	+0.78	+0.35
75% (vol.) aq. CH₃OH	−0.18	+0.28	I	−1.88	−1.38	+0.50	−0.28
	—CO₂H	—CH₂CO₂H	—OH	—CH₂OH	—CH(CH₃)OH	—B(OH)₂	—SH
H.C.[a]	−0.88	−0.15	−1.43[c]	+0.18	+0.10	I	−0.78
CH₃OH	−0.35	+0.53	−0.43	+0.55	+0.38	+1.25	−0.60
75% (vol.) aq. CH₃OH	−0.68	+0.43	−0.68	+0.43	+0.25	+1.13	−0.75

[a] Mean of shielding parameters observed in carbon tetrachloride and in cyclohexane; precision of the mean, 0.10 ppm or less.

[b] Shielding parameter for dimethylformamide.

[c] *Note added in proof:* Using the dispersion rather than the absorption mode, M. G. Schwartz has obtained the shielding parameter for 1% *meta*-fluorophenol in CCl₄ of −1.20. This value apparently is attributable to largely monomeric phenol—cf. G. C. Pimentel and A. L. McClellan, *The Hydrogen Bond*, Freeman, San Francisco, Calif., 1960, pp. 149, 369, 370, 376, 377, 383.

TABLE A5.IX

Effects of Hydrogen Bonding on $\int_{H}^{m-X}$

Solvent	meta Substituent							
	—N(CH$_3$)$_3$$^+$	—NH$_3$$^+$	—CH$_2$NH$_3$$^+$	—SO$_3$$^-$	—CH$_2$CO$_2$$^-$	—O$^-$	—CO$_2$$^-$	—B(OH)$_3$$^-$
CH$_3$OH	−5.95	−3.63	−1.25	−1.03	+1.23	+1.90	+1.05	+3.33
75% (vol.) aq. CH$_3$OH	−6.18	−3.40	−1.13	−1.30	+1.03	+1.63	+0.80	+2.93

shielding parameters for both the $-NH_3^+$ and $-CH_2NH_3^+$ substituents are at higher field strengths in aqueous methanol than methanol. This behavior is exceptional to that observed for uncharged proton-donating substituents. The result perhaps follows from greater ion pair association in methanol or from preferential solvation of $-NH_3^+$ by water over methanol. The origin of this effect may be either the smaller steric requirements of water and/or its greater capacity to delocalize the positive charge through hydrogen bonding to second and third solvation shells (which becomes involved with the higher solvation energy for the charged substituent).

The present qualitative survey of hydrogen bonding effects on the ^{19}F shielding parameters serves as a useful preliminary investigation which will be utilized as a background to work in progress in quantitative equilibrium studies of hydrogen bonding utilizing this new analytical technique.

2. Solvent Polarity Effects on Amide and Ester Functions

A second type of behavior was encountered with the two ester substituents $-CO_2C_2H_5$ and $SO_3C_2H_5$ and the two amide substituents $CONH_2$ and SO_2NH_2 (cf. Table A5.II). In nonhydroxylic solvents, the shielding for these groups is appreciably shifted to lower field strengths (implying greater electron withdrawal) in strongly polar than in nonpolar solvents. The latter solvents include cyclohexane and carbon tetrachloride, the former solvents include, for example, dimethyl sulfoxide, acetonitrile, nitromethane and dimethylformamide. Shielding in benzene, diethyl ether, and methylene iodide is intermediate between that for these two groups of solvents. Table A5.X lists some typical results illustrating this behavior.

TABLE A5.X

Effects of Solvent Polarity on Shielding Parameters, $\int_H^{m-x}$, for Amide Groups

Solvent	meta Substituent			
	$-SO_2NH_2$	$-CONH_2$	$SO_3C_2H_5$	$-CO_2C_2H_5$
Cyclohexane	I	I	-2.83	-0.15
Diethyl ether	I	I	-3.15	-0.33
Diethyl maleate	-2.20	—	-3.60	-0.65
Acetone	—	-0.15	-3.48	-0.58
$C_6H_5NO_2$	-2.70	—	-3.48	-0.38
CH_3CN	—	-0.50	—	-0.65

These substituents have important resonance stabilization, i.e.

$$-C\underset{X}{\overset{O}{\Big\langle}} \longleftrightarrow -\overset{+}{C}\underset{X}{\overset{O^-}{\Big\langle}} \longleftrightarrow -C\underset{X^+}{\overset{O^-}{\Big\langle}}$$

I   II   III

Apparently polar solvents increase the contributions of both the ionic forms II and III to the resonance hybrid so that the given substituent as a whole becomes more electron attracting (contrast, however, subsequent discussion).

3. Carbonyl-Addition Compounds

The shielding parameters for the relatively poorly resonance-stabilized electron-deficient carbonyl substituents, CHO, $COCF_3$, and COCN, are very strongly shifted to higher field strengths in methanol solution. In the case of the benzaldehyde two signals are observed, one of which is reasonable for the normal carbonyl compound. The second weaker signal at 1.74 ppm higher field strength is apparently due to the hemiacetal (80) in equilibrium with the benzaldehyde. It is probable that the shielding observed for the $COCF_3$ and COCN substituents in methanol is due to the complete formation of the methanol–carbonyl addition compounds since the shielding parameter for the single observed signal is in both instances 2.4 ppm greater in methanol than cyclohexane solutions. This assignment is in reasonable agreement with the fact that the shielding parameter for the $-CH_2CN$ substituent is 2.2 ppm greater than for the $-COCN$ substituent in cyclohexane solution.

In dimethylformamide and dimethyl sulfoxide solutions of the $m\text{-}COCF_3$ compound, two signals were observed for both the fluorobenzene and CF_3 group fluorine atoms. The higher field signal of the former occurs at essentially the same position in these two solvents ($\int_H^{m-X} = +0.64$ ppm). Although molecular complexes between solute and solvent are possible, we suspect that these results and that for pyridine may be due to the hydrate–carbonyl addition compound. In 75% aqueous (vol.) methanol a single fluorine signal at -0.08 ppm is observed. Attempts to remove the last traces of water from dimethylformamide by heating with CaH_2, however, did not alter the result.

4. Effects of Interactions of Groups Having Lewis-Acidic and Basic Character

The fourth type of solvent effect encountered is a shift to higher field strengths (implying smaller inductive electron withdrawal) which is consistently found (Table A5.XI) for substituents which may be characterized as having an atom (generally the first atom of the group) with an electronic deficiency and a partially open orbital. These Lewis-acidic substituents give rise to higher-field resonance in basic solvents, e.g., dioxane and dimethyl sulfoxide. It is implied that as the result of solute–solvent interaction the electron deficiency of the first atom is partially removed. The direction of the effect is that expected by the changes in formal changes which take place on the formation of a Lewis acid–base complex, i.e., $A + B: \rightleftharpoons {}^{(-)}A:B^{(+)}$. However, the magnitude of the effect is much smaller than that expected for the formation of a stable electron pair bond between a solute and solvent molecule (81). It is thus further implied that the interaction is weak (perhaps intermediate between chemical bonding and physical interactions) and perhaps characterizable by relatively small equilibrium constants for formation (82).

The effect involved appears to bear the analogous relationship to Lewis-acid-base complex formation that hydrogen bonding bears to proton transfer equilibria. That is, the former involves a much weaker interaction which is very much less sensitive to variation in acid and base strength (75). Because we believe the analogy to be useful, we shall hereafter refer to the effect as arising from a Lewis-acid bonding solute–solvent interaction. The Lewis-acid bonding interaction may be considered to mimic Lewis-acid–base coordinate covalency and potentially to prevail when the acid and base are of such nature that the chemical potential for formation of the conventional covalency is inadequate or the potential barrier for reaction sufficient to ensure slow rates of formation of the conventional complex.

Several clear trends are apparent in Table A5.XI for the relationship between structure of the substituent and the effect on shielding of the Lewis-acid bonding. The upfield shift in dioxane compared to normal shielding increases in the order CHO < C≡N, $COCF_3$ < COCN, or −NO < −NO_2. This is the order of decreased shielding in normal solvents and, apparently, of increased charge deficiency on the common first atom (C or N) of the substituent. Over the carbon series of substituents, the effect of Lewis-acid bonding by dioxane increases from 0.2 to 1.0 ppm. Thus, the larger of these solvent effects is of the same order of magnitude, for example, as the effect on intramolecular shielding produced in changing the —HC=CH_2 to the —CH_3 substituent.

For the substituents –$SO_3C_2H_5$, $CO_2C_2H_5$, and $CONH_2$ there is

TABLE A5.XIA

Effects of Lewis-Acid Bonding on $\int_H^{m-x}$

Solvent	—CONH$_2$	—CO$_2$C$_2$H$_5$	—SO$_3$C$_2$H$_5$	meta Substituent —COCH$_3$	—CHO	—NO	—COCl	—COF
Normal[a]	I	−0.25[b]	−3.00[b]	−0.68	−1.18	−1.80	−2.10[d]	−2.15
Acetonitrile	−0.50	−0.65	—	−0.73	−1.23	−1.78	—	−2.08
Tetrahydrofuran	−0.03	−0.40	−3.20	−0.58	−1.08	−1.83	−2.05	—
Pyridine	−0.08	−0.43	R	−0.60	−1.13	−1.68	−1.93	−1.98
Diglyme	—	—	−3.25	−0.63	−1.03	—	−2.00	—
Dimethyl sulfoxide	−0.03	−0.70	R	−0.60	−1.15	−1.88	—	−1.90
Dioxane	−0.15	−0.30	−3.03	−0.48	−0.98	−1.55	−1.73	−1.83

TABLE A5.XIB

Effects of Lewis-Acid Bonding on $\int_H^{m-x}$

Solvent	—CN	—CH=CHNO₂	—SO₂CH₃	—SOCH₃	—NO₂	—COCF₃	—COCN
				meta Substituent			
Normal	−2.78	−1.13[c]	−3.28[e]	−2.90[e]	−3.38	−2.62	−3.33[b]
Acetonitrile	−2.68	−1.18	—	—	−3.33	−2.60	−2.78
Tetrahydrofuran	−2.75	—	—	—	−3.25	—	—
Pyridine	−2.78	−1.08	—	—	−3.23	—	−2.55
Diglyme	−2.63	—	—	—	−3.18	—	—
Dimethyl sulfoxide	−2.58	−0.95	—	−2.40	−3.15	−2.35	−2.13
Dioxane	−2.43	−1.00	−2.98	−2.48	−2.90	−2.28	−2.35

[a] Mean value from Table A5.IV unless otherwise designated.

[b] Mean of shielding parameters observed in carbon tetrachloride and in cyclohexane; precision of the mean, 0.10 ppm or less.

[c] Mean of shielding parameters observed in benzene, ethyl acetate, diethyl maleate, acetone, pyridine, dimethylformamide, nitrobenzene, nitromethane and acetonitrile; precision of the mean, ±0.08 ppm.

[d] Mean of shielding parameters observed in cyclohexane, carbon tetrachloride, diethyl ether, tetrahydrofuran, diethyl maleate, acetic anhydride and nitrobenzene; precision of the mean, ±0.05 ppm.

[e] Mean of shielding parameters observed in carbon tetrachloride and in nitromethane; precision of the mean, ±0.05 ppm.

little apparent effect of Lewis-acid bonding, although the situation may be somewhat confused by the solvent polarity effect (cf. Table A5.X). These substituents are all characterized as strongly resonance stabilized by interactions within the group (below) (cf. subsequent discussion).

Apparently, the Lewis-acid bonding interaction is not sufficiently strong to off-set resonance stabilization which is lost as the result of such an interaction, so that the effect tends to be reduced or to be absent for strongly resonance-stabilized substituents (82). By this criterion, for example, the results in Table A5.XI suggest that the $-SOCH_3$ substituent is less resonance stabilized than the $-SO_2CH_3$ substituent as is reasonable on other grounds (both of these substituents give approximately the same normal shifts but the former is more affected by dioxane).

The order of the effects of Lewis-acid bonding among the solvents of Table A5.XI apparently is not the same for all substituents. For substituents showing the smaller efiects the order quite generally is acetonitrile < pyridine, diglyme < dimethyl sulfoxide < dioxane. However, for COCN the order changes to dioxane < dimethyl sulfoxide (83). Changes in order may be associated with orientation, steric effects and other group specificities. However, the effects of Lewis-acid bonding which are apparent for the $-CF_3$ and $-SF_5$ substituents in dioxane (cf. Table A5.III) suggest that the interaction is a relatively long range one and is not as susceptible to steric factors as conventional chemical equilibria.

From the variation of shielding parameters with substituent under "normal" conditions is it apparent that ^{19}F intramolecular shielding is a relatively sensitive probe to changes in electron withdrawal (presumably inductively) from the *meta* position of the benzene ring. Solvent–substituent group interactions which involve acceptor–donor complexing (84) presumably would appreciably alter electron withdrawal. The effects of Lewis-acid bonding (84,85) appear to relate to the extent of net transfer of electronic charge from solvent to substituent and the extent to which this charge transfer is transmitted to the benzene ring (the transmission factor from the *meta* position to the fluorine nucleus apparently is essentially a constant for any *meta*-substituted fluorobenzene in any solvent). The smaller solvent shift in dioxane for the β-nitrovinyl than for the nitro group is reasonable in terms of attenuated transmission through the —C=C— system of the charge transfer which occurs between the similar nitro groups and dioxane.

5. Effects of Solvent on σ_I Values

The empirical correlation (10) between intramolecular shielding by *meta* substituents in fluorobenzenes and the inductive substituent parameter σ_I can be examined with present results much more critically than was previously possible and the implied dependence of σ_I values on solvent can be evaluated.

The results for the chemically inert substituent groups (Table A5.III) give plots of $\int_H^{m-X}$ vs. σ_I which, to a precision on the order of the experimental error, are linear and for which the slope is independent of medium. In Figure A5.1, the closed circles give the mean shielding

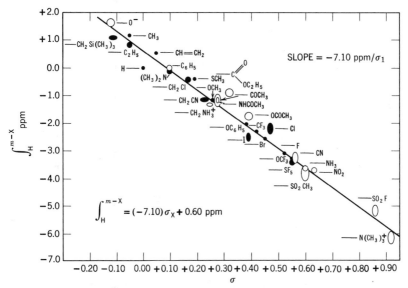

Fig. A5.1. Correlation of fluorine NMR shielding parameters for *meta*-substituted fluorobenzenes with inductive constant, σ_I; standard of reference is internal fluorobenzene: (●) "normal" shielding, cf. Tables A5.III and A5.IV; (○) mean shielding in protonic solvents no more acidic than formic acid.

parameters (Table A5.III) for the chemically inert *meta* substituents (and their precision measures) plotted vs. σ_I values. The only deviation which appears outside the combined experimental errors of $\int_{H}^{m-X}$ and σ_I is that for the unsubstituted fluorobenzene. The deviation for the hydrogen substituent is unexplained, although several observations may be made in this connection. The molecular symmetry of fluorobenzene differs from that of any *meta*-substituted fluorobenzene which may give rise to an intramolecular shielding term (86). In the correlations of inductive effects in aliphatic series reactivities, the hydrogen atom is a rather frequent deviator (87). Since in aromatic series reactivities, however, the σ_I value for hydrogen has found wide applicability (88), it appears less likely that the deviation is due to a dependence of σ_I on the type of system involved.

The important consequence of the results of Table A5.III is the implied constancy in nearly any solvent of the σ_I values for the chemically inert substituents. Generality in ground electronic states is implied since a remarkably activating derivative of benzene would be required to convert the chemically inert *meta* substituent of the fluorobenzene to a chemically active substituent. This conclusion may also be extended to additional substituents, for example: alkyl groups; alternant hydrocarbon groups; halogens (Cl and perhaps I); $-SiX_3$ (X = halogen); $-SCX_3$ (X = halogen); $-NX_2$ (X = halogen). If solvents giving rise to either hydrogen or Lewis-acid bonding are excluded, the additional substituents of Table A5.IV (and structurally similar substituents) may be added to this category.

$$\overset{\displaystyle O}{\overset{\displaystyle \|}{}}$$

Inasmuch as the σ_I values for groups such as $-\overset{\|}{C}CH_3$, $-CN$, $-NO_2$, O^-, NH_3^+ etc., have been determined in solvents in which hydrogen bonding effects are implied (Tables A5.V–A5.IX), it is not appropriate to utilize the mean shielding parameters for these substituents (Table A5.IV) in the $\int_{H}^{m-X}$ vs. σ_I plot. Instead, the shielding parameter appropriate to the solvent conditions from which the σ_I value is derived is apparently required. Although the results of Tables A5.V–A5.IX imply that σ_I values for such substituents are measurably variable among different protonic solvents, mean shielding parameters for these substituents in any protonic solvent with acidity equal or less than formic acid (including various aqueous organic mixed solvents and non-hydroxylic solvents; e.g., $HCCl_3$ and H_2CCl_2) can be given which apparently have useful precision. The open circles in Figure A5.1 give the mean shielding parameters (and their precision measures) obtained in this manner from the results in Tables A5.V–A5.IX.

Since the σ_I values for the —CH=CH$_2$, —SCH$_3$, —OCH$_3$, and —OC$_6$H$_5$ substituents are apparently appropriate to the solvents on which the mean values of Table A5.IV are based, these shielding parameters have been plotted in Figure A5.1. For the substituents –CH$_2$OH, –NHNH$_2$, –N(CH$_3$)$_2$ and –CH$_2$NH$_2$ the shielding parameters plotted in Figure A5.1 are mean values from the results in cyclohexane and carbon tetrachloride. For the –NHNH$_2$ and the CH$_2$X substituents the σ_I values have been calculated from the relationship (89): for $-AB$, $\sigma_I = (\sigma_I)_{AH} + (\sigma_I)_B/2.0$. The σ_I values plotted for the substituents –O$^-$, –OCF$_3$, SF$_5$ and SO$_2$F in Figure A5.1 have been obtained from at least two aromatic series reactivities by the method of Taft and Lewis (90), but have not been confirmed in aliphatic series reactivities. The applicability to the latter is especially questionable for the charged substituents (11).

The correlation in Figure A5.1 of the intramolecular shielding produced by *meta* substituents covers a truly remarkable range of structures with relatively good precision, according to eq. (1): $\int_{H}^{m-X} = (-7.10)\sigma_I + 0.60$. The fit of both closed and open circles with essentially equal precision indicates that the slope of the regression line, -7.10 ppm/σ_I, is within its precision completely independent of solvent, and offers confirmation of the interpretation that variation of shielding for a given substituent with solvent is due to modification in the σ_I values resulting from solvent–substituent interaction.

The fit of the shielding for –CH$_2$X substituents to the same line as for $-X$ substituents offers confirmation of the interpretation of shielding for the latter. The interposed methylene essentially destroys any direct conjugative interactions of X with the benzene ring (leading potentially to σ_R contributions) so that the effects of CH$_2$X substituents must apparently be necessarily inductive. It is further significant that the fall-off factor for shielding parameters is closely the same as that observed in reactivities (89) and that there is essentially the same insensitivity of the shielding parameters for CH$_2$X as for X substituents to solvent dielectric constant.

The small but real deviations from eq. (1) may be caused by neighboring group anisotropies and by small σ_R contributions which are not recognizable due to the contributions from the former (and other second-order) effects. It is quite plain, however, that any σ_R contributions are of little practical consequence and this contribution, if present, is relatively unimportant, for example, compared to σ_R contributions to Hammett *meta-σ-values* ($\sigma_{(m)} = \sigma_I + 0.50\sigma_R^0$) (10).

Equation 1 may be utilized for two practical purposes: *1.* for the calculation of (σ_I and σ^*) values for new substituent groups and *2.* for determination of the variation of σ_I with solvent. The result may apparently be utilized with some confidence in view of the nature of Figure A5.1.

Table A5.XII lists values of σ_I calculated (column two) for the present work for 50 substituents in solvents in which no hydrogen- or Lewis-acid bonding effects are observed. Table A5.XII also lists (column three) the σ_I values for the substituents calculated for dioxane solutions as an illustration of the effects on σ_I of Lewis-acid bonding. Further, the average σ_I values for protonic solvents which are no more acidic than formic acid are listed in column four of Table A5.XII and the σ_I values for the acidic solvent trifluoroacetic acid are listed in column five. These results illustrate the effects of hydrogen bonding on σ_I values (in trifluoroacetic acid substantial proton transfer equilibria are probably involved with the $-CONH_2$ and $-SOCH_3$ substituents). Finally, Table A5.XII lists the σ_I values based on chemical reactivities (87) (column 1) which have been used in Figure A5.1.

The general applicability of the NMR derived σ_I values in solvents which involve hydrogen and Lewis-acid bonding effects is a matter of much interest. The results in Table A5.VIII for the $-NH_2$ and $-CH_2NH_2$ substituents clearly establish that hydrogen-bonding solvent effects will not always be the same for a given functional group, independent of its molecular environment. As noted previously, solvent effects are observed to be in opposite directions for the NH_2 group directly attached to the benzene ring as compared to that separated by an interposed methylene group. The environmental effect in this particular instance is, of course, quite severe since direct conjugation of $-NH_2$ with phenyl produces a large decrease in base strength and presumably a substantial increase in acid strength. On the other hand, the same structural change for the $-OH$ group, which produces similar effects on acid and base strength, gives results in methanol compared to hydrocarbon solvents for the $-OH$ and $-CH_2OH$ substituents which apparently imply a very nearly constant solvent effect on the σ_I value for an OH group (i.e., either aliphatic or aromatic). Unfortunately, these results are not unambiguous, however, for they have not been corrected for the unknown effects of polymerization in hydrocarbon media.

The question of whether one will observe the former or the latter behavior in a given instance therefore depends not only upon the nature of the structural change in the solute but is also determined by the nature of the change involved in the hydrogen-bonding donor and acceptor capacities of the solvent. It does not appear unreasonable to sometimes utilize σ_I values which are characteristic of *a given substituent in a given solvent* (or solvent class) as a useful approximation. Such a procedure, however, must be used with appropriate caution for it is not presently possible to define the conditions under which results such as those illustrated by the $-NH_2$ group may be expected to invalidate such a treatment.

The comparison of solvent effects in σ_I values obtained from the fluorine shielding effects of *meta*-substituted fluorobenzenes and those obtained from the acidities of 4-substituted [2,2,2]bicyclooctanecarboxylic acids is of special interest. Unfortunately, only one substituent, $-Br$, investigated by Ritchie and Lewis falls into the present category of a generally chemically inert substituent. It is implied by present results that the accurate evaluation of solvent effects on σ_I values from reactivities must be based upon reaction constants (ρ or ρ^*) which are determined from the results for chemically inert substituent groups (i.e., $\rho = \log (k/k_0)/\sigma_I$ for such substituents). With ρ values based upon the single substitution of 4-Br for H– (which therefore must be regarded as tentative and rather uncertain ρ values), σ_I values may be calculated from the data reported by Ritchie and Lewis. Table A5.XIII lists values

TABLE A5.XIII
Comparison of Solvent Effects on σ_I Values From
Reactivities and Shielding Parameters

Substituent	Acetone		CH$_3$OH		Dimethyl sulfoxide	
	Acidity	NMR	Acidity	NMR	Acidity	NMR
OH	+0.15	+0.16[a]	+0.24	+0.15	+0.03	+0.16[a]
CO$_2$C$_2$H$_5$	+0.21	+0.17	+0.27	+0.18	+0.36	+0.18
CN	+0.51	+0.48	+0.53	+0.51	+0.43	+0.45

[a] Based upon shielding parameters obtained by M. G. Schwartz (unpublished).

of σ_I for the uncharged CO$_2$Et, CN and OH substituents in the pure solvents, acetone, dimethylsulfoxide and methanol, which were obtained in this manner and corresponding values obtained from the shielding parameters. Considering the combined uncertainties of the ρ and $\log (K/K_0)$ values, the agreement is fair (the agreement for the CN

TABLE A5.XII. σ_I Values

Substituent	Reactivities — Weakly protonic solvents[e]	"Normal" solvents	^{19}F Shielding[f] — Dioxane	^{19}F Shielding[f] — Weakly protonic solvents	^{19}F Shielding[f] — Trifluoroacetic acid
1 —N(CH₃)₃⁺	+0.92	—	—	+0.93 ± 0.02	—
2 —SO₂Cl	—	+0.80	—	+0.86 ± .01	—
3 —SO₂F	+0.86[a]	+0.75	—	+0.81 ± 0.01	—
4 —NO₂	+0.63	+0.56	+0.49	+0.60 ± 0.02	+0.80
5 —SF₅	+0.55[a]	+0.56	+0.54	+0.56 ± 0.01	+0.57
6 —SO₂CH₃	+0.60	+0.55	+0.50	+0.62 ± 0.03	—
7 —NH₃⁺	+0.60	—	—	+0.58 ± 0.02	—
8 —COCN	—	+0.55[c]	+0.42	+0.55 ± 0.03	+0.73
9 —OCF₃	+0.55[a]	+0.55	+0.55	+0.55 ± 0.01	+0.54
10 —SO₃C₂H₅	—	+0.50[c]	+0.51	+0.63 ± 0.02	+0.88
11 —CN	+0.56	+0.48	+0.43	+0.53 ± 0.02	+0.74
12 —F	+0.52	+0.52	+0.52	+0.52 ± 0.01	+0.54
13 —SOCH₃	+0.52	+0.49	+0.43	+0.62 ± 0.04	+1.00
14 —COCF₃	—	+0.45	+0.41	+0.48 ± 0.02	+0.61
15 —SO₂NH₂	—	—	+0.38	+0.46 ± 0.03	—
16 —Br	+0.45	+0.44	+0.43	+0.44 ± 0.01	+0.44
17 —CF₃	+0.41	+0.41	+0.38	+0.41 ± 0.01	+0.41
18 —COF	—	+0.39	+0.34	+0.42 ± 0.01	+0.56
19 —OC₆H₅	+0.38	+0.37	+0.37	+0.37 ± 0.01	+0.50
20 —NO	—	+0.34	+0.30	+0.37 ± 0.01	—
21 —OH	+0.25	+0.29[c,g,h]	—	+0.16 ± 0.04	—
22 —OCOCH₃	+0.39	+0.27[c]	—	+0.33 ± 0.02	—
23 —CH₂NH₃⁺	+0.25[b]	—	—	+0.25 ± 0.01	—
24 —CHO	—	+0.25	+0.22	+0.31 ± 0.02	—
25 —OCH₃	+0.25	+0.25	+0.26	+0.29 ± 0.03	+0.53
26 —NHCOCH₃	+0.28	+0.20[d,g]	—	+0.24 ± 0.01	+0.51

27 —COCH₃	+0.28	+0.18	+0.15	+0.23 ± 0.02	+0.46
28 —CH=CHNO₂	+0.38	+0.24	+0.23	+0.27 ± 0.01	—
29 —SO₃⁻	—	—	—	+0.25 ± 0.02	—
30 —CH₂CN	+0.23ᵇ	+0.24ᶜ	—	+0.24 ± 0.02	—
31 —N=NC₆H₅	—	+0.19ᶜ	—	+0.25	—
32 —SH	+0.25	+0.19ᶜ	—	+0.18 ± 0.01	+0.65
33 —CONH₂	—	—	+0.11ᵍ	+0.21 ± 0.02	+0.34
34 —SCH₃	+0.19	+0.14	+0.15	+0.14 ± 0.01	—
35 —NHNH₂	+0.15ᵇ	+0.14ᶜ,ᵍ	—	+0.10 ± 0.02	—
36 —CH₂Cl	+0.17	+0.14ᶜ	—	+0.14 ± 0.02	+0.35
37 —CO₂C₂H₅	+0.30	+0.11ᶜ	+0.13	+0.21 ± 0.02	—
38 —N(CH₃)₂	+0.10	+0.10ᶜ	—	+0.10 ± 0.01	+0.10
39 —C₆H₅	+0.10	+0.08	+0.08	+0.08 ± 0.01	—
40 —CH₂OH	+0.10ᵇ	+0.06ᶜ,ᵍ	—	+0.01 ± 0.02	—
41 —NH₂	+0.10	+0.01ᶜ,ᵍ	—	+0.05 ± 0.08	—
42 —CH=CH₂	+0.05	+0.01	+0.01	+0.01 ± 0.02	—
43 —CH₂—NH₂	0.00ᵇ	0.00ᶜ,ᵍ	—	+0.04 ± 0.01	—
44 —CH₂CH₃	-0.05	-0.03	—	-0.03 ± 0.01	—
45 —CO₂⁻	—	—	—	-0.35 ± 0.02	—
46 —CH₂Si(CH₃)₃	-0.11ᵇ	-0.07ᶜ	—	-0.07 ± 0.01	—
47 —CH₃	-0.05	-0.08	-0.07	-0.08 ± 0.01	-0.06
48 —B(OH)₂	—	—	—	-0.08 ± 0.01	—
49 —O⁻	-0.12ᵃ	—	—	-0.16 ± 0.01	—
50 —B(OH)₃⁻	—	—	—	-0.36 ± 0.03	—

ᵃ Calculated by the method of Taft and Lewis (ref. 90) from aromatic series reactivities. For the SO₂F substituent, unpublished results of Taft and Davis on the ionization of ArCO₂H and ArNH₃⁺, H₂O, 25°, have been used. For the OCF₃ and SF₅ substituents, the results of W. A. Sheppard, *J. Am. Chem. Soc.*, 83, 4860 (1961), and 84, 3072 (1962), have been used. For the O⁻ substituent, σ-values given by J. Hine, *ibid.*, 82, 4880 (1960), have been used.   ᵇ Obtained by eq. (2).   ᶜ Obtained from average shielding in carbon tetrachloride and cyclohexane solutions.   ᵈ Measured in dimethylformamide solution only.   ᵉ Precision ±0.03.   ᶠ Precision ±0.01 unless otherwise indicated; $\sigma_I = (J^m_{H-x} - 0.60)/7.10$.   ᵍ Uncorrected for polymeric association.   ʰ The corrected value for OH is approximately +0.25, cf. footnote c, Table A5.VIII.

substituent is actually quite satisfactory). These results indicate that the solvent effects on shielding parameters have promising applicability to appropriate reactivity systems, but the test may not yet be regarded as critical for even the favorable examples.

5. Some Comments on the Effects of Structure on σ_I Values

A more detailed treatment of this subject will be considered in a subsequent publication. In the present discussion several of the more important and obvious aspects of this subject deserve additional comment to that given previously (91).

The effect of a well-localized positive charge on the first atom of the substituent group is to increase σ_I by approximately $+0.80$ unit. It is clear that the effect is readily dissipated by delocalization of charge within the substituent and by interactions with the solvent. The comparable σ_I values for $-SOCH_3$, $-SO_2CH_3$, $-NO_2$, but the much smaller value for $-COCH_3$ indicate, in accord with other evidence (92), that the $-\overset{+}{S}-\overset{-}{O}$ bond is much more ionic than the $-\overset{|}{C}=O$ bond.

The distribution of charge within the substituent group, especially as it affects the charge of the first atom, is a primary factor determining the inductive effect of the substituent as a whole. Thus, we may estimate that the $-C\equiv N$ substituent would have a σ_I value of perhaps 0.1–0.2 unit more positive than the $-CH_2NH_2$ substituent (i.e., $\sigma_I \simeq +0.2$) if the triple bond were nearly nonpolar. The estimated increase is due to the greater electronegativities of the sp than sp^3 valence states (93). Therefore about 0.3 unit of the observed "normal" σ_I value is due to the polarization, $-C\equiv N \leftrightarrow -\overset{+}{C}=\overset{-}{N}$. Partial neutralization of the negative charge by hydrogen bonding with trifluoroacetic acid raises this figure to about 0.45 σ_I unit.

The systematic variation of structure carried out in the present work permits an especially instructive analysis of the effect on σ_I of charge distribution within substituents of the general formulas $-COX$ and $-SO_2X$. In the absence of any appreciable charge delocalization interactions between the X and $-C=O$ or $-SO_2$ groups, one would anticipate a relationship [eq. (2)] similar to that noted earlier for $-CH_2X$ substituents, i.e.,

$$(\sigma_I)_{-COX} = (\sigma_I)_{-CHO} + [(\sigma_I)_X/2.0] \tag{2a}$$

and

$$(\sigma_I)_{-SO_2X} = (\sigma)_{-SO_2H} + [(\sigma_I)_X/2.0] \tag{2b}$$

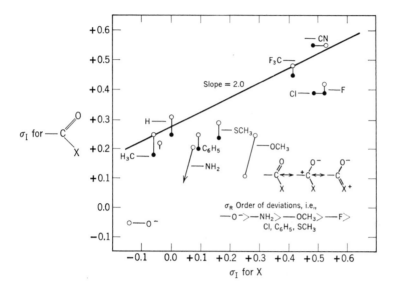

Fig. A5.2. Effect of charge delocalization within substituent on σ_I; COX substituents (X is as indicated): (●) σ_I value for "normal" aprotic solvents; (○) mean σ_I value for hydroxylic solvents. The results for the —COC$_6$H$_5$ and —COSCH$_3$ substituents were obtained by Dr. Y. Tsuno, unpublished results.

Figures A5.2 and A5.3 illustrate that these relationships do hold roughly. For COX substituents, a plot of $(\sigma_I)_{COX}$ vs. $(\sigma_I)_X$ gives an approximately linear relationship of slope 12.0 for the substituents X = –CN, –CF$_3$, –H, and –CH$_3$. These substituents are of such character that no appreciable delocalization interaction is anticipated (a weak hyperconjugative interaction probably occurs for the –CH$_3$ substituent). On the other hand, the relationship fails for the –R substituents, X = O$^-$, –NH$_2$, –OCH$_3$, –F, –Cl, –C$_6$H$_5$, for which there is delocalization of negative charge into the carbonyl group. These substituents give σ_I values (for the –COX substituent as a whole) which are less than anticipated by the above relationship. The magnitude of the deviation decreases in the order of substituents given above, and corresponds to correlation with σ_R values (10). That is, this is the qualitative order expected for decreasing delocalization of charge from X into —C=O. The charge distribution effect on the σ_I value may be

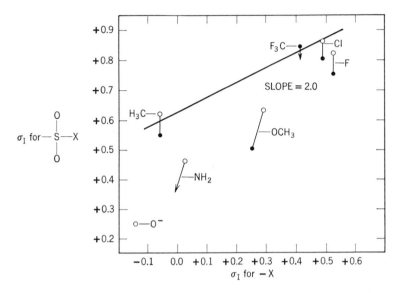

Fig. A5.3. Effect of charge delocalization within substituent on σ_I; SO$_2$X substituents (X is as indicated): ($\bullet$) σ_I value for "normal" aprotic solvents; ($\bigcirc$) mean σ_I value for hydroxylic solvents.

interpreted in terms of the valence bond structures

as indicating that as structure III is increased in importance, structure II decreases in importance (i.e., negative charge from X accumulates on both C and O). This result is in accord with charge density relationships obtained by HMO calculations (94).

Similar results are obtained for –SO$_2$X substituents. In Figure A5.3 a line of slope 2.0 is drawn between the two substituents CH$_3$ and CF$_3$ (95), for which eq. (2b) is expected to hold approximately. Deviations for X = – R substituents are apparent in Figure A5.3. It is of interest that the magnitudes of the deviations for a given X in –COX and –SO$_2$X are similar. A consequence is the fact that corresponding σ_I values are quite generally about +0.40 more positive in the latter series (implying the equivalent of about 0.5 electronic unit more positive charge on S than C).

The present results illustrate in an apparently unambiguous manner how a given (but arbitrary) position in the molecule must be specified in order that the terms resonance and inductive effect of substituents can have even an approximate utility.

References and Footnotes to Appendix 5

1. This work was supported in part by the Office of Naval Research, Project NRO55-328.
2. Deceased; NSF Cooperative Graduate Fellow, 1959–1960.
3. Public Health Service Postdoctoral Research Fellow, 1959–1960.
4. D. F. Evans, *Proc. Chem. Soc.*, *1958*, 115; *J. Chem. Soc.*, *1960*, 877.
5. R. E. Glick and S. J. Ehrenson, *J. Phys. Chem.*, *62*, 1599 (1958).
6. B. J. Alder, *Ann. Rev. Phys. Chem.*, *12*, 211 (1961), and references listed therein.
7. J. A. Pople, W. G. Schneider, and H. J. Bernstein, *High-Resolution Nuclear Magnetic Resonance*, McGraw-Hill, New York, 1959, Chapter 7.
8. (a) F. Prosser and L. Goodman, *J. Chem. Phys.*, *38* (1963); (b) R. W. Taft, F. Prosser, L. Goodman, and G. T. Davis, *ibid.*, *38* (1963).
9. M. Karplus and T. P. Das, *J. Chem. Phys.*, *34*, 1683 (1961).
10. R. W. Taft, *J. Phys. Chem.*, *64*, 1805 (1960).
11. C. D. Ritchie and E. S. Lewis, *J. Am. Chem. Soc.*, *84*, 581 (1962).
55. Reference 7, pp. 80–82.
56. A. D. Buckingham, T. Schaefer, and W. G. Schneider, *J. Chem. Phys.*, *32*, 1227 (1960).
57. A. D. Buckingham, T. Schaefer, and W. G. Schneider, *J. Chem. Phys.*, *32*, 1227 (1960).
58. E. G. McRae, *J. Phys. Chem.*, *61*, 562 (1957); *Spectrochim. Acta*, *12*, 192 (1958).
59. Reference 7, Chapt. 15.
60. A. A. Bothner-By and R. E. Glick, *J. Chem. Phys.*, *26*, 1651 (1957); J. R. Zimmerman and M. R. Foster, *J. Phys. Chem.*, *61*, 282 (1957).
61. L. Onsager, *J. Am. Chem. Soc.*, *58*, 1486 (1936).
62. J. G. Kirkwood, *J. Chem. Phys.*, *2*, 351 (1934).
63. A. A. Maryott and E. R. Smith, "Table of Dielectric Constants of Pure Liquids," *Natl. Bur. Std. Circ.*, *514* (1951).
64. S. G. Smith, A. H. Fainberg, and S. Winstein, *J. Am. Chem. Soc.*, *83*, 618 (1961).
65. G. R. Leader and J. F. Gormley, *J. Am. Chem. Soc.*, *73*, 5731 (1951).
66. L. P. Hammett, *Physical Organic Chemistry*, McGraw-Hill, New York, 1940, pp. 80–87, 184–204; M. N. Davis and H. B. Hetzer, *J. Res. Natl. Bur. Std.*, *60*, 569 (1958).
67. F. H. Westheimer and J. G. Kirkwood, *J. Chem. Phys.*, *6*, 506, 613 (1938); J. N. Sarmousakis, *ibid.*, *12*, 277 (1944).
68. Cf. E. A. Moelwyn-Hughes, *States of Matter*, Interscience, New York, 1961, p. 5.
69. A detailed study, similar to the present one, on solvent effects for fluorine

NMR shielding of *p*-substituted fluorobenzenes has been carried out and will be reported in the near future. A preliminary communication of these results has appeared; R. W. Taft, R. E. Glick, I. C. Lewis, I. R. Fox and S. Ehrenson, *J. Am. Chem. Soc.*, *82*, 756 (1960). In general, the magnitude of solvent effects on the shielding parameters for $-R$ *para*-substituted fluorobenzenes (relative to internal fluorobenzene) is comparable to that reported here for *meta*-substituted fluorobenzenes. However, for many $+R$ *para*-substituted fluorobenzenes the solvent effects are an order of magnitude greater.

70. Estimated from data of R. Stewart and K. Yates, *J. Am. Chem. Soc.*, *80*, 6355 (1958).
71. G. V. Tiers, *J. Am. Chem. Soc.*, *78*, 4165 (1956).
72. R. W. Taft and P. L. Levins, *Anal. Chem.*, *34*, 436 (1962), and unpublished results.
73. G. C. Pimentel and A. L. McClellan, *The Hydrogen Bond*, Freeman, San Francisco, 1960, Chapter 3.
74. J. R. Gordon, *J. Org. Chem.*, *26*, 738 (1961).
75. The fluorine shielding parameters for a series of 1:1 hydrogen bonded complexes of *p*-fluorophenol with a variety of oxygen and nitrogen bases in carbon tetrachloride solutions have been found to correlate with the standard enthalpy of complex formation, unpublished results of R. W. Taft and M. G. Schwartz.
76. P. Ballinger and F. A. Long, *J. Am. Chem. Soc.*, *82*, 795 (1960).
77. G. E. K. Branch and M. Calvin, *The Theory of Organic Chemistry*, Prentice-Hall, New York, 1941, p. 229.
78. A. V. Willi, *Z. Physik. Chem.*, *27*, 233 (1961).
79. R. W. Taft, *J. Am. Chem. Soc.*, *82*, 2965 (1960).
80. G. W. Meadows and B. deB. Darwent, *Can. J. Chem.*, *30*, 501 (1952); *Trans. Faraday Soc.*, *48*, 1015 (1952).
81. For example, the formation of a Lewis acid–base complex between *m*-fluorophenyl methyl sulfide and BCl_3 in H_2CCl_2 solution results in a downfield shift of the fluorine signal of 4.1 ppm; unpublished results of J. Carten and R. W. Taft.
82. C. D. Ritchie has kindly communicated results of a study (*J. Am. Chem. Soc.*, in press) of the integrated intensity of the infrared bands of benzonitrile in binary solutions of carbon tetrachloride and several polar solvents. The results are interpreted in terms of an equilibrium constant for the formation of a weak 1:1 complex. The apparent equilibrium constants for formation are in the order: dimethyl sulfoxide > dimethylformamide > pyridine. Further, there is no detectable effect of *meta* and *para* substituents on the formation constant for benzonitrile–dimethyl sulfoxide complex, indicating that the apparent Lewis-acid bonding has a low sensitivity to Lewis acidity of the nitrile. Finally, no evidence was found for a definite complex between ethyl benzoate and dimethyl sulfoxide. It is of interest to note that entirely similar trends are obtained in the solvent shielding effects of Tables A5.II and A5.XI.
83. The Lewis acid bonding effect with this substituent is apparently sufficiently large to discriminate among other solvents as follows (cf. Table A5.II): dimethyl sulfoxide > dioxane > dimethylformamide > pyridine > acetone, benzene, acetonitrile, nitromethane, nitrobenzene > diethyl maleate > diethyl ether.

84. Cf. R. S. Mulliken and W. B. Person, *Ann. Rev. Phys. Chem.*, *13*, 107 (1962).
85. Molecular self-association through Lewis-acid bonding can potentially affect physical properties in an analogous manner to that commonly accepted for hydrogen bonding. As possible examples, we note that aliphatic nitriles have boiling points 35–65° higher than corresponding primary amines. The lactones β-propiolactone and γ-butyrolactone have boiling points 70–120° higher than nonhydrogen bonded liquids of comparable molecular weight. (We are indebted to Prof. N. C. Deno for this observation.)
86. W. S. Brey, Jr., and K. D. Lawson, Abstracts, Am. Chem. Soc. Natl. Meeting, Chicago, Sept., 1961, p. 4-T.
87. R. W. Taft and I. C. Lewis, unpublished summary.
88. R. W. Taft and I. C. Lewis, *J. Am. Chem. Soc.*, *81*, 5343 (1959).
89. G. E. K. Branch and M. Calvin, "The Theory of Organic Chemistry," Prentice-Hall, New York, 1941, Chapter VI. The Branch and Calvin Scheme employs a fall-off factor of 1/2.8. However, J. C. McGowan, *J. Appl. Chem.*, *1960*, *312*, and references therein, has pointed out that the value 1/2.0 frequently better fits available data. A recent survey (87) indicates that there is no general single precise value of the fall-off factor. Values from 1/1.8 to 1/3.0 are found dependent upon both substituent and reactivity type. In the present application the McGowan factor serves as a useful approximation.
90. R. W. Taft and I. C. Lewis, *J. Am. Chem. Soc.*, *80*, 2436 (1958); *81*, 5343 (1959).
91. R. W. Taft, *Steric Effects in Organic Chemistry*, M. S. Newman, Ed., Wiley, New York, 1956, Chapter 13.
92. C. C. Price, private communication.
93. R. W. Taft, *J. Chem. Phys.*, *26*, 93 (1957).
94. Reference 8b and unpublished results.
95. The σ_I value of $+0.84$ for the $-SO_2CF_3$ substituent has been communicated to the authors by Dr. W. A. Sheppard.

XI. APPENDIX 6. FLUORINE NUCLEAR MAGNETIC RESONANCE SHIELDING IN *para*-SUBSTITUTED FLUOROBENZENES. THE INFLUENCE OF STRUCTURE AND SOLVENT ON RESONANCE EFFECTS (1)*†‡

Intramolecular ^{19}F NMR shielding in the special case of *para*-substituted fluorobenzenes apparently can be directly related, at least approximately, to the π-electron charge density on the fluorine atom (4,5) or its bonded carbon atom (5). Extensive studies (6) of the solvent effects on the fluorine shielding in *meta*-substituted fluorobenzenes

* By Robert W. Taft, Elton Price, Irwin R. Fox (2), Irwin C. Lewis, K. K. Andersen (3), and George T. Davis, College of Chemistry and Physics, the Pennsylvania State University, University Park, Pennsylvania.

† Reprinted from the *Journal of the American Chemical Society*, 85, 3146 (1963). Copyright 1963 by the American Chemical Society and reprinted by permission of the copyright owner.

‡ References and footnotes for this appendix will be found on pp. 320–322.

suggests that the intramolecular shielding effect of a *para* substituent in fluorobenzene can be obtained at high dilution in any solvent from the measurement of shielding referred to the standard, internal fluoro-benzene. The substituted and unsubstituted fluorobenzene present to their environment fluorine atoms which are sufficiently similar that intermolecular shielding cancels with considerable precision in the shielding parameter of the substituted relative to the unsubstituted internal fluorobenzene.

Fluorine shielding is so sensitive to the very small (absolute) intramolecular perturbations in the fluorine atom π-charge density produced by a distant *para* substituent that this measurement ranks as the most sensitive probe (relative to the experimental error) currently available to investigate such interactions (4–7). Using this distant but very sensitive probe, new and highly instructive investigations of both the effects of substituent structure and of solvent–substituent inter-actions on the intramolecular shielding effect of *para* substituents are made possible. In the present paper we report the results of extensive studies of both types.

Solvent–substituent interactions previously investigated by this technique include proton transfer equilibria (8), Lewis acid–base equilibria (6), hydrogen bonding (6), and polar interactions (6). This paper reports additional results of the last two kinds, especially the latter. The results can apparently be interpreted in terms of the effects of polar solvents on the relative contributions of various types of dipolar resonance forms to the ground electronic state of the fluorobenzene.

The effects of structure on the intramolecular shielding of $-R$ *para* substituents have been correlated with the σ_R^0 resonance effect parameter from chemical reactivities (7b). The correlation is extended with the present results and is utilized both to obtain σ_R^0 values for new substituent groups and to assess the magnitude of solvent effects on σ_R values. Comparison of the present results with the previously reported corresponding *meta* substituent effect shielding parameters (6) provides new insight into relationships between the inductive and resonance effects of substituents.

A. Experimental (omitted)

B. Results

Table A6.I lists the results of a nearly complete survey of the shielding parameters for eight $-R$ *para*-substituted fluorobenzenes relative to fluorobenzene as the internal standard in 20 widely varying

TABLE A6.I

Shielding Parameters, $\int_H^{p-x}$, for −R para-Substituted Fluorobenzenes[a]

Solvent	para Substituent							
	OCH_3	OC_6H_5	F	CH_3	SCH_3	C_6H_5	Br	OCF_3
Cyclohexane	11.70	7.45	6.80	5.40	4.40	3.00	2.60	2.25
Benzene	11.45	7.15	6.60	5.40	—	—	2.50	2.10
Carbon tetrachloride	11.50	7.40	6.80	5.40	4.30	2.90	2.50	2.10
Diethyl ether	11.65	7.25	6.65	5.45	4.80	—	2.35	1.85
Tetrahydrofuran	—	7.20	6.40	5.40	4.75	—	2.20	—
Dioxane	11.45	7.25	6.60	5.50	4.95	2.90	2.50	2.10
Diglyme	11.40	6.95	6.35	5.30	4.75	—	2.15	—
Diethyl maleate	11.50	7.25	6.30	5.40	4.75	—	2.00	—
Acetic anhydride	11.40	6.95	6.25	5.45	4.70	—	2.00	1.55
Acetone	11.45	6.90	6.30	—	—	2.70	2.05	1.60
Pyridine	—	7.00	6.35	5.45	4.75	—	2.25	1.75
Dimethylformamide	11.50	—	6.20	—	4.95	—	2.00	—
Methylformamide	11.55	—	6.35	—	—	—	—	—
Benzonitrile	11.45	—	6.35	—	—	—	—	—
Nitrobenzene	11.55	7.05	6.30	5.55	4.80	2.10	2.20	1.85
Nitromethane	11.40	—	6.25	—	—	—	—	c
Dimethyl sulfoxide	11.35	6.85	6.15	5.30	—	—	2.00	—
Methanol	11.45	7.25	6.40	5.55	4.50	2.80	2.10	1.55
75% (vol.) aq. methanol	11.20	I[b]	6.40	5.60	4.40	I	2.10	—
Formic acid	10.70[d]	—	6.35	—	—	—	—	—
Trifluoroacetic acid	6.95[d]	5.20[d]	6.40	—	—	—	2.25	2.00
Mean value	11.45	7.15	6.40	5.45	4.70	2.75	2.25	1.85
Av. dev.	±0.08	±0.16	±0.13	±0.07	±0.16	±0.22	±0.16	±0.21

[a] In ppm relative to fluorobenzene; exptl. error = ±0.08.   [b] I designates insufficient solubility.   [c] 1.55 in acetonitrile.   [d] Excluded in obtaining mean value.

TABLE A6.II

Shielding Parameters, $-\int_{\mathrm{H}}^{p-\mathrm{x}}$, for $+\mathrm{R}$ *para*-Substituted Fluorobenzenes[a]

Solvent					*para* Substituent					
	CF₃	SF₅	—CH=CHNO₂	COCH₃	CN	NO₂	NO	COF	COCF₃	SO₂F
3-Methylpentane	4.95	5.20	—	6.05	8.65	9.00	—	—	—	—
Cyclohexane	5.05	5.35	—	6.10	8.95	9.20	10.50	11.15	12.00	12.20
Benzene	5.15	—	5.80	6.50	9.05	9.45	11.50	11.35	12.45	12.35
Carbon tetrachloride	5.15	5.50	—	6.60	9.20	9.55	11.10	11.40	12.35	12.50
Dioxane	5.15	5.50	5.40	6.15	8.95	9.10	11.60	—	11.80	12.20
Diethyl ether	5.45	5.90	—	6.40	9.20	9.65	11.35	—	12.65	12.70
Tetrahydrofuran	5.65	6.00	—	6.35	9.45	9.75	12.00	—	—	12.90
Ethyl acetate	—	—	5.80	6.50	9.45	9.85	12.00	11.70	12.85	13.00
Diglyme	5.75	—	—	6.55	9.45	10.00	—	—	—	—
Chlorobenzene	5.50	5.95	—	6.70	9.60	10.05	11.95	—	—	—
Acetic acid	—	—	—	—	—	10.00	12.25	11.70	13.00	13.05
Diethyl maleate	5.85	6.35	6.10	6.75	9.70	10.25	12.40	12.00	13.25	13.40
Methanol	5.90	6.35	6.20	7.65	10.25	10.35	12.45	12.10	R	13.50
Acetone	5.80	6.35	5.95	6.55	9.70	10.10	12.45	11.85	12.95	13.35
Acetic anhydride	5.80	—	—	6.80	9.75	10.35	12.65	11.95	13.15	13.45
Pyridine	5.70	6.15	6.10	6.80	9.80	10.25	12.55	11.85	12.90	13.55

Solvent										
Dimethylformamide	5.90	6.40	5.95	6.70	9.80	10.30	12.85	11.95	12.95	13.50
Monomethylformamide	5.70	6.25	—	7.00	9.90	10.30	12.75	—	—	13.40
Benzonitrile	—	—	—	6.90	9.90	10.45	12.65	—	—	—
Nitrobenzene	5.70	6.25	6.40	6.85	9.90	10.50	12.65	12.05	13.20	13.45
Acetonitrile	5.90	—	6.15	—	9.90	10.35	12.80	12.00	13.10	13.55
Nitromethane	5.75	6.35	6.30	6.95	9.95	10.55	12.90	12.15	13.15	13.60
Dimethyl sulfoxide	5.90	6.40	5.95	6.80	9.85	10.30	13.20	R	12.85	13.55
Methylene chloride	—	—	—	—	—	10.85	12.90	—	—	13.60
Chloroform	—	—	—	7.65	10.45	11.05	12.95	12.50	13.45	13.70
Formamide	—	—	—	9.55	11.15	11.00	13.05	—	—	13.85
Formic acid	5.35	6.00	6.80	8.50	11.00	11.00	R[b]	12.40	13.55	13.60
75% (vol.) aq. methanol	6.05	I[b]	—	—	—	11.20	13.70	R	—	14.25
2,2,3,3-Tetrafluoropropanol	5.45	6.05	—	9.30	11.60	11.85	14.80	13.05	—	14.20
Trifluoroacetic acid	5.35	5.95	—	12.80	13.75	14.05	R	15.15	16.80	15.25

[a] In ppm relative to fluorobenzene; expt. error = ±0.08.
[b] I designates insufficient solubility; R designates that a fast reaction occurs.

pure solvents at high dilution. Table A6.II gives similar results for ten + R *para*-substituted fluorobenzenes in 30 solvents. Shielding parameters for 23 additional − R *para* substituents in four solvents are given in Table A6.III. Tables A6.IV and A6.V lists shielding parameters for 22 additional + R *para* substituents in a variety of solvents. All shielding parameters have been rounded to the nearest 0.05 ppm (exptl. error = ± 0.08 ppm).

TABLE A6.III

Shielding Parameters, $\int_{H}^{p-x}$, for − R *para*-Substituted Fluorobenzenes[a]

	Solvent			
Substituent	Cyclohexane	Carbon tetrachloride	Methanol	75% (vol.) aq. methanol
O^-	—	—	19.50	19.30
$N(CH_3)_2$	15.90	15.65	15.05	13.75
NH_2	14.40	14.20	14.05	13.40
$NHNH_2$	I[b]	—	12.25	13.10
OH	10.65	10.85(11.40)[f]	12.95	12.35
$CH_2Si(CH_3)_3$	6.95	7.00	7.15	I
C_2H_5	5.05	5.00	5.10	5.10
$NHCOCH_3$	I	I[c]	5.15	4.90
$CH_2CO_2^-$	I	I	4.65	4.55
$OCOCH_3$	4.80	4.55	4.15	3.95
CH_2NH_2	3.75	3.60	3.00	3.00
SH	3.50	3.50	4.35	4.20
Cl	3.20	3.10	2.70	2.75
$CH(OH)CH_3$	2.75	2.55	2.85	2.80
CH_2CO_2H	I	2.30	2.75	2.85
CH_2OH	2.15	2.05	2.50	2.40
I	1.70	1.55	1.35	1.35
$OCOCF_3$	1.60	1.50	R[d]	R[e]
$CH=CH_2$	1.45	1.40	1.30	1.30
CH_2CN	1.30	1.20	1.35	1.35
CH_2Cl	0.50	0.35	0.45	0.35
NCS	—	−0.55	—	—
$CH_2NH_3^+$	—	—	−0.85	−1.00

[a] In ppm relative to fluorobenzene; exptl. error = ± 0.08.
[b] I designates insufficient solubility; R designates that a fast reaction occurs.
[c] 6.85 in dimethylformamide.
[d] 1.60 in CH_3NO_2.
[e] 0.20 in CF_3CO_2H.
[f] In 1% soln. (unpublished results of Mr. M. G. Schwartz).

In the previous paper of this series (6) it was found that the shielding of fluorobenzene relative to a fixed external standard varies widely with solvent, in part, apparently due to intermolecular dispersion force interactions between the fluorine atom and the solvent. However, the shielding parameters for a wide variety of chemically inert *meta*-substituted fluorobenzenes relative to the standard internal fluorobenzene were found to be solvent invariant to a precision of the same order as the experimental error. This result indicates the considerable precision with which intermolecular shielding cancels in the shielding parameter of the *meta*-substituted fluorobenzene relative to internal fluorobenzene, providing a measure of the intramolecular shielding effect of the *meta* substituent. Fluorobenzene as an internal reference serves an additional function. Variable physical interactions (or even weak complexing) between the solute aromatic ring and solvents of widely varying internal pressure should be essentially equal for substituted and unsubstituted fluorobenzene. Thus little or no effect arising from this source should enter the substituent shielding effects ($\int_H^X$ parameters) based upon the internal fluorobenzene standard.

The results in Tables A6.I and A6.III for $-R$ *para*-substituted fluorobenzenes indicate that a similar conclusion can be drawn (although in general somewhat less precisely) regarding the shielding parameters for such compounds relative to internal fluorobenzene.

The *meta* substituents CH_3, C_6H_5, H, Br, F, and OCF_3 were found to be very generally chemically inert in interactions with the solvent. The shielding parameters for these substituents in the *para* position of fluorobenzene vary in a given solvent over a range of approximately 5 ppm. However, the shielding parameters for the given substituent in a wide variety of solvents at high dilution define mean values precise to approximately ± 0.2 ppm (cf. Table A6.I). Further, the results for the OCH_3 and OC_6H_5 substituents, which extend the range of shielding an additional 5 ppm, show a similar solvent insensitivity except in formic and trifluoroacetic acid solutions. It was previously noted that these two substituents (in the *meta* position) act as bases toward these solvents, giving a similar result to that indicated by Table A6.I, namely that resonance occurs at lower field strengths than normal. Table A6.III provides several additional examples of expected variations in shielding parameters resulting from hydrogen bonding interactions between *para* substituent and solvent.

In contrast, the results in Tables A6.II, A6.IV, and A6.V for $+R$ *para*-substituted fluorobenzenes show the shielding parameters for

TABLE A6.IV

Shielding Parameters, $-\int_{H}^{p-x}$, for $+R$ para Substituted Fluorobenzenes[a]

Solvent	para Substituent									
	Si(CH₃)₃	C≡C—H	SOCH₃	CONH₂	SO₂NH₂	CO₂C₂H₅	CO₂H	SO₂CH₃	SO₃C₂H₅	CHO
3-Methylpentane	0.50	—	—	I	I	5.75	I	—	7.70	8.75
Cyclohexane	0.50	2.35	I[b]	I	I	5.90	I	I	8.10	9.15
Carbon tetrachloride	0.50	2.50	3.00	I	I	6.20	I	8.00	8.30	9.40
Dioxane	—	2.50	2.45	3.50	5.10	6.20	6.20	7.30	8.05	8.90
Diethyl ether	—	—	—	—	—	6.25	—	—	—	9.10
Tetrahydrofuran	—	2.80	—	3.40	—	6.45	6.00	—	—	9.15
Diethyl maleate	—	—	—	—	5.65	6.65	—	—	8.90	9.50
Methanol	0.65	2.70	4.80	5.00	6.00	7.15	—	8.85	9.45	R
Acetone	—	—	—	3.80	—	6.60	—	—	—	9.30
Pyridine	—	—	—	3.65	—	6.70	5.55	—	—	9.55
Dimethyl-formamide	0.60	—	—	—	—	—	6.05	—	—	9.45
Monomethyl-formamide	0.50	—	—	4.15	—	6.70	—	—	—	9.65
Nitrobenzene	—	2.85	—	I	6.45	6.60	—	—	9.15	9.55
Acetonitrile	—	—	—	4.25	—	6.70	—	—	—	9.65
Nitromethane	0.55	—	3.35	—	—	6.70	—	—	—	—
Dimethyl sulfoxide	—	—	2.75	3.45	—	6.90	—	—	—	9.50
75% (vol.) aq. methanol	I	2.85	5.25	5.55	6.80	7.65	—	9.65	10.15	10.50
Trifluoroacetic acid	R[c]	—	—	13.95	—	10.35	11.80	—	12.75	16.10

[a] In ppm relative to fluorobenzene; exptl. error = ±0.08.
[b] I denotes insufficient solubility.
[c] R denotes that a fast reaction occurs.

TABLE A6.V

Shielding Parameters, $-\int_{H}^{p-x}$, for $+R$ *para*-Substituted Fluorobenzenes[a]

	Solvent			
Substituent	Cyclohexane	Carbon tetrachloride	Methanol	75% (vol.) aq. methanol
$CO_2^- Na^+$	I	I	I	2.30
$B(OH)_2$	I	I	2.45	3.05
$SO_3^- H^+$	I	I	3.00	3.30
$C_6H_5N{=}N$	3.10	3.25	3.90	I
$COSCH_3$	—	—	6.0	I
C_6H_5CO	5.75	6.05	7.25	8.00
$CO_2C_6H_5$	7.45	7.70	8.65	I
$COCl$	11.20	11.40	R	R
SO_2Cl	11.90	12.20	13.35	13.85
N_2^+	I	27.80[b]	29.80[c]	—

[a] In ppm relative to fluorobenzene; exptl. error $= \pm 0.08$.
[b] In acetone.
[c] In acetonitrile.

these compounds are generally markedly solvent dependent. The nature of the solvent dependence (to be discussed in further detail in subsequent sections), however, shows no resemblance to that for the shielding of internal fluorobenzene relative to a fixed external standard. Thus, for example, the shielding parameters from Table A6.II for one of the *para*-substituted fluorobenzenes plotted vs. the shielding parameter for fluorobenzene relative to a fixed external standard (6) leads to a wide scattering of points. Since the solvent insensitive shielding parameters for *meta*-substituted fluorobenzenes and for $-R$ *para*-substituted fluorobenzene cover as wide a range as for most $+R$ *para*-substituted fluorobenzenes[positively charged $+R$ *para* substituents are exceptions (8)], it may be reasoned that the shielding parameters for $+R$ *para*-substituted fluorobenzenes relative to internal fluorobenzene also involve essentially complete cancellation of intermolecular shielding. The variation of the latter shielding parameters with solvent is then also to be attributed to modification of the intramolecular substituent effect through substituent–solvent interactions [as previously found (6), on a much reduced scale, for certain *meta*-substituted fluorobenzenes in certain solvents]. The nature of the solvent dependence indeed supports this interpretation (cf. subsequent discussion).

1. Correlation of Shielding Parameters

The following relatively precise correlation of F NMR shielding parameters for $-R$ *meta-* and *para-*substituted fluorobenzenes has been reported (7c)

$$\int_{H}^{p-X} = -29.5\sigma_R^0 + \int_{H}^{m-X} = -29.5\sigma_R^0 - 7.1\sigma_I + 0.60 \quad (1)$$

where $\int_{H}^{p-X}$ and $\int_{H}^{m-X}$ are the shielding parameters for the *para-*substituted

TABLE A6.VI
"Normal" σ_R^0 Values from Fluorine Shielding Parameters

Subst.	$-\sigma_R^0$ (reactivity)	$-\sigma_R^0$ (shielding)
O^-	0.66^a	0.60^c
$N(CH_3)_2$	0.52	0.539 ± 0.006
NH_2	0.48	0.481 ± 0.006
$NHNH_2$	—	0.43^c
OCH_3	0.41	0.429 ± 0.003
OH	0.40	0.427 ± 0.003^d
F	0.35	0.322 ± 0.006
OC_6H_5	—	0.312 ± 0.006
$OCOCH_3$	—	0.21
$CH_2Si(CH_3)_3$	—	0.20
$OCOCF_3$	—	0.19
Cl	0.20	0.18
OCF_3	0.17^b	0.176 ± 0.008
SCH_3	—	0.173 ± 0.006
Br	0.19	0.163 ± 0.006
SH	—	0.15
CH_2NH_2	—	0.15
CH_3	0.10	0.146 ± 0.003
C_2H_5	0.09	0.14
I	0.12	0.14
C_6H_5	0.10	0.093 ± 0.008
CH_2CN	—	0.08
$-NCS$	—	0.06
CH_2Cl	—	0.03
$CH=CH_2$	—	0.03
$CH_2NH_3^+$	—	0.00^c

[a] Based upon σ-values given by J. Hine, *J. Am. Chem. Soc.*, *82*, 4880 (1960).

[b] W. A. Sheppard, *J. Am. Chem. Soc.*, *83*, 4860 (1961).

[c] In 75% (vol.) aq. CH_3OH solution.

[d] In 1% CCl_4 solution.

fluorobenzene and its *meta* isomer, respectively, relative to internal fluorobenzene (in ppm). The resonance effect parameter, σ_R^0, is obtained by the method of Taft from σ^0 reactivities (7b).

The correlation is illustrated in Table A6.VI by comparison of σ_R^0 values obtained from reactivities and "normal" values calculated by eq. (1) from the shielding parameters (Tables A6.I and A6.III) in solvents (including hydrocarbon solvents) which define relatively precise mean values of both $\int_H^{p-X}$ and $\int_H^{m-X}$.

In Table A6.VII are listed σ_R^0 values obtained from shielding parameters which show the hydrogen bonding solvent effects of 75% (vol.) aqueous methanol and of trifluoroacetic acid on the $N(CH_3)_2$, OCH_3, OH, OC_6H_5, and SH substituents. The results for the chemically

TABLE A6.VII
Hydrogen Bonding Solvent Effects on σ_R^0 Values
from Fluorine Shielding[a]

Subst.	Normal	$-\sigma_R^0$ value 75% aq. MeOH	CF_3CO_2H
$N(CH_3)_2$	0.54	0.47	—
OCH_3	0.43	0.42	0.34
OH	0.43	0.45	—
F	0.32	0.32	0.33
OC_6H_5	0.31	—	0.27
OCF_3	0.18	—	0.18
Br	0.16	0.16	0.16
SH	0.15	0.17	—

[a] Exptl. error $= \pm 0.01$.

inert substituents, F, Br, and OCF_3, are included for comparison. Although the indicated solvent effects on the σ_R^0 value obtained from the shielding parameters may not be precisely applicable [as previously noted (6) for the inductive parameters, σ_I] to other systems, it is very probable that they do establish expected general trends and magnitudes.

The σ_R^0 values of Table A6.VI may be combined with σ_I values of the previous paper (6) to obtain σ^0 values. These values are potentially applicable via the modified Hammett equation, $\log (k/k_0) = \sigma^0\rho$, to the side-chain reactivities of benzene which are uncomplicated by the effects of direct interaction structures (between substituent and side-chain function) and polarization effects (7b, 7c).

$$\sigma^0(meta) = \sigma_I + 0.50\sigma_R^0$$
$$\sigma^0(para) = \sigma_I + \sigma_R^0$$

A discussion of the relationship of σ_R^0 and structure is presented in the following sections.

2. π-Electron Charge Densities

Approximate localized MO π-electron density changes produced by the various *para* substituents may be estimated from the σ_R^0 values given in Table A6.VI through relationships derived previously (5a, 5b) between charge density changes and the quantity $\int_H^{p-X} - \int_H^{m-X}$. The resulting relationships are

$$\text{for the fluorine atom, } \Delta q_{(F)} = 0.025\sigma_R^0 \qquad (2)$$

$$\text{for the attached (\textit{para}) carbon atom, } \Delta q_{(C)} = 0.133\sigma_R^0$$

where $\Delta q_{(F)}$ is the difference in the π-electron charge density of the fluorine atom of fluorobenzene and that for the *para*-substituted fluorobenzene, and $\Delta q_{(C)}$ is the corresponding difference in the π-electron charge density at the aromatic carbon to which the fluorine atom is bonded. These relationships, in addition to other uncertainties, are based upon the assumption of a fixed ΔE parameter (5a) for all *para*-substituted fluorobenzenes, and consequently must be accepted with reserve (cf., however, subsequent discussion).

We have made (at the suggestion of Professor Lionel Goodman) an independent test of eq. (2), which is based upon the steric inhibition of resonance effect of a 2-methyl substituent on the *para*-dimethylamino substituent (27a). In cyclohexane solution, the shielding parameter for 3-methyl-4-dimethylaminofluorobenzene relative to internal *meta*-methylfluorobenzene is $+7.05$ ppm (we are indebted to Mr. John Carten for this determination). From Table A6.III the normal *para*-dimethylamino shielding effect is found to be $+15.90$ ppm, indicating that the 2-methyl substituent produces a 56% steric inhibition of the resonance effect (note that $\int_H^{p-N(CH_3)_2}$ is essentially equal to $\int_H^{p-N(CH_3)_2} - \int_H^{m-N(CH_3)_2}$; cf. Table A6.VIII). This is essentially the same figure as reported earlier (27b) based upon σ_R values and upon extinction coefficients for 3-methyl-4-dimethylaminoethyl benzoate.

From eq. (1), one obtains $\sigma_R^0 = -0.24$ for the twisted *para*-dimethylamino substituent. From eq. (2) then we obtain for the *para* carbon atom a π-electron charge density 0.040 greater in the *para*-dimethylaminofluorobenzene. This is essentially the same figure (0.046) obtained by Lauterbur (27c) from the ^{13}C NMR shieldings of

TABLE A6.VIII
Typical Fit of Data to Eq. (3) for UAFPD Substituents

UAFPD substituent	σ_I	F NMR Shielding Parameters f_H^{p-x} ppm Calc.	Obsd.	f_H^{m-x} ppm Calc.	Obsd.	C-1 ioniz. ArOH, H_2O, 25° $\log(K^m/K_0)$ Calc.	Obsd.	A-15 sapon. rate, $ArCO_2Et$, aq. acetone, 25° $\log(k^m/k_0)$ Calc.	Obsd.	A-13 ion-pair formation, $ArCO_2H$, DPG, benzene, 25° $\log(K^m/K_0)$ Calc.	Obsd.	$\log(K^p/K_0)$ Calc.	Obsd.	Ioniz. 6X-quinolines, H_2O, 25° $\log(K/K_0)$ Calc.	Obsd.
O^-	-0.15[a]	19.4	19.3[a]	$+1.7$	$+1.6$[a]	-0.31	—	-0.83	—	-0.90	—	-1.57	—	-1.54	—
NMe_2	$+0.05$	15.6	15.8	$+0.2$	-0.1	$+0.03$	$+0.10$	-0.36	-0.35	-0.42	-0.37	-0.98	—	-0.88	—
NH_2	$+0.10$	14.6	14.1	-0.1	$+0.3$	$+0.11$	$+0.11$	-0.25	-0.24	-0.30	-0.33	-0.84	-0.81	-0.70	-0.69
$NHNH_2$	$+0.15$	13.7	13.1[a]	-0.5	-0.4	$+0.20$	—	-0.13	—	-0.18	—	-0.69	—	-0.53	—
OCH_3	$+0.26$	11.6	11.6	-1.3	-1.2	$+0.38$	$+0.35$	$+0.13$	$+0.13$	$+0.09$	$+0.12$	-0.36	-0.34	-0.16	—
OH	$+0.27$	11.4	11.4	-1.3	-1.2	$+0.40$	$+0.42$	$+0.15$	—	$+0.11$	$+0.09$	-0.34	-0.32	-0.13	-0.24
F	$+0.51$	6.8	6.8	-3.0	-3.0	$+0.80$	$+0.79$	$+0.71$	$+0.72$	$+0.69$	$+0.68$	$+0.37$	$+0.35$	$+0.68$	$+0.69$
C-1		-19.10		-7.10		$+1.69$		$+2.32$		$+2.40$		$+2.94$		$+3.36$	
C-2		$+16.55$		$+0.60$		-0.06		-0.48		-0.54		-1.13		-1.04	

[a] In 75% (vol.) aq. CH_3OH.

the *para* carbon atoms in *N*,*N*-dimethylaniline and *N*,*N*-dimethyl-*o*-toluidine and is in satisfactory agreement with the theoretically calculated value of McRae and Goodman (27d) of 0.031. Also in quite satisfactory agreement are $\Delta q_{(C)}$ values calculated from eq. (2) for the *p*-NH$_2$ and *p*-N(CH$_3$)$_2$ substituents and corresponding values listed by Lauterbur (27c) obtained from ^{13}C NMR shieldings and from theoretical calculations.

From the largest $-\sigma_R^0$ value (0.60 for O$^-$) of Table A6.VI, we find the greatest increase in fluorine atom π-charge density is on the order of 0.02 electron, supporting the earlier assumption that the substituted and unsubstituted compounds present to the solvent nearly identical fluorine atoms. The *para* carbon atom of the phenoxide ion (in aq. CH$_3$OH) has an increased π-electron density on the order of 0.1 electron, sufficient to account for the highly activating influence of the –O$^-$ substituent in electrophilic substitution reactions at the *ortho* and *para* positions.

Equation (2) provides a basis for direct association of the reactivity resonance effect parameter, σ_R^0, with the π-charge density localized at the *para*-carbon atom in the ground electronic state of a benzene derivative [of the σ^0 type (7b)]. Since σ_R^0 parameters came from measurements of chemical equilibria, a linear relationship (or an approximate one) between the localized *para*-carbon atom π-charge density and the change in this charge density between initial and final electronic ground states (of the σ^0 type) is implied. Equation (2) serves as an empirical relationship which demonstrates that the molecular orbital theory concept of localized π-charge density is a useful one in the interpretation of observed effects of *para* substituents on chemical reactivities.

3. Regarding σ_I–σ_R Relationships

The inductive parameter σ_I and the resonance parameter σ_R have been proposed as independent parameters, i.e., parameters which *in general* are not directly related to one another (7c). Recently, Bauld (28a) and McDaniel (28b) have made limited comparisons of σ_I and σ_R and have suggested that linear relationships exist between these parameters for all substituents having a common first atom.

Since σ_I is linearly related to $\int_H^{m-X}$ and σ_R^0 is linearly related to $\int_H^{p-X} - \int_H^{m-X}$ (for –R substituents), plots of $\int_H^{p-X}$ vs. $\int_H^{m-X}$ are highly instructive regarding σ_I–σ_R relationships (7b, 7c). In Figure A6.1 such a plot has been constructed from all of the available data in cyclohexane

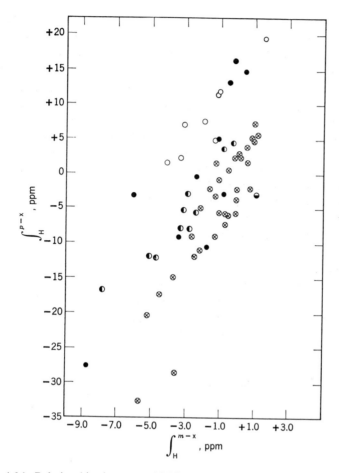

Fig. A6.1. Relationship between shielding parameters for *meta*- and *para*-substituted fluorobenzenes. First atom of substituent is as designated: ($\bigcirc$) oxygen, ($\pmb{\Phi}$) sulfur, ($\bullet$) nitrogen, ($\otimes$) carbon, ($\pmb{\ominus}$) boron.

solution (a few data in other solvents for substituents not soluble in cyclohexane are also included).

It is abundantly clear from Figure A6.1 that *in general* there is no direct relationship between σ_I and σ_R, even for substituents having a common first atom (designated by the set of distinguishing symbols listed under Figure A6.1). This result holds for both $-R$ and $+R$ substituents. Rather similar scatter patterns are obtained in plots of substituent effects on aromatic series reactivities, i.e., plots of log (k/k_0) vs. σ_I; cf., for example, Figure 3, ref. 29. It will be noted from Figure A6.1, however, that for substituents with a common first atom there are generally rough trends for σ_R (or $-\int_H^{p-X}$) to increase (become more positive) as σ_I (or $-\int_H^{m-X}$) increases. In general, such trends are very crude, however, and frequent exceptions are to be noted. For example, σ_I is substantially greater for the NO_2 group [normal value (6), $+0.56$] than for the NO group [normal value (6), $+0.34$], while $\bar{\sigma}_R$ is appreciably greater for NO (in cyclohexane solution, $+0.30$) than for NO_2 [in cyclohexane solution, $+0.20$ (30)]. Several additional examples of pairs of substituents which show this inverse order behavior include: OC_6H_5 and $OCOCH_3$; $SOCH_3$ and SO_2NH_2; C_6H_5 and $CH{=}CH_2$; CF_3 and CH_3CO; CN and COF; NCS and $N{=}NC_6H_5$.

If one restricts comparison to series of closely related structures, relatively precise linear relationships between $\int_H^{p-X}$ and $\int_H^{m-X}$ (and therefore between σ_R and σ_I or between σ_p and σ_m) do appear. Five such series are illustrated in Figures A6.2–A6.4. Figure A6.2 gives the plot of $\int_H^{p-X}$ vs. $\int_H^{m-X}$ for substituents of the formulas COX and SO_2X, both of which generate (separately) relatively precise linear relationships. Points for $SOCH_3$ and SF_5 are included to illustrate that these structures do not qualify for the SO_2X line. Similarly it is illustrated in Figure A6.2

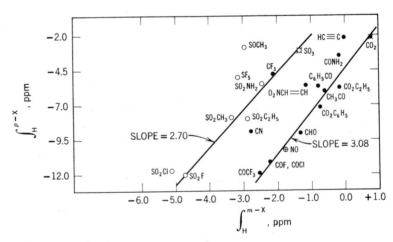

Fig. A6.2. Linear $\int_H^{p-X}$ vs. $\int_H^{m-X}$ relationships for $-COX$ and $-SO_2X$ substituents. (●) Carbon, (○) sulfur, (⌀) denotes 74% (vol.) aq. CH_3OH.

that the points for CF_3, $HC{\equiv}C$, $C{\equiv}N$, and $(NO_2)CH{=}CH$ do not follow the linear relationship defined by COX substituents. Relatively wide variation in the structure of X is apparently permitted, the approximate relationships encompassing $X = O^-$, NH_2, OC_2H_5, C_6H_5, CH_3, OC_6H_5, H, F, Cl, and CF_3.

Figure A6.3 illustrates linear relationships between $\int_H^{p-X}$ and $\int_H^{m-X}$ for substituents of the formula CH_2X and, also for an especially interesting series constituted by the substituents O^-, $N(CH_3)_2$, $NHNH_2$, OH, OCH_3, and F. Included in the latter plot are unpublished data of Mr. M. G. Schwartz for the phenols in 1% CCl_4 solutions in the essentially unassociated form (6) and as 1:1 complexes with several bases (as indicated). A point for the OCH_3 group complexed with trifluoroacetic acid (in this solvent) is also included. Whereas the first three series of linear $\int_H^{p-X} - \int_H^{m-X}$ (or $\sigma_R-\sigma_I$) relationships do involve substituents with a common first atom, this last series illustrates the fact that this feature is not a necessary condition for such a relationship. (The fact that the NO substituent fits the line for COX substituents is a further illustration.) It is of further interest to note in Figure A6.3 that the substituents OC_6H_5, $OCOCH_3$, $NHCOCH_3$, OCF_3, CH_3, $CH{=}CH_2$ SCH_2, Br, and H, for example, do not qualify for this linear relationship.

The substituents which do qualify for the linear $\sigma_R^0-\sigma_I$ relationship include only relatively simple structures in which the first atom of the substituent is from the first row of the periodic table (thus excluding, for example Cl, Br, I, SH, and SCH_3). Further, this first atom bears at least one unshared electron pair (thus excluding, for example, H, CH_3, and $CH{=}CH_2$) in simple π (p-p) conjugation with the benzene ring. This interaction must also be essentially unfettered by further conjugative interactions with the groups bonded to this atom (i.e., the other atoms of groups attached to the first atom are essentially active only by their inductive effect, thus excluding, for example, the substituents OC_6H_5, $OCOCH_3$, $NHCOCH_3$, and OCF_3). Quite plainly the substituents which do qualify are those for which the united atom approximation of the substituent in $\pi(p-p)$ conjugation with the ring is most appropriate.

It has been previously demonstrated that the π-electron shielding effect for this simple class of *para* substituent is directly related for corresponding X and COX substituents (5b). This relationship is anticipated by HMO theory using the united atom model (5b) (the relationship shows the same structural restrictions given above, which is more restrictive than for the $\int_H^{p-X}$ versus $\int_H^{m-X}$ relationship for COX

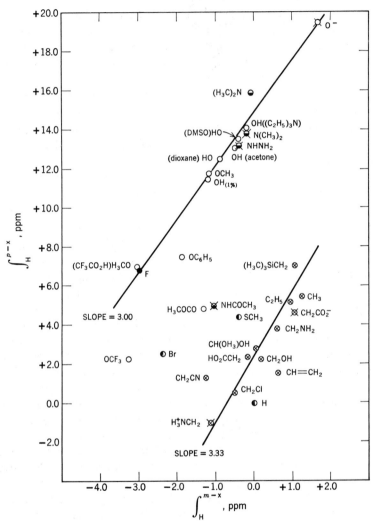

Fig. A6.3. Linear $\int_H^{p-X}$ vs. $\int_H^{m-X}$ relationships for UAFPD substituents and for —CH$_2$X substituents. (●) Fluorine, (○) oxygen, (◖) nitrogen, (⊗) carbon, (◗) others, (◯◯) denotes 75% (vol.) aq. CH$_3$OH.

given in Figure A6.2. Employing a similar model, Ehrenson (31) has shown the HMO π-electron energy changes which are presumably involved in the resonance effects of these substituents in σ⁰ chemical reactivities are directly related to the effective coulomb integral of the united atom substituent. It seems abundantly clear for this very special class of substituent that both the inductive and resonance effect parameters σ_I and σ_R^0 are directly controlled by the effective nuclear charge of the united atom-like substituent. Because of the importance and apparently well defined character of this class of substituent, we propose for them the designation UAFPD (united atomlike first row pair donor) substituents.

The ensuing linearity between σ_I and σ_R^0 ($\sigma_R^0 = 0.40\sigma_I - 0.54$) for the UAFPD substituents has important consequences. Consider the very general system X–A–Y, whose appropriate chemical or physical property, P, is measured at Y (X is the UAFPD substituent, A is an intervening molecular cavity, of essentially fixed structure, and Y is some functionality). To the approximation that inductive and resonance effects originating at the X–A bond and transmitted inductively at the A–Y bond can be given by the relationship (32)

$$P = \sigma_I\rho_I + \sigma_R^0\rho_R^0$$

(where ρ_I and ρ_R are series constants which are dependent upon A and conditions of solvent and temperature and which measure the relative effectiveness of propagation of the inductive and resonance effects of X, respectively, and σ_I and σ_R^0 are the substituent parameters for X), the property P becomes a function of a single substituent parameter, i.e.

$$\text{(for UAFPD substituents) } P = C_1\sigma_I + C_2 \tag{3}$$

where C_1 and C_2 are constants.

Equation (3) has previously been demonstrated (7c,33) for saturated A systems, in which case it is not limited to only the UAFPD substituents. Table A6.VIII lists some typical reactivities (also $\int_H^{p-X}$ and $\int_H^{m-X}$ values) illustrating the applicability of eq. (3) to systems in which A is unsaturated. The listed reactions are designated by the numbering system employed previously by Taft and Lewis (34). Although eq. (3) is of limited utility (for A unsaturated) because of the relatively few available UAFPD substituents, it is clear from Table A6.VIII that when it is applicable the relationship holds with very high precision (higher, for example, than the Hammett equation, in the original or modified forms). The precision of eq. (3) is all the more remarkable when it is recognized that although the Hammett equation is also a single substituent parameter equation, different values of the substituent parameter are required for the *meta* and *para* position of benzene. The substituent parameter in eq. (3) depends only upon the character of the substituent and is apparently independent of position of substitution, or of the nature of the molecular cavity A (with the limitation noted below).

Equation (3) can apparently be used as a moderately critical diagnostic tool for detection of the effects of direct conjugative interaction between X and Y (A permitting). Plots of log (k/k_0) vs. σ_I for the UAFPD substituents are distinctly nonlinear curves [i.e., log (k/k_0) is a nonlinear function of σ_I] for the following reactivities (which in each instance include data for N, O, and F substituents); for example: for *para* substituents, A-1, A-14, A-15, A-16, A-19, D-1, D-10, E-3; for *meta* substituents (34), E-3; for the 7 position, the ionization constants of quinolines (35), and the rates of methoxy-dechlorination of 4-chloroquinolines (36); for the 2 position, the rates of sodium borohydride reduction of fluorenone (37). For each of these reaction series, direct interaction valence bond forms (between X and Y) can be written and effects of the observed reactivities due to this charge delocalization interaction are anticipated from the nature of the reactivities involved.

Although the number of UAFPD substituents is quite limited, several additional substituents which are presumably of this class could conceivably be investigated under especially favorable circumstances, including $^-:CH_2$, $^-:NH$, $NHOCH_3$, $N(CH_3)F$, OCl, etc.

Figure A6.3 (or A6.1 and A6.2) is unique in regard to the structural distinctions which it makes. The familiar inductive order of charge withdrawal $O^- < H, N(CH_3)_2 < SCH_3 < OCH_3 < Br < F$ is clearly demonstrated by the $\int_H^{m-X}$ values (for the latter four substituents $\int_H^{m-X}$ is substantially negative while corresponding $\int_H^{p-X}$ values are substantially positive). Yet, in the linear $\int_H^{p-X} - \int_H^{m-X}$ relationships it is quite clear that it is the H atom, the conventional standard substituent, which occupies an especially unique position. Perhaps no previous measurements of substituent effects have so clearly placed the hydrogen substituent in a position appropriate to its truly unique structure.

Figure A6.4 illustrates a linear $\int_H^{p-X} - \int_H^{m-X}$ relationship for "triple bond" substituents. Again it is clear that the linear relationship is not limited to substituents with a common first atom.

It is not possible at present to specify the detailed structural features which give rise to linear $\sigma_R - \sigma_I$ relationships. Clearly, the nuclear charge of the first atom of the substituent can be varied without destruction of these relationships. It appears that a nearly fixed hybridization state of the first atom of the substituent is a necessary

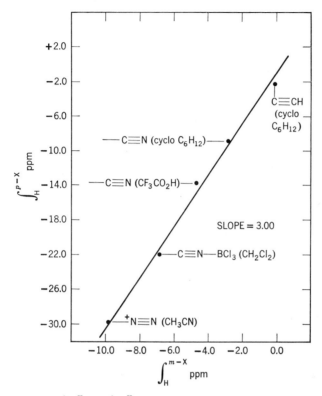

Fig. A6.4. Linear $\int_H^{p-X}$ vs. $\int_H^{m-X}$ relationship for "triple bond" substituents; solvent indicated.

(but not a sufficient) condition. In addition, there must be a fixed number of delocalized electrons and atomic positions within the substituent. If this former criterion is correct, it would mean a nearly common hybridization state for the UAFPD substituents. It should be noted in this connection that data for the NH_2 substituent have not been included (to prevent crowding) in Figure A6.3. The data for this substituent including the unpublished results of Mr. M. G. Schwartz for the aniline complexed with dimethylacetamide in 1% CCl_4 solution define a line of essentially parallel slope to that shown in Figure A6.3 but apparently displaced by a small amount (1.8 ppm) to the left. This displacement is unexplained, although conceivably it could represent the effect of some small difference in the necessary conditions given above.

As a further criterion of the fundamental character of all five of the linear $\int_H^{p-X} - \int_H^{m-X}$ relationships illustrated in Figures A6.2–A6.4, it is significant that essentially the same slope, 2.70–3.33 (with a mean value of 3.0), is obtained.

The nearly identical slopes may be used as an argument that the ΔE "excitation energy" parameter (5a) must be approximately constant for most *meta*- and *para*-substituted fluorobenzenes. For example, if ΔE were constant for only a given family of closely related substituents in either the *meta* or *para* position, then linear $\int_H^{p-X} - \int_H^{m-X}$ relationships of essentially the same slope would not result (as actually observed) among families of substituents of widely different structure, but rather lines of variable slope would be observed. It is, of course, conceivable that deviations from the linear $\int_H^{p-X} - \int_H^{m-X}$ relationships may arise from variable ΔE, but in view of the nature of structures which do and do not follow the linear relationships this does not appear to be generally the case. The matter, however, does need further theoretical analysis.

For the CH_2X substituents, the implied linear $\sigma_R^0-\sigma_I$ relationship leading to the applicability of eq. (3) has been critically tested by Exner (38). The ionization constants for an extensive series of

$$XCH_2-\langle\bigcirc\rangle-CO_2H$$

acids were shown to follow eq. (3).

Our solvent effect studies provide additional evidence that linear $\int_H^{p-X} - \int_H^{m-X}$ relationships are not general. Figure A6.5 gives a plot of $\int_H^{p-NO_2}$ vs. $\int_H^{m-NO_2}$ in a wide variety of solvents. Similar results have been obtained with most $+R$ substituents. The unlabeled points, which define a vertical line (a trivial $\int_H^{p-X} - \int_H^{m-X}$ relationship) to the precision (essentially the experimental error) indicated by the line breadth shown, correspond to nonprotonic solvents of varying polarity (with increasing polarity $\int_H^{p-NO_2}$ decreases but $\int_H^{m-X}$ is unaffected). It will be recognized that all of the labeled solvents which lie to the right of this line are protonic solvents which hydrogen bond with the nitro group (6). For these, a very crude linear $\int_H^{p-NO_2}$ vs. $\int_H^{m-NO_2}$ trend does occur. The solvents dioxane, tetrahydrofuran, pyridine, and dimethyl sulfoxide, which lie to the left of the line, are presumably complexed by Lewis acid bonding (6) to the nitro group.

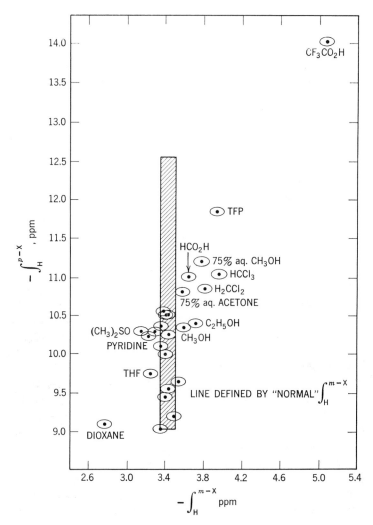

Fig. A6.5. Solvent effects on shielding parameters for *meta*- and *para*-nitro-fluorobenzene.

4. Qualitative Interpretation of Solvent Effects

The results of the present and the previous study of solvents effects (6) can apparently be interpreted conceptually in terms of the solvent susceptibility of the contribution of various (physically insignificant) valence bond structures to the ground (hybrid) electronic state. The

very slight (if any) effect of solvent polarity (excluding effects of hydrogen bonding and Lewis acid bonding) on the intramolecular shielding
effect of most *meta* substituents indicates that there is little or no
practical change with solvent polarity in the contribution to the ground
electronic state of the right hand ionic valence bond structure such as

Solvent variation from hydrocarbons to *N*-methylformamide,
encompassing a twofold variation in the bulk dielectric constant of the
solvent, is accompanied by no practical change in intramolecular
shielding of fluorine (6). Comparison of the shielding parameters for
various *meta* substituents, as well as other properties such as dipole
moments and chemical reactivities, leaves no doubt that the ionic
structures do make important contributions to the ground electronic
state. The contribution simply shows no practical solvent dependence
at room temperature.

Similarly, the results for $-R$ *para*-substituted fluorobenzenes
(excluding again specific solvent–substituent hydrogen and Lewis acid
bonding interactions) indicate that with similar gross changes in solvent
polarity there is at room temperature little practical change in the
contribution to the normal state of the right-hand ionic valence bond
structures, such as

This conclusion follows from the fact that only relatively small changes in the intramolecular shieldings of $-R$ para substituents with substantial changes in solvent polarity are observed [the effects in general are only slightly beyond the experimental error (39)]. Again, this result cannot be attributed to the fact that the ionic structures make no measurable contributions to the ground electronic states. Comparison of the shielding parameters for various para-substituted fluorobenzenes (Table A6.VI) leaves no doubt that the substantial variations in shielding observed can be interpreted in terms of small but measurable contributions to the ground electronic state from the dipolar forms (5).

With the sole exception of the p-Si(CH$_3$)$_3$ substituent, all of the various $+R$ (electron acceptor) para-substituted fluorobenzenes investigated have shielding parameters which are increasingly shifted to lower field strengths with increasing solvent polarity. The present study offers a rigorous confirmation of the early communication of this result (40). The measurable variation with solvent of the contribution to the normal state of valence bond structures such as the following is indicated. Further, the magnitude of this effect is such as to indicate in general quite practical consequences.

This interpretation is made more certain when it is borne in mind that the shielding parameters for the *meta* isomers (for which such structures cannot be written) are generally solvent insensitive (cf. Fig. A6.5). The ability of solvent to support the charge-separated direct-interaction paraquinoid forms will be considered in greater detail in a subsequent publication.

The only structural feature in common among the $+R$ para substituents for which shielding parameters are substantially shifted to lower field strengths by polar solvents is the fact that the dipolar direct-interaction resonance forms involve atomic centers of opposite formal charge *both* of which are on the periphery of the molecule. The dipolar resonance forms for $-$R *para* substituents do not meet this requirement. Instead, in these structures at least one of the centers of formal charge is buried within the molecular cavity (i.e., is not on an end atom of the molecule). A similar situation prevails for the dipolar forms given above for *meta*-substituted fluorobenzenes (with perhaps the special exception of the hyperconjugative form for the CF_3 substituent).

It is apparently not a sufficient condition for the observation of such solvent effects that the substituent be a $+$R *para* substituent. The point is well illustrated by the p-$Si(CH_3)_3$ substituent. The shielding parameter for this substituent is negative (that for the *meta* isomer is positive) thus classifying it (consistent with reactivity and dipolement results) (41) as a $+$R substituent. There is, however, no measurable solvent dependence of the shielding parameter for this substituent (cf. Table A6.IV). Inspection of the dipolar direct-interaction resonance form readily confirms the fact that this form does not meet our necessary condition since the formal charge on the Si atom is buried within the molecular cavity.

These results on the relative susceptibilities of the various valence bond forms to polar solvent effects are qualitatively consistent with the Kirkwood model of electrostatic solvation energy of a dipole in a spherical cavity (42). The contribution of a dipolar valence bond form to the resonance hybrid will be enhanced by the lowering of its energy state (relative to that of the other contributing structures) through solvation. The solvation energy given by the Kirkwood model is

$$\frac{1}{2}\left(\frac{D_i - D}{2D + D_i}\right)\left(\frac{\mu^2}{b^3}\right)$$

where D_i is the internal dielectric constant of the cavity, D is the dielectric constant of the bulk solvent, μ is the dipole moment, and b is the radius of the molecular cavity. For a given increase in dielectric constant of the medium, the greater the distance of separation of the charges in the dipolar resonance form relative to the diameter of the molecular cavity, the greater will be the expected lowering of the energy state of this form and therefore the greater the unshielding of the fluorine atom. This corresponds qualitatively to the observations. Also, if essentially identical dipolar centers are kept at the end atoms and the diameter of the molecular cavity is increased (μ^2/b^3 decreases), the solvation energy of the dipolar form will be decreased. Essentially such a situation prevails in the comparison of the direct interaction forms of *para*-fluoronitrobenzene and *para*-fluorophenyl-*trans*-nitroethylene

In accord with the expected weaker solvation of the latter form, solvent effects on $\int_H^{p-X}$ are appreciably smaller (on either an absolute or a percentage basis) for the latter than the former compound (cf. Table A6.II).

Although the Kirkwood model accounts qualitatively for the observed solvent effects, the shielding parameter data indicate that the bulk dielectric constant is an inadequate *quantitative* measure of the ability of the solvent to support charge separations in solute molecules. A more detailed consideration of this subject will be presented in a subsequent paper.

5. Applications to Reactivities

The important consequences on reactivities (7c,43), barriers to rotation (44a), and infrared frequencies (44b) of increasing the contribution of dipolar direct interaction forms to the ground electronic state in solvents of increasing polarity have been previously pointed out. We wish to note a further example of a reactivity effect, which apparently has not been previously recognized. The effects of select *meta* substituent on the rate of reaction of aniline with benzoyl chloride in benzene (45), 25°, follows the Hammett equation ($\rho = -3.20$) with excellent precision:

meta Substituent	log $(k/k^0)_{obsd}$	$\sigma\rho$
CH_3	$+0.27$	$+0.22$
Cl	-1.23	-1.18
NO_2	-2.20	-2.24

However, for the p-NO_2 substituent $\sigma\rho = -4.06$, whereas log $(k/k_0)_{obsd} = -3.25$. The discrepancy can be attributed to the fact that

$\sigma = +1.27$ for the $p\text{-NO}_2$ is applicable only to aqueous solutions (46). In benzene solution the direct interaction form

contributes less to the normal state of the reactant molecule than in aqueous solutions. This leads therefore to a reduced $\bar{\sigma}$ value of $+1.02$, applicable in this solvent. The resulting solvent effect on reactivity amounts to the appreciable decrease in free energy of activation (in benzene relative to water) of approximately 1.0 kcal/mole.

Equation (1) may be rearranged to the following form, which may be utilized to obtain the effective $\bar{\sigma}_R$ values for the $+R$ substituents in various solvents from the shielding parameters listed in Tables A6.II, A6.IV, and A6.V of this paper and Tables A6.II–A6.IX of ref. 6.

$$\bar{\sigma}_R = (-0.0339)\left(\int_H^{p-X} - \int_H^{m-X}\right) \qquad (4)$$

The solvent effects on $\bar{\sigma}_R$ values obtained from eq. (4) may be taken to indicate the expected magnitudes of solvent effects on the reactivity effects of these substituents, but the precise quantative applicability of these substituent parameters is questionable.

In a subsequent paper linear relationships generated by plots of $\int_H^{p-X}$ vs. $\int_H^{p-X}$ for pairs of $+R$ substituents (from the data of Tables A6.II and A6.IV) are considered in detail. The effect of solvent on the contribution of dipolar direct interaction forms is used as a model to obtain a generalized scale of solvent polarity; i.e., $\int_H^{p-X}$ values for $+R$ substituents are utilized as measures of the ability of solvent to support charge separation in simple solute molecules. Applications to the effect of solvent on the rates of ion-forming reactions are made.

References and Footnotes to Appendix 6

1. This work was supported in part by the National Science Foundation.
2. Deceased; NSF Cooperative Graduate Fellow, 1959–1960. Taken in part from the Ph.D. Thesis of Irwin R. Fox, Pennsylvania State University, August, 1961.
3. Public Health Service Postdoctoral Research Fellow, 1959–1960.
4. M. Karplus and T. P. Das, *J. Chem. Phys.*, *34*, 1683 (1961).
5. (a) F. Prosser and L. Goodman, *J. Chem. Phys.*, *38*, 374 (1963); (b) R. W. Taft, F. Prosser, L. Goodman, and G. T. Davis, *ibid.*, *38*, 380 (1963); (c) P. C. Lauterbur, *Tetrahedron Letters*, *8*, 274 (1961); (d) H. Spiesecke and W. G.

Schneider, *J. Chem. Phys.*, *35*, 731 (1961); (e) K. Ito, K. Inukai, and Isobe, *Bull. Chem. Soc. Japan.*, *33*, 315 (1960).

6. R. W. Taft, E. Price, I. R. Fox, I. C. Lewis, K. K. Andersen, and G. T. Davis, *J. Am. Chem. Soc.*, *85*, 709 (1963).

7. (a) H. S. Gutowsky, D. W. McCall, B. R. McGarvey, and L. H. Meyer, *J. Am. Chem. Soc.*, *74*, 4809 (1952); (b) R. W. Taft, S. Ehrenson, I. C. Lewis, and R. E. Glick, *ibid.*, *81*, 5253 (1959); (c) R. W. Taft, *J. Phys. Chem.*, *64*, 1805 (1960).

8. (a) I. R. Fox, P. L. Levins, and R. W. Taft, *Tetrahedron Letters*, *7*, 249 (1961); (b) R. W. Taft and P. L. Levins, *Anal. Chem.*, *34*, 436 (1962).

27a. B. M. Wepster, *Progress in Stereochemistry*, Vol. 2, W. Klyne and P. D. D. de la Mare, Eds., Butterworths, London, 1958.

27b. R. W. Taft and H. D. Evans, *J. Chem. Phys.*, *27*, 1427 (1957).

27c. P. C. Lauterbur, *J. Chem. Phys.*, *38*, 1415, 1432 (1963).

27d. E. G. McRae and L. Goodman, *J. Chem. Phys.*, *29*, 334 (1958).

28. (a) N. L. Bauld, Abstracts, 139th National Meeting, American Chemical Society, March, 1961; (b) D. H. McDaniel, *J. Org. Chem.*, *26*, 4692 (1961).

29. R. W. Taft and I. C. Lewis, *Tetrahedron*, *5*, 210 (1959).

30. The larger σ_I value for NO_2 is in accord with the more positive nitrogen atom of this substituent and the predominant role of the first atom in determining the σ_I parameter for the substituent as a whole (6). In spite of the larger electron-attracting power of the first atom, σ_R for NO_2 is dramatically less that that for NO. This is apparently to be attributed to the effect of electron repulsion, which is unfavorable to the conjugative accumulation of negative charge on the two oxygen atoms of the NO_2 group. The NO substituent is unencumbered by such a structural feature. It is apparent from the discussion which follows that among structurally similar substituents such factors either do not enter or adjustments within the series occur so that linear $\sigma_I - \sigma_R$ relationships appear. In fact, it may be anticipated that the same structural features which lead to the decay of the linear $\int_H^{p-X} - \int_H^{m-X}$ relationships of the following discussion are also responsible for limitations of direct linear free energy relationships. That is, these structural changes produce reactivity effects which require for description two or more substituent parameters, e.g., σ_I and σ_R^0 or σ_R^+, etc., rather than the single parameter, σ_I.

31. S. Ehrenson, unpublished results referred to in ref. 7c.

32. R. W. Taft, *J. Am. Chem. Soc.*, *79*, 1045 (1957).

33. (a) J. D. Roberts and W. T. Moreland, Jr., *J. Am. Chem. Soc.*, *75*, 2167 (1953); (b) S. Siegel and J. M. Komarmy, *ibid.*, *82*, 4547 (1960); (c) H. Kwart and L. J. Miller, *ibid.*, *83*, 4552 (1961).

34. R. W. Taft, I. R. Fox, and I. C. Lewis, *J. Am. Chem. Soc.*, *83*, 3349 (1961).

35. S. B. Knight, R. H. Wallick, and J. Bowen, *J. Am. Chem. Soc.*, *76*, 3780 (1954); S. B. Knight, R. H. Wallick, and C. Balch, *ibid.*, *77*, 2577 (1955).

36. E. Baciocchi, G. Illuminati, and G. Marino, *J. Am. Chem. Soc.*, *80*, 2270 (1958).

37. G. G. Smith and R. P. Bayer, *Tetrahedron*, *18*, 323 (1962).

38. O. Exner and J. Jones, *Collection Czech. Chem. Commun.*, *27*, 2296 (1962).

39. The small decreases observed in $\int_H^{p-f}$ (and other) values between hydrocarbon and highly polar solvents (cf. Table A6.I) apparently are to be attributed

largely to a small increase in polar solvents in the contribution of resonance forms, e.g.,

of the unsubstituted fluorobenzene reference.

40. R. W. Taft, R. E. Glick, I. C. Lewis, I. Fox, and S. Ehrenson, *J. Am. Chem. Soc.*, *82*, 756 (1960).
41. Cf. R. W. Taft, *Steric Effects in Organic Chemistry*, M. S. Newman, Ed., Wiley, New York, 1956, p. 596.
42. J. G. Kirkwood, *J. Chem. Phys.*, *2*, 351 (1934).
43. (a) B. Gutbezahl and E. Grunwald, *J. Am. Chem. Soc.*, *75*, 559 (1953); (b) M. M. Davis and H. B. Hetzer, *J. Res. Natl. Bur. Std.*, *60*, 569 (1958).
44. (a) J. C. Woodrey and M. T. Rogers, *J. Am. Chem. Soc.*, *84*, 13 (1962); (b) L. J. Bellamy et al., *Trans. Faraday Soc.*, *1959*, 1677.
45. F. J. Stubbs and C. Hinshelwood, *J. Chem. Soc.*, *1949*, 571.
46. L. P. Hammett, *Physical Organic Chemistry*, McGraw-Hill, New York, 1940, p. 188.

Chemistry of Radical-Ions

M. Szwarc

*Department of Chemistry, State University College of Forestry
at Syracuse University, Syracuse, New York*

I. RADICAL-IONS

A suitable molecule A may accept an extra electron and form a new species which we shall denote by $A^{\cdot-}$. The negative sign appearing in this symbol indicates the presence of a negative charge characteristic for an ion, and the dot indicates that the new species possesses an odd number of electrons and therefore a radical nature. Consequently, the term radical-anion seems to be appropriate for such an entity.

A reverse process, ionization through which a molecule A is deprived of one of its electrons, produces a positive radical-ion. The symbol $A^{\cdot+}$ is proposed for such an entity. Positive radical-ions are observed in mass spectrographs, in which case they are referred to as the parent ions of the analyzed gas. They are usually short lived, because the vertical electron ejection process leaves them in vibrationally excited states. A more gentle removal of an electron from a neutral molecule, e.g., through a suitable chemical or electrochemical reaction, yields positive radical-ions in their ground state. Such positive radical-ions are intrinsically stable, although their binding energy may be negative if they can decompose into fragments which gain additional stability, say by delocalization of their electrons.

A relatively stable negative radical-ion may be formed if the parent molecule possesses a sufficiently low-lying empty orbital. Otherwise, electron capture leads to dissociation. For example, in the course of an electron-transfer process H_2O or CH_3Br decompose into $H + OH^-$ and $CH_3 + Br^-$, respectively, because the interactions of the pairs $H–OH^-$ and $CH_3–Br^-$ are repulsive.

Unsaturated compounds such as aromatic hydrocarbons, some substituted olefins, dienes, substituted acetylenes, ketones, etc., are particularly suitable materials from which stable negative, as well as positive, radical-ions may be formed. Molecules of such compounds are conveniently described in terms of π orbitals, half of which are empty and, therefore, capable of accommodating additional electrons. Let us remark in passing that these empty orbitals are designated as antibonding ones. The term originates from the calculation which shows that an electron placed into such an orbital contributes less to the binding energy than an electron in a localized orbital associated with a particular atom. Nevertheless, energy levels of antibonding orbitals may lie *below* the ionization potential of $A^{\cdot-}$, i.e., the electron-capture process is then exothermic and the respective parent molecule has, therefore, a positive electron affinity.

Many radical-anions generated from unsaturated compounds are perfectly stable and, under favorable experimental conditions, may exist indefinitely. Even more stable, in respect to their decomposition, are the corresponding positive radical-ions formed from the parent molecule by removing an electron from the highest occupied π orbital.

However, the greater reactivity of radical-cations, as compared with the respective radical-anions, calls for more stringent precautions in their preparation. Their annihilation caused by interaction with the surrounding species is rather rapid, because the potential energy barriers of various reactions in which they may participate are, on the whole, lower for the positive than for the negative radical-ions.

II. REVIEW OF EARLIER STUDIES OF RADICAL-IONS

It has been known for about one hundred years that aromatic hydrocarbons may react with alkali metals. For example, in 1867 Berthelot (1) described the formation of a black addition product on fusing metallic potassium with naphthalene in a closed tube. The first real comprehension of such processes (expressed, however, in terms which differ from our modern notation) should be attributed to Schlenk. As early as 1914 he described (2) the reaction of alkali metals with anthracene in ether solution and reported the formation of two distinct compounds: a one-to-one adduct (sodium anthracene) and a two-to-one adduct (disodium anthracene). These two species were characterized by their visible spectra and by chemical analysis. Although the modern concepts of radicals and radical-ions were not yet developed in those days, Schlenk's description of sodium anthracene and of ketyls, which were extensively studied in his laboratory (28), is remarkably close to our present interpretation. Using the language of his day, Schlenk reported his findings in terms which closely correspond to our modern notation of electron-transfer processes involving carbanions, radicals, and radical-ions.

In the following years, the radical nature of such adducts was explicitly stressed, e.g., in a paper by Willstätter (3), and this interpretation was thoroughly developed, particularly by Schlenk and Bergmann (4). Unfortunately, the emphasis on the radical nature of the alkali adduct distracted the attention of the earlier workers from the ionic properties of these compounds. For example, sodium naphthalene was described by the formula which implies that its molecule is electrically neutral. Had this been the case, one should not expect a significant role of solvent in the process of its formation. However, studies of Scott and his colleagues (5) demonstrated that in some specific solvents the reaction of metallic sodium with naphthalene produced the adduct, although in others no reaction was observed. For example, the characteristic green color of naphthalenide rapidly appeared when the interaction took place in tetrahydrofuran or dimethyl ether but not in diethyl ether or benzene. Moreover, addition of benzene to the green solution of sodium naphthalene in tetrahydrofuran followed by the removal of the latter solvent through distillation led to the reverse reaction—sodium naphthalene decomposed into naphthalene and sodium dust. All these observations indicated that sodium naphthalene and, therefore, the other analogous alkali adducts had an ionic character. Indeed, Scott et al. (5) remarked that the green solution showed electric conductance, but they did not consider further the implications of their own observations.

The first objection to Willstätter's formulation of the alkali adducts was raised by Hückel and Bretschneider (6). They noted that the reduction by an alkaline earth metal, such as calcium, should closely resemble the reaction with sodium and, indeed, both metals reduce naphthalene to 1,4-dihydronapthalene in liquid ammonia. However, had the structure proposed by Willstätter been correct, the calcium adduct should be represented by the improbable structure,

and should, therefore, exhibit different properties than the sodium adduct. This is contrary to the observations. Consequently they proposed that the reduction involves an electron transfer, viz:

$$\text{Na(metal)} + \text{naphthalene} \rightleftharpoons$$

producing a heteropolar C^-,Na^+ bond. Here, for the first time, the idea of an electron-transfer process was explicitly expressed in interpreting this class of reactions.

The conclusive evidence proving the radical-anion nature of the alkali adducts was furnished by the pioneering studies of Lipkin and Weissman (7), who reported and extensively discussed the paramagnetic properties of these compounds. Moreover, examination of their ESR spectra demonstrated that the extra electron is delocalized and occupies the lowest antibonding π orbital. Numerous studies published by various workers in the following years fully confirmed these conclusions.

III. FORMATION OF NEGATIVE AND POSITIVE RADICAL-IONS

Negative radical-ions may be formed in the gas phase by electron attachment. This subject is fully discussed in the following section. The attachment of solvated electrons with formation of the respective radical-ions may be studied by radiolysis (Sec. XIII) or by photolysis (Sec. XVI). Radical-anions are also formed by reduction of suitable electron acceptors by alkali and alkali-earth metals—a subject discussed in Sec. XII. Alternatively, the reduction may be performed electro-chemically, and this subject is dealt with in Sections X and XXI. In this connection one should mention the interesting polarographic technique developed by Maki and Geske (179) which permits the formation of radical-ions in the cavity of an ESR spectrometer.

The formation of a radical-anion by electron transfer involving another radical-ion is discussed in Sec. IX, while the disproportionation producing radical-anions from dianions is discussed in Sec. XVII. Electron transfer from carbanions to acceptors forming radical-ions is reviewed in Sec. XX.

Finally, processes involving charge-transfer complexes are discussed in Sec. XXII. These form simultaneously the positive and negative radical-ions. The positive radical-ions may be also formed electrolytically (Secs. X and XXI) and by radiolysis (Sec. XIV) or photolysis (Sec. XVI). Oxidation by chemical means is discussed in Sec. XXII. This involves the reaction of electron donors with various Lewis acids.

IV. ELECTRON CAPTURE IN THE GAS PHASE AND ELECTRON AFFINITY

There are two ways in which stable negative ions may be formed in the gas phase following collision between an electron and a neutral molecule. Either the electron is captured and a negative parent radical-ion is formed, or the collision breaks up the molecule into a positive and a negative fragment. It is the former alternative which interests us.

Electron capture by a neutral molecule may be regarded as a transition taking place between two electronic levels of the negative radical-ion. In the initial state one of the electrons of the radical-ion occupies an unbound orbital, the potential energy curve of the system being identical to that of the neutral molecule in its ground state. The final state is, of course, the electronic ground state of the radical-anion. The electron is captured if its kinetic energy is equal to the potential energy of the negative radical-ion in its electronic ground state, the negative radical-ion having, however, a configuration of the neutral molecule in its lowest vibrational state. The transition then takes place and produces a vibrationally excited radical-ion. The relevant relations are clarified by inspection of the potential energy curves of diatomic molecules shown in Figure 1. In case (a) the potential energy of $AB\cdot^-$ is repulsive and electron capture results in the dissociation of AB into $A\cdot + B^-$. The capture is efficient if the kinetic energy of the electron is greater than E_2 and smaller than E_1; electrons with energies higher than E_1 or lower than E_2 will be scattered but not captured. In case (b) electrons having kinetic energies higher than E_2 but lower than E_3 form on capture $AB\cdot^-$ radical-ions which may be stabilized by subsequent collision. Denoting the

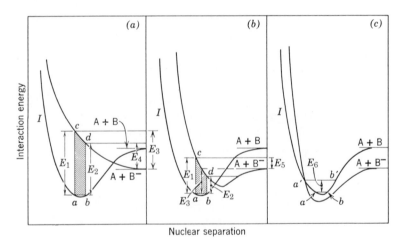

Fig. 1. Potential energy curves illustrating three possible ways in which negative ions may be formed from a molecule AB by electron capture.

relaxation time of spontaneous decomposition of a nonstabilized $AB\cdot^-$ into AB + electron by θ, and the average time between collisions by τ, one finds that the probability of formation of a stable radical-ion arising from an electron-capture process is $\breve{\omega} = \theta/(\theta + \tau)$. Since $1/\tau$ is proportional to the gas pressure, p, while θ is independent of it, $\breve{\omega} = p/(p + p')$ where p' is the pressure at which $\tau = \theta$.

Case (c) represents a situation in which only a vibrationally excited molecule AB has a high probability of electron capture. Of course, left up to itself the radical-ion will lose its extra electron through a reverse process, but on collision with a gas molecule, it may be stabilized like the radical-ion of case (b).

Recapitulating, diatomic molecules in the gas phase may capture electrons if their kinetic energies are within well specified limits. Vibrationally excited negative radical-ions are produced if the potential energy curves are favorably interrelated. These ions lose their extra electron in about 10^{-8} sec, but they become stable if a favorable collision with a gas molecule has taken place during this time. Hence, at atmospheric pressure, when $\tau \sim 10^{-9}$ to 10^{-10} sec, stable radical-ions may be formed through an electron-capture process involving suitable molecules.

Electron capture by polyatomic molecules resembles the process described for diatomic molecules. However, a polyatomic molecule has many degrees of freedom and a large number of available energy levels. Consequently, the restrictions imposed on the kinetic energy of the captured electron are greatly relaxed.

The ionization potential of a negative radical-ion is, by definition, identical with the electron affinity of its parent molecule. Hence, the latter may be determined by investigating thermal ionization of the radical-ions. Interesting studies based on this principle were reported by Becker, Wentworth, and their associates. The method utilizes an electron-capture detector, designed by Lovelock (8) for gas-chromatographic analyses. Such a detector operates at a low electric field and utilizes low energy electrons. The vapor to be investigated is mixed with a very large excess of a suitable carrier gas which flows, at atmospheric pressure, through the device. A mixture of argon containing 10% of methane is most convenient to use for this purpose. The ionization is achieved by irradiating the gas with β rays of tritium. The primary fast electrons are rapidly slowed down, their kinetic energy being reduced to the thermal level through inelastic collisions with the molecules of the carrier; and in the course of this process secondary electrons and positive ions, as well as neutral radicals, are produced. The stationary concentrations of the latter two species remain constant if the operating conditions are standardized. In particular, they are not affected by the presence of the investigated vapor, because its partial pressure is negligible when compared with that of argon and methane. On the other hand, only the molecules of the investigated vapor are capable of capturing *thermal* electrons; neither methane nor argon can undergo an electron-capture process. Therefore, the presence of the vapor reduces the concentration of thermal electrons in the plasma. The use of chromatographic column ascertains the purity of the investigated compound.

By applying a short-duration positive potential to the collecting electrode of such a device, one can measure the concentration of thermal electrons. In the argon–methane mixture the current drawn from the electrode was found to be proportional to the electron density in plasma, its value being independent of the pulse duration (0.5–5 μsec) and of the applied voltage (10–80 V).

The following model (9) was proposed to account for the events occurring within the electron-capture cell during the pulse sampling period. (*1*) The rate of formation of thermal electrons is assumed to be constant and not affected by the presence of the investigated vapor. (*2*) The pulse is assumed to remove most of the electrons and leave a large excess of positive ions, $\oplus$, and of radicals, R, in the plasma, i.e., $[\oplus] \gg [e^-]$ and $[R] \gg [e^-]$. (*3*) Thermal electrons are supposed to be lost by three reactions,

$$e^- + \oplus \longrightarrow \text{products}, \qquad k_N$$

$$e^- + R \longrightarrow R^- \qquad k_R$$

$$e^- + A \underset{k_{-1}}{\overset{k_{-1}}{\rightleftarrows}} A^{\cdot-}$$

Hence, in the absence of the investigated compound, A,

$$d[e^-]/dt = k_p R_\beta - \{k_N[\oplus] + k_R[R]\}[e^-]$$

and in its presence,

$$d[e^-]/dt = k_p R_\beta - \{k_N[\oplus] + k_R[R]\}[e^-] - k_1[A][e^-] + k_{-1}[A^{\cdot-}]$$

Here $k_p R_\beta$ is the rate of formation of thermal electrons through the action of β rays.

The stationary concentration of radical-ions, $[A^{\cdot-}]$, is determined by four processes: their formation, their thermal ionization, their destruction by the positive ions, and their destruction by the radicals. Thus,

$$d[A^{\cdot-}]/dt = k_1[A][e^-] - k_{-1}[A^{\cdot-}] - \{k_{N_1}[\oplus] + k_{R_1}[R]\}[A^{\cdot-}]$$

Denoting the sums $\{k_N[\oplus] + k_R[R]\}$ and $\{k_{N_1}[\oplus] + k_{R_1}[R]\}$ by k_D and k_L, respectively, we may rewrite the above equations, viz.,

$$d[e^-]/dt = k_p R_\beta - k_D[e^-] - \text{in the absence of A, and in its presence,}$$
$$d[e^-]/dt = k_p R_\beta - k_D[e^-] - k_1[A][e^-] + k_{-1}[A^{\cdot-}]$$

and

$$d[A^{\cdot-}]/dt = k_1[A][e^-] - k_{-1}[A^{\cdot-}] - k_L[A^{\cdot-}]$$

At a sufficiently long time ($t = \infty$), the simultaneous solution of these equations gives the stationary concentrations of electrons, $[e^-]_0$, in the absence of A, and their stationary concentration in the presence A, viz., $[e^-]_A$. The electron capture coefficient, K, defined as $\{[e^-]_0 - [e^-]_A\}/[A][e^-]_A$, may be therefore calculated:

$$K = k_L k_1/k_D(k_L + k_{-1})$$

The pseudo-first-order constants k_L and k_D refer to the diffusion-controlled reactions and, hence, their ratio is essentially temperature independent, whereas k_{-1} increases exponentially with temperature. Thus, at sufficiently high temperatures $k_{-1} \gg k_L$ and then $K = (k_L/k_D)K_{eq}$, where K_{eq} is the equilibrium constant of the reaction,

$$A + e^- \rightleftharpoons A^{\cdot-}$$

At low temperatures $k_{-1} \gg k_L$ and then $K = k_1/k_D$ becomes virtually temperature independent because k_1 and k_D are rate constants of reactions controlled by diffusion.*

At high temperatures $\ln K = \ln(k_L/k_D) + \ln(\gamma/T^{3/2}) + \epsilon/kT$, where ϵ is the electron affinity of A, and $\gamma/T^{3/2} = f(A^{\cdot-})/f(A) \cdot f(e^-)$. The symbols $f(\ldots)$ denote the partition functions of $A^{\cdot-}$, A, and of the free electron, respectively. In this high temperature region a plot of $\ln(KT^{3/2})$ versus $1/T$ should be linear, the slope being given by ϵ/k and the intercept by $\ln(k_L/k_D) + \ln f(A^{\cdot-})/f(A) + \text{constant}$. The ratio k_L/k_D is expected to be only slightly affected by the nature of the investigated hydrocarbon and the ratio $f(A^{\cdot-})/f(A)$ should be close to 2 due to the spin factor, whereas the other factors are nearly cancelled. In spite of the cancellation the ratio of partition functions is not exactly constant, and it may

---

* A similar model which accounts for dissociative electron-capture processes has been developed recently by Wentworth, Becker, and Tung (123).

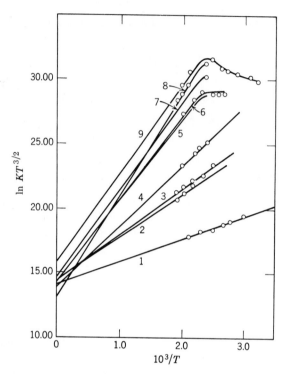

Fig. 2. Plot of ln ($KT^{3/2}$) versus $1/T$. K is the electron-capture coefficient for the gaseous reaction electron + aromatic hydrocarbon $\rightleftarrows$ (aromatic hydrocarbon)$^{\cdot-}$. *1*, naphthalene; *2*, triphenylene; *3*, phenanthrene; *4*, chrysene; *5*, benzo(*c*)phenanthrene; *6*, anthracene; *7*, pyrene; *8*, benzanthracene; *9*, azulene.

vary within a factor of 1.5–2. Therefore, the intercepts of the lines obtained for various aromatic hydrocarbons are not exactly identical—their values may vary slightly, probably within ± 1.

The results obtained for some aromatic hydrocarbons (9) are shown in Figure 2, which indicates only minor variations in the intercepts. Nevertheless, the assumption of a *constant* intercept does not seem to be justified, and it is doubtful whether the accuracy of the data could be genuinely improved by choosing a fixed value for the intercept (9). The electron affinities of 14 aromatic hydrocarbons calculated from the slopes of the respective lines are reported in refs. 9 and 10, and these values are listed in Table I. Only one electron affinity value seems to be in doubt, namely that for anthracene; the magnitude of the error and its cause will be discussed later. Included in Table I are electron affinities

TABLE I

Electron Affinities[a]

Compound	Electron affinity, eV	Intercept
Aromatic Hydrocarbons in the Gas Phase		
Naphthalene	0.152 ± 0.016	14.21
Triphenylene	0.284 ± 0.020	14.38
Phenanthrene	0.308 ± 0.024	14.33
Chrysene	0.419 ± 0.036	13.98
Benzo-(e)-pyrene	0.486 ± 0.155	—
Picene	0.490 ± 0.110	—
Benzo-(c)-phenanthrene	0.542 ± 0.040	14.44
Anthracene	0.552 ± 0.061 (?)	14.44
Pyrene	0.579 ± 0.064	14.66
Dibenz-(a,h)-anthracene	0.676 ± 0.122	—
Dibenz-(a,j)-anthracene	0.686 ± 0.155	—
Benzo-(a)-anthracene	0.696 ± 0.045	—
Benzo-(a)-pyrene	0.829 ± 0.121	12.94
Azulene	0.587 ± 0.065	16.06
Some Aromatic Aldehydes and Ketones		
Acetophenone	0.334 ± 0.004	14.70
Benzaldehyde	0.42 ± 0.010	15.89
Naphthaldehyde-2	0.62 ± 0.040	14.46
Naphthaldehyde-1	0.745 ± 0.070	12.61
Phenanthrene aldehyde-9	0.655 ± 0.14	14.92

[a] Wentworth, Chen, and Lovelock, *J. Phys. Chem.*, *70*, 445 (1966); Becker and Chen, *J. Chem. Phys.*, *45*, 2403 (1966); Wentworth and Chen, *J. Phys. Chem.*, *71*, 1929 (1967).

of some aromatic aldehydes and ketones determined by the same technique (122,123).

The method described above applies to nondissociative electron-capture processes. This limitation was emphatically stressed by Wentworth and Becker (11). Furthermore, the variation of the electron-capture coefficients with temperature is a necessary condition for a reliable determination of the electron affinities. Disregard of these restrictions may lead to erroneous values (12).

In the low temperature region the electron capture coefficient, K, remains constant and its value is given by the ratio k_1/k_D. The pseudo-constant k_D may be estimated from the dependence of the current on the duration of the pulse. Thus, an approximate evaluation of k_1's is possible and their values, which are given in ref. 9, were found to range from 3.5–25.10^{12} liter/mole-sec. This nearly tenfold variation of the rate constants of the diffusion controlled processes again indicates that the assumption of a fixed intercept is questionable (see also Table XII in Sec. XIII).

The results obtained for azulene are most interesting. The rate constant, k_1, of the electron-capture process for this hydrocarbon appears to involve an activation energy of about 0.15 eV, whereas no activation energy was anticipated or observed for other hydrocarbons. This might indicate that only a vibrationally excited molecule of azulene may capture an electron, i.e., the relevant process corresponds to case (c) of Figure 1.

Although the technique developed by Becker, Wentworth, and Lovelock appears to be the most reliable and admirably applicable to studies of electron affinities of polyatomic molecules in the gas phase, other experimental methods may, nonetheless, deserve a brief discussion. In principle, one needs to determine the equilibrium constant of a system containing thermal electrons, molecules A, and their radical-ions, $A^{\cdot-}$. This calls for techniques by which one may analyze plasma for $A^{\cdot-}$ and e^-. Equilibrium between these species is attained when thermal electrons drift under the influence of a low and uniform potential gradient through an investigated gas towards an anode. The observed current gives a value proportional to the sum of $[e^-] + [A^{\cdot-}]$. Now, if an "electron filter" is placed in front of the anode, the electrons may be deflected and then only the heavy ions reach the electrode. Hence, the fraction of $[A^{\cdot-}]$ in the swarm may be determined. Early use of this technique is illustrated by Bradbury's investigation of electron affinity of oxygen molecules (13). This method was applied in other studies which showed that CO, NH_3, CO_2, N_2O, H_2O, and H_2S have no affinity for electrons, whereas SO_2, NO, BF_3, Cl_2, Br_2, and I_2 have positive electron affinities. However, determination of electron affinities by this technique is rather unreliable.

Equilibria established in flames were investigated originally by Rolla and Piccardi (14). A fine wire inserted into the flame served as a cathode and a source of electrons. The distinction between electrons and negative ions was possible because their mobilities differ greatly. This technique led to the first estimates of electron affinities of I and Br atoms and subsequently to the determination of electron affinities of SO_2, SeO_2, and MoO_3 (14b–d). The method was refined by Sutton and Mayer (15), who used a device based on "magnetron effect" (driving electrons into circular paths and preventing them from reaching the anode) in order to discriminate between ions and electrons. Subsequently, this device was applied with moderate success to other systems. Its further modification was described by Glockler and Calvin (16).

Recently the magnetron technique has been greatly improved by Page and his associates (17) and successfully used in studies of electron affinities of quinone and chloranil (18). Thus, a value of 32 ± 2 kcal/mole was derived for the electron affinity of quinone and 57 ± 6 kcal/mole for that of chloranil. Further developments (19) made it possible to determine the electron affinity of the OH radical. A value of 43.6 ± 3 kcal/mole was deduced (20), indicating that the previously reported high values are apparently in error.

Dissociation of alkali halides at high temperature (1800° K) gives, among others, M^+ and X^- ions. From their concentrations the equilibrium constant of the reaction

$$MX \rightleftharpoons M^+ + X^-$$

was determined (21). These data permitted Mayer to calculate the electron affinity of X atoms, because the bond dissociation energy of MX and the ionization potential of M were known.

Injection of alkali metals, or their salts, into flames provides a convenient source of thermal electrons. Their concentration may be determined from attenuation of microwaves traversing the flame. In the presence of a suitable electron acceptor, A, the following equilibria are established,

$$Na \rightleftharpoons Na^+ + e^- \qquad K_1$$

and

$$A + e^- \rightleftharpoons A^{\cdot-} \qquad K_2$$

Hence, K_2 may be determined if the concentrations of Na and A are known. This approach was extensively used by Sudgen and his co-workers.

Electron affinities may be also deduced from some spectroscopic data. Recent studies of Person (22), who calculated the values for electron affinities of halogen molecules, provide a good example of such an investigation.

V. QUANTUM-MECHANICAL CALCULATION OF ELECTRON AFFINITIES AND THEIR CORRELATION WITH IONIZATION POTENTIALS

The simplest quantum mechanical calculation of electron affinities of π systems, e.g., of aromatic hydrocarbons, is based on Hückel's approach. In his treatment the electron affinity is given by the negative energy of the lowest unoccupied π orbital, its value being $-(\alpha + \chi_{N+1}\beta)$. Unfortunately, this method is unreliable because the repulsion between electrons is not accounted for. Approximate calculations of electron affinities of alternant aromatic hydrocarbons and radicals, in which the Hückel self-consistent field approach was used, were reported by Hush and Pople (23). In their treatment, the energy of an orbital depends on whether other orbitals are occupied or empty. Similar calculations were subsequently performed by other workers, a different degree of sophistication being introduced in their approach. For the sake of illustration, the results of Hedges and Matsen (24), of Hoyland and Goodman (25), and of Becker and Chen (10) are listed and compared in Table II. Its

TABLE II

Calculated Electron-Affinities of Aromatic Hydrocarbons in eV

Hydrocarbon	Hedges and Matsen (24)	Hoyland and Goodman (25)	Becker and Chen (10)	Experimental
Naphthalene	−0.38	−0.21	0.17	0.15
Phenanthrene	−0.20	−0.25	0.31	0.31
Triphenylene	−0.28	—	0.29	0.28
Chrysene	0.04	—	0.49	0.42
Pyrene	0.68	0.55	0.55	0.58
Benzo(e)pyrene	—	—	0.54	0.49
Anthracene	0.49	0.61	0.55	0.55
Benzo(a)pyrene	—	—	0.71	0.83
Naphthacene	0.82	—	—	—

inspection shows that on the whole the theory is still unsatisfactory, and the answers depend, to a great extent, on the choice of parameters used in computations. The calculations of Becker and Chen were based on the $\check{\omega}$-method (26) with $\check{\omega} = 3.75$, $\alpha_0 = 5.98$, and $\beta_0 = 1.23$. With this choice of parameters a very good agreement between calculated and observed values of electron affinities was obtained. However, the rather high value of $\check{\omega}$ may be questioned.

Electron affinities and ionization potentials of alternant hydrocarbons were first correlated by Hush and Pople (23) who concluded that their sum should be constant. In addition, Hedges and Matsen (24) stressed that the Hückel and the ASMOH theories suggest that the electron affinities and the corresponding ionization potentials should be symmetrical in respect to work function of graphite. The experimental data now available seem to confirm these predictions.

According to Mulliken's definition, the electronegativity of an atom or a molecule is the average of its ionization potential and electron affinity. Hence, the results of the calculations given above indicate that electronegativities of alternant aromatic hydrocarbons are constant, their most probable value being estimated at 4.1 or 4.2 eV. On this basis, the ionization potentials of an aromatic hydrocarbon may be predicted if its electron affinities are known, or vice-versa.

The semiempirical calculations of electron affinities of gaseous radicals have been reported recently by Gaines and Page (29). On the whole, their results agree well with the experimental data.

VI. ELECTRON AFFINITIES IN SOLUTION

In solution the addition of an electron to a suitable acceptor, A, differs in two respects from the reaction proceeding in the gas phase: (a) Dissipation of energy arising from the exothermicity of the process is ascertained by the interaction of the radical-ions with solvent molecules. (b) The heat of reaction is substantially larger than that in the gas phase because the solvation energy of the radical-ion substantially contributes to the exothermicity of the overall process. The heat of solvation may be calculated by the conventional Born approach, $\Delta H_{solv} = (e^2/2r)(1 - 1/D)$, if the ions are assumed to be spherical (42). Certainly, this is a gross oversimplification. A more refined approach was proposed by Hush and Blackledge (27), who calculated the distribution of charge over all the atoms of a radical-ion, attributed then an effective radius, r_j, to each atom, and finally summed up all the Born-type interaction terms. Thus,

$$\Delta H_{solv} = \sum_j (q_j^2/2r_j)(1 - 1/D)$$

The results of their computations are listed in Table III. The heat of solvation decreases on increasing the area of the radical-ion, and for sufficiently large radical-ions their values should be nearly constant. The calculations of Hush exaggerate, however, the exothermicity of solvation because the "effective" dielectric constant is smaller than its bulk value.

An interesting attempt to determine the solvation energy of aromatic radical-ions was reported by Prock et al. (269). They studied

TABLE III

Heats of Solvation of Aromatic
Radical-Ions in Tetrahydrofuran

Parent hydrocarbon	ΔH, eV
Benzene	1.8
Naphthalene	1.1
Phenanthrene	0.8
Anthracene	0.8

the photoinjection of electrons into liquids. These electrons were produced by irradiating with visible light a negatively charged rhodium electrode immersed in a benzene solution of aromatic hydrocarbons such as naphthalene, phenanthrene, anthracene, and pyrene. The photocurrent ($\sim 10^{-11}$ Å) far exceeded the dark current, and the decreased threshold, when compared with the photoelectric effect observed in the gas phase, was due to two factors: (1) electron affinity of the aromatic hydrocarbon which acquired an electron and became the current carrier; and (2) solvation energy of the radical-ion. Indeed, the process may be represented by three steps,

$$\text{Rh (sol)} \rightarrow \text{Rh (sol)}^+ + e^- \text{ (g)} - \text{work function } \phi$$
$$e^- \text{ (g)} + \text{A (g)} \rightarrow \text{A}\cdot^- \text{ (g)} - \text{electron affinity, } \epsilon, \text{ in the gas phase}$$
$$\text{A}\cdot^- \text{ (g)} + \text{A (sol)} \rightarrow \text{A}\cdot^- \text{ (sol)} + \text{A (g)} - \text{solvation energy, } \Delta E_{\text{sol}}$$

Hence, the wave threshold, $h\nu$, is given by the equation $h\nu = \phi - \epsilon + \Delta E_{\text{sol}}$ in which only the last term is unknown. Thus, the solvation energy is calculated from the experimental data.

Although the method is undoubtedly original and interesting, it remains to be seen how reliable it is. For example, the drop of potential between the electrode and the adjacent layer of liquid should be added to the left side of the equation, the omission of this term making $-\Delta E_{\text{sol}}$ too large. This may account for some surprising conclusions drawn by the authors.

The negative radical-ions are not specifically coordinated with molecules of ethereal solvents, whereas alkali ions such as Li^+ or Na^+ interact strongly with ethers (see, e.g., ref. 47). The lack of coordination of anions with ethers is evident from studies of mobilities of radical-anions (37). The respective self-diffusion constants, calculated from such data, are only slightly smaller than the diffusion constants of the parent hydrocarbons.

"Electron affinities" in solution were investigated by several workers who applied potentiometric, polarographic, and spectrophoto-metric methods in their studies. The details of these techniques and the significance of the derived results will be discussed now.

VII. POTENTIOMETRIC STUDIES OF ELECTRON AFFINITIES IN SOLUTION

The first reported potentiometric studies involving aromatic com-pounds were those of Bent and Keevil (31). They measured the potential established between the mercury amalgam electrode and a bright platinum electrode which was in contact with a saturated solution of, e.g., triphenylmethyl sodium in ether. Although the potential was shown to obey the Nernst dilution equation, certainly for triphenylmethyl sodium, the results were not too conclusive. The solvation energies and the dissociation constants of ion pairs were not known. In fact, Keevil and Bent (32) attempted to determine the dissociation constant of $C(Ph)_3{}^-, Na^+$ in diethyl ether, but their results are questionable because the study was performed in solutions which were too concentrated. Subsequently, the dissociation constant was determined by Swift (50) and, on this basis, the electron affinity of triphenylmethyl radical was calculated to be -48 ± 5 kcal/mole. The potentiometric technique was used in subsequent work to determine electron affinities of benzo-phenone, fluorenone, bis-biphenylyl ketone, and of some aromatic hydrocarbons such as anthracene, stilbene, and tetraphenylethylene in diethyl ether solutions. Thus, the change in free energy for the process anthracene + sodium $\rightarrow$ sodium-anthracene was estimated as -12 to -13 kcal/mole.

A most successful approach to potentiometric studies was devel-oped by Hoijtink and his associates (33). These workers measured the potential developed between two platinum electrodes, one placed in a standard solution of biphenyl and its radical-anion, the other in contact with a solution of the investigated hydrocarbon and its radical anion. The procedure resembled a potentiometric titration. The standard solution of the mixture, biphenyl (B) and biphenyl radical-anion (B·⁻), was gradually added to the solution of the investigated hydrocarbon, A. On addition of B·⁻ a virtually instantaneous electron transfer converted A into A·⁻ and hence, the second electrode responds to the redox system A–A·⁻.

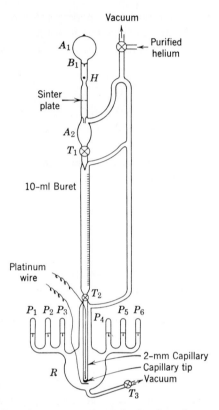

Fig. 3. Apparatus used in the potentiometric titration of aromatic hydrocarbons
with sodium biphenyl (34).

The original apparatus of Hoijtink was slightly modified in the
writer's laboratory (34), and its new version is shown in Figure 3. The
buret containing the B + B·⁻ solution was terminated by a 2-mm bore
capillary, its lower tip being sealed and then punched with a needle to
form six or seven tiny parallel capillaries. This arrangement considerably
slowed down the diffusion of the liquid from reactor R to the upper
platinum electrode which was touching the sealed tip. The resistance of
the unit, when filled with a $0.016M$ THF solution of sodium biphenyl,
was about 10 megohms.

The potential between electrodes was measured by a valve voltmeter which was described by Scroggie (35). It functioned as a very
stable impedance converter by means of which an input dc voltage in a

high resistance circuit was converted into an identical output dc voltage generated in a very low resistance circuit. The latter was then measured by any conventional voltmeter with a resistance exceeding 400 ohms.

All the operations were performed in helium atmosphere, the gas having been purified and freed from any traces of oxygen or moisture. The investigated hydrocarbons were introduced from ampoules P into the reactor by crushing the respective breakseals. After the titration was complete, the titrated solution was sucked out without exposing the electrodes to air, a new solution introduced, and another titration performed. A typical titration curve is shown in Figure 4.

Results of Hoijtink et al. (33) are given in Table IV where, for the sake of comparison, the values reported by Chaudhuri, Jagur-Grodzinski, and Szwarc (36) are also listed. In spite of some differences, both sets of data are approximately linearly related. Several factors could contribute to the observed divergencies, and some of them will be considered now.

In his discussion of the potentiometric method, Hoijtink assumed tacitly that the dimethoxyethane (DME) solutions of the investigated radical-ions are virtually completely ionized, although he stressed the incomplete dissociation of $A^{2-},2Na^+$. Therefore, although he was

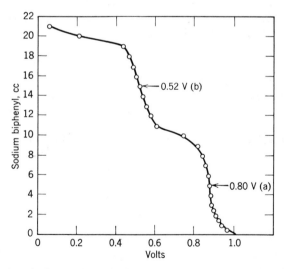

Fig. 4. A typical potentiometric titration curve. Perylene titrated with sodium biphenyl. The arrows denote the points corresponding to the addition of (a) $\frac{1}{2}$ and (b) $\frac{3}{2}$ equivalents of sodium biphenyl. The respective potentials are ϵ_1 and ϵ_2.

TABLE IV

Aromatic hydrocarbon	Original results of Hoijtink et al. in DME	Corrected results of Szwarc et al. in THF
1. Biphenyl	(0.0)	(0.0)
2. Naphthalene	0.09	0.043 ± 0.02
3. Triphenylene	0.19	0.132 ± 0.01[a]
4. Phenanthrene	0.17	0.142 ± 0.01
5. Benzo(c)pyrene	0.58	0.484 ± 0.02
6. Pyrene	0.60	0.529 ± 0.01
7. Benzo(e)anthracene	0.76	0.590 ± 0.02
8. 9,10-Dimethylanthracene	—	0.616 ± 0.01
9. Anthracene	0.78	0.642 ± 0.01
10. Benzo(a)pyrene	0.90	0.760 ± 0.02
11. Acenaphthylene	1.12	0.880 ± 0.03
12. Fluoranthrene	0.94	0.820 ± 0.02
13. Perylene	1.09	0.965 ± 0.01
14. Napthacene	1.28	1.058 ± 0.02

[a] The original value given by Slates and Szwarc (37) was 0.128; after re-checking, the correct value was found to be 0.132.

concerned with the problem of ion solvation, he neglected, in that early study, the effect of ion pairing. The latter problem was considered by Jagur-Grodzinski et al. (34), who showed that the potential determined at the middle point of titration curve is given by

$$\epsilon = \epsilon_0 + 0.03 \log (K_B/K_A) + 0.03 \log ([A^{\cdot-},Na^+]_t/[B^{\cdot-},Na^+]_t)$$

Here ϵ_0 represents the correct standard potential corresponding to the ionic equilibrium

$$B^{\cdot-} + A \rightleftarrows A^{\cdot-} + B$$

while K_B and K_A refer to the equilibrium constants of ion-pair dissociation, viz.,

$$B^{\cdot-},Na^+ \rightleftarrows B^{\cdot-} + Na^+$$

and

$$A^{\cdot-},Na^+ \rightleftarrows A^{\cdot-} + Na^+$$

respectively. The last term of the equation giving ϵ accounts for the difference in the *total* concentration of $B^{\cdot-},Na^+$ in the buret and of $A^{\cdot-},Na^+$ in the reactor. At the middle point of titration the concentration of $A^{\cdot-},Na^+$ in the titrated solution is usually 3–4 times lower than that of $B^{\cdot-},Na^+$ in the buret, and hence the above correction term amounts to less than 0.02 V. The correction terms, $0.03 \log K_B/K_A$, were also expected to be small. The required dissociation constants were determined (37). Thus, the proper ϵ_0 values were calculated and these are listed in the last column of Table IV.

It has been shown recently (268) that sodium salts of radical-ions are virtually completely dissociated in hexamethyl phosphoramide (HMPA). Therefore, it is advantageous to carry out the potentiometric titrations in this solvent since the results give ϵ_0 directly. Moreover, some disturbing reactions which take place, e.g., in the course of titration of heteroaromatics in tetrahydrofuran, are avoided in HMPA.

Fortunately, no corrections are necessary to account for the difference in dissociation of the various ion pairs if the titration is carried out in dimethoxyethane (the solvent used in Hoijtink's studies). Chang, Slates, and Szwarc (38) showed that in DME the relevant dissociation constants are closely similar for all the pertinent aromatic$\cdot^-$,Na$^+$ ion pairs. Therefore, the discrepancy between the two sets of data which are given in Table IV cannot be explained by this omission. The reliability of both sets of data will be discussed later, but first we wish to consider in some detail the dissociation of A$\cdot^-$,Alkali$^+$ ion pairs into free ions.

VIII. DISSOCIATION OF SALTS OF RADICAL-ANIONS IN ETHEREAL SOLUTION

Dissociation of sodium salts of naphthalene$\cdot^-$, biphenyl$\cdot^-$, pyrene$\cdot^-$, anthracene$\cdot^-$, triphenylene$\cdot^-$, tetracene$\cdot^-$, and perylene$\cdot^-$ in tetrahydrofuran (THF) was studied conductometrically (37,38) over a wide temperature range. Some results are shown graphically in Figure 5 giving the plots of log K_{diss} versus $1/T$. At room temperature the dissociation constant of sodium salts of naphthalene$\cdot^-$ differs by more than two powers of ten from that of perylene$\cdot^-$ or naphthacene$\cdot^-$. On the other hand, their dissociation constants are nearly identical at $-70°C$. All these dissociation processes are exothermic. At 25°C the exothermicity is high for naphthalene$\cdot^-$,Na$^+$ (8.2 kcal/mole) but low for perylene$\cdot^-$,Na$^+$ (2.2 kcal/mole). At the lowest temperatures (about $-70°C$) the heat of dissociation of all these salts approaches 0. These findings indicate that small radical-anions, e.g., napthalene$\cdot^-$, form contact ion pairs in THF at 25°C, whereas the larger the radical-anion the weaker its attraction for the counterion and, therefore, the more extensive the coordination of Na$^+$ ion of the pair with THF molecules.

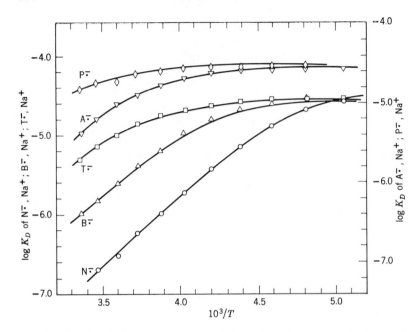

Fig. 5. Dissociation constants, K_D, of ion pairs of aromatic radical-ions at different temperatures (range +25 down to −70°C). Naphthalene·⁻, Na⁺ (○); biphenyl·⁻, Na⁺ (△); triphenylene·⁻, Na⁺ (□); anthracene·⁻, Na⁺ (▽); perylene·⁻, Na⁺ (◇). Solvent, tetrahydrofuran; counterion, sodium. Note the striking difference in the heat of dissociation of naphthalene·⁻, Na⁺ and perylene·⁻, Na⁺. The scale for A·⁻ and P·⁻ was shifted by 0.2 units. This means that all the lines nearly coincide with the lowest temperatures.

The coordination of the counterion with the solvent becomes more extensive at lower temperatures, i.e., in spite of the partial separation of the ions which takes place in the process the conversion of ion pairs from a less solvated to a more solvated form is exothermic.

Dissociation of the sodium salts was also investigated in dimethoxyethane (DME) (38). The results indicated that this more powerfully solvating ether is coordinated with the sodium ion of all the investigated ion pairs, even at 25°C. Consequently, the dissociation constants are relatively high, but the exothermicities of these processes are rather low. This accounts for the relative insensitivity of the dissociation constants on the nature of radical-ion, as is clearly demonstrated by the data given in Table V. It is instructive to compare the behavior of these ion pairs in DME and in THF and, therefore, the required data are included

TABLE V

Dissociation Constants and Heats of Dissociation of Sodium Salts of Aromatic Radical Anions

	Biphenyl$^{\cdot-}$	Triphenylene$^{\cdot-}$	Perylene$^{\cdot-}$
In DME			
$10^6 K_{\text{diss}}$ at 25°C, M	4.6	5.6	6.0
$10^6 K_{\text{diss}}$ at -55°C, M	17.0	14.8	18.8
ΔH_{diss} at 20°C, kcal/mole	-2.1	-2.4	-2.5
ΔS_{diss} at 20°C, eu	-31.5	-32.5	-32.5
ΔH_{diss} at -55°C, kcal/mole	0	0	0
ΔS_{diss} at -55°C	-22.0	-22.0	-21.0
In THF			
$10^6 K_{\text{diss}}$ at 25°C, M	1.0	5.2	15.5
$10^6 K_{\text{diss}}$ at -55°C, M	24.5	28.8	28.5
ΔH_{diss} at 20°C, kcal/mole	-7.3	-5.2	-2.2
ΔS_{diss} at 20°C, eu	-52	-42	-29
ΔH_{diss} at -55°C, kcal/mole	-1.6	-0.4	0.0
ΔS_{diss} at -55°C, eu	-28	-23	-21

in this table. Attention should be paid to the ΔS values. The solvation of free ions produced by the dissociation of the noncoordinated contact ion pairs leads to a high negative ΔS of dissociation, e.g., its value for biphenyl$^{\cdot-}$,Na$^+$ in THF at 20°C is -52 eu. Much lower negative values, of the order 20–30 eu, are found for the dissociation of solvated and partially separated ion pairs. This is a very general observation, as may be seen, e.g., from data collected in Table VII of ref. 67.

It is also interesting to note that the dissociation of *solvent separated* ion pairs is higher in THF than in DME, if conditions are identical. This is due to the smaller size of Na$^+$ ion when coordinated with DME, its Stokes radius then being ~ 3.5 Å, whereas the Na$^+$ ion coordinated with THF has a Stokes radius of ~ 4.2 Å (47).

Our knowledge of salts of other alkali metals is less extensive. Some data became available recently for the lithium salts. Contrary to the previous report (115), it was found that the Stokes radii in THF are identical for the Li$^+$ and Na$^+$ ions (39). The evidence is provided by the observation that the limiting conductance, Λ_0, is the same for the sodium or lithium salts of anthracene$^{\cdot-}$ or perylene$^{\cdot-}$. The lithium salts are, however, more solvated than the sodium. This is shown by the low heat of dissociation found for the lithium derivatives. Nevertheless, such lithium salts are often less dissociated than the respective *solvent-separated* sodium salts. This indicates that the lithium salts are solvated to a large extent on the periphery of the pair and, therefore, the tendency to separate the ions is less pronounced than that of the solvent-separated sodium salts. Thus, the small lithium cation, even when solvated, may be more strongly attached to the anion than the solvent coordinated sodium ion. Moreover, covalent character of the $\bar{C}$—Li^+ bond may also contribute to this bias. Some data collected in Table VI

TABLE VI

Λ_0 of Lithium and Sodium Salts of Aromatic Radical-Ions in THF

T°C	Perylene·⁻		Anthracene·⁻		Naphthalene·⁻	
	$\Lambda_0(Li^+)$	$\Lambda_0(Na^+)$	$\Lambda_0(Li^+)$	$\Lambda_0(Na^+)$	$\Lambda_0(Li^+)$	$\Lambda_0(Na^+)$
25	107	106	121	120	129	128
5	86.4	85.6	97.5	95.2	104	102
−15	67.0	66.5	75.7	75.1	80.6	79.5
−35	48.9	49.5	56.4	55.8	60.0	59.0
−55	35.2	34.9	39.8	39.4	42.3	42.0
Dissociation constants × 10⁶ in M						
25	4.5	15.5	4.5	4.3	3.1	0.14
5	5.4	20.3	5.4	9.7	4.3	0.33
−15	6.1	25.3	6.3	16.8	6.0	1.07
−35	6.9	28.2	7.1	23.9	7.4	3.9
−55	7.0	28.5	7.4	28.7	9.0	13.1
Heat of dissociation, ΔH kcal/mole at 25°C						
	−1.4	−2.2	−1.9	−6.1	−2.8	−8.2

illustrate these relations. There is no doubt, however, that the free energy of formation of lithium salts from gaesous ion is greater than that of sodium salts (see Sec. XII).

Finally, a few words of caution may be not out of place. In a most interesting paper Hoijtink et al. (40) discussed the dissociation of salts of aromatic radical-ions in tetrahydrofuran and in methyl tetrahydrofuran. The conductance of each salt was investigated at *one* concentration ($10^{-4}M$) although over a wide temperature range, and the degree of dissociation was estimated from the temperature dependence of the conductance. This method may often be misleading. For example, Hoijtink considered a salt to be completely dissociated if the relevant conductance versus temperature curve showed no maximum and monotonically decreased on lowering the temperature. This led to the conclusion that $10^{-4}M$ solution of anthracene$\cdot^-$,Na$^+$ at 25°C in THF is virtually completely dissociated, while according to the data derived from Figure 5, its dissociation amounts to only about 8% under these conditions. Similarly, the lithium salt was considered to be completely dissociated at room temperature, whereas its dissociation is even slightly lower than that of the sodium salt. The dissociation of the lithium salt increases only insignificantly at −50°C.

Three factors are important in determining the temperature dependence of conductance at constant salt concentration: (1) "activation energy" of the solvent's viscosity; (2) the heat of dissociation of the investigated ion pair; (3) the effect of temperature on the Stokes radii of the ions. Whenever the "activation energy" of solvent's viscosity is higher than one-half of the exothermicity of dissociation of the investigated ion pair, the conductance diminishes monotonically with decreasing temperature. This, undoubtedly, is the case for the system anthracene$\cdot^-$,Li$^+$ in THF, because the heat of dissociation is only −1.9 kcal/mole (39). Apparently a similar situation exists for the sodium salt. The heat of dissociation of anthracene$\cdot^-$,Na$^+$ rapidly decreases from −6 kcal/mole at 20°C to 0 kcal/mole at −60°C, whereas the "activation energy" of THF viscosity is about 1.8 kcal/mole (47).

Dissociation of salts of aromatic radical-ions was investigated by Weissman and his co-workers by means of ESR spectroscopy. The results and the significance of their studies will be discussed later.

IX. EQUILIBRIA BETWEEN RADICAL-IONS AND THEIR PARENT MOLECULES

We are now in position to consider the equilibria established between two radical-ions, A$\cdot^-$ and B$\cdot^-$, and their parent molecules A and B. Thus,

$$A\cdot^- + B \rightleftharpoons B\cdot^- + A \qquad K_{AB,i}$$

The equilibrium constant may be obtained from spectrophotometric studies because the spectra of A$\cdot^-$ and B$\cdot^-$, which extend to the visible or near IR region, are often sufficiently different to permit their distinction and quantitative determination of their concentrations. Under most common experimental conditions, e.g., in tetrahydrofuran at concentrations $10^{-4}M$ or higher, the radical-ions exist virtually as

ion pairs and, therefore, the spectrophotometrically observed equilibrium refers to the reaction

$$A^{\cdot-},M^+ + B \rightleftharpoons B^{\cdot-},M^+ + A, \qquad K_{AB,p}$$

There is a simple relation between $K_{AB,i}$ and $K_{AB,p}$, namely,

$$K_{AB,p} = K_{AB,i} \cdot K_{diss,A^{\cdot-},M^+} / K_{diss,B^{\cdot-},M^+}$$

where $K_{diss,A^{\cdot-},M^+}$ and K_{diss,B^-,M^+} refer to the dissociation constants
of $A^{\cdot-},M^+$ and $B^{\cdot-},M^+$ ion pairs, respectively. Hence, if the latter
constants are known, the determination of $K_{AB,p}$ leads to the value of
$K_{AB,i}$. The equilibrium constant $K_{A,B,i}$ gives the difference of electron
affinities of A and B in solution. The latter differs insignificantly from
the $\epsilon_A - \epsilon_B$ determined in the gas phase (36), because the variations in
the solvation energies of radical-anions in tetrahydrofuran, or more
correctly the differences $\Delta H_{sol(A^{\cdot-})} - \Delta H_{sol(B^{\cdot-})} + \Delta H_{sol(B)} - \Delta H_{sol(A)}$,
are usually small.

The above-discussed equilibria were investigated by Paul, Lipkin,
and Weissman (41), whose pioneering work demonstrated the feasibility
of the method and established the general trend of electron affinities of
the aromatic hydrocarbons. Unfortunately, various technical difficulties
involved in such studies were not yet appreciated at that time. In addition, some confusion arose because it was not fully realized that the
conventional method of preparation of aromatic radical-ions, viz.,
reduction with alkali metals, may produce dialkali adducts as well as
monoalkali adducts. Consequently, the quantitative findings reported
in this important investigation were somewhat in error. For example,
the equilibrium constant of the reaction

$$\text{anthracene}^{\cdot-} + \text{naphthacene} \rightleftharpoons \text{naphthacene}^{\cdot-} + \text{anthracene}$$

was found to be of the order of unity, while the potentiometric data
(33,34,37) indicated an equilibrium constant of the order 10^6 to 10^7.

Studies of Weissman's group were, therefore, repeated in this
writer's laboratory. The equilibria were investigated in THF at room
temperature and the systems studied were chosen on the basis of the
following criteria. (1) The difference in the reduction potentials of the
investigated hydrocarbons should not exceed 0.15 V, i.e., the determined equilibrium constants should not be larger than 250. Otherwise,
a reliable determination of the concentration of one of the radical-ions
calls for an enormous excess of the other hydrocarbon. This, in turn,
could lead to partial destruction of some radical-ions and, therefore, to

erroneous results. Indeed, such a difficulty vitiated Weissman's study of the system naphthalene$\cdot^-$ + anthracene $\rightleftarrows$ anthracene$\cdot^-$ + naphthalene. (2) The spectra of the investigated radical-ions should not overlap too closely. A successful spectrophotometric study obviously requires such a condition.

The following systems were eventually selected for the spectrophotometric studies: tetracene–perylene, anthracene–pyrene, 9,10-dimethylanthracene–pyrene, and triphenylene–naphthalene. The results are given in Table VII. The agreement between the difference of the

TABLE VII

Equilibria A$\cdot^-$,Na$^+$ + B $\rightleftarrows$ B$\cdot^-$,Na$^+$ + A in THF at 25°C

Investigated pair	$K_{AB,p}$	$K_{AB,i}$	$\Delta\epsilon$ from $K_{AB,i}$	$\Delta\epsilon^a$ observed
Perylene (A)–Tetracene (B)	52	34	0.092	0.093
Pyrene (A)–Anthracene (B)	111	74.5	0.112	0.113
Pyrene (A)–9,10-dimethyl-anthracene (B)	91	29.5	0.088	0.087
Naphthalene (A)–triphenylene (B)	3	73	0.111	0.089

^{a} $\Delta\epsilon$ observed taken from ref. 36, in volts.

observed potentials and that calculated from the respective equilibrium constant is remarkably good.

The equilibrium method is particularly valuable when small differences of electron affinities had to be determined. An alternative method, in which the same equilibrium constants are derived from kinetic studies, will be described later (see Sec. XX).

X. POLAROGRAPHIC DETERMINATION OF ELECTRON AFFINITIES

The polarographic technique of determining electron affinities is closely related to the potentiometric method. The reduction on a dropping mercury electrode has to be performed in a medium of low proton activity to avoid rapid protonation of the primary product. The 75% and 96% aqueous dioxanes 2-methoxyl-1-ethanol (cellosolve) and dimethylformamide have been used as the most suitable solvents, with tetramethyl- or tetrabutylammonium iodide as the supporting electrolyte. Depending on the reduced substrate, one or two polarographic waves were observed. The first is attributed to the formation of A·⁻, and the second is associated with the formation of A^{2-}.

This interesting field was opened through the pioneering work of Laitinen and Wawzonek (43–45), who initially investigated the polarographic reductions of styrene, α-methylstyrene, and 1,1-diphenylethylene and then extended their studies to polarography of aromatic hydrocarbons. To account for the independence of the recorded potential of the pH of the solution, the reduction was assumed to proceed through the step

$$A + e \rightleftarrows A^{\cdot-} \text{ (reversible and potential determining)}$$

This is followed by

$$A^{\cdot-} + e \longrightarrow A^{2-}$$

if a second wave is observed at higher potentials. The resulting product of the latter process is rapidly destroyed by protonation, viz.,

$$A^{2-} + \text{proton} \longrightarrow AH^- \text{ (irreversible)} \longrightarrow AH_2 \text{ (irreversible)}$$

We should inquire whether the first step could be represented by an alternative equation,

$$A + X^+ + e \rightleftarrows A^{\cdot-},X^+$$

demanding that $X^+ = NMe_4^+$ or NBu_4^+ be involved in the energy controlling step. Such a step may appear plausible in view of the presence of a supporting electrolyte, and it could be expected in media of low dielectric constant, e.g., in 96 or 75% dioxane. However, the halfwave potentials observed in 75% aqueous dioxane and in the 96% dioxane differed only by 0.02 V (46). The change in the proportion of water decreases the dielectric constant of the medium from about 13 to 3.5 and, therefore, the dissociation constant of the respective ion pairs is greatly reduced. For example, the dissociation constant of tetrabutylammonium nitrate decreases by about 10^6 as the proportion of water in the solvent is lowered from 25% to 4%. Hence, had ion pairing been involved in the actual electron transfer, a larger variation in the potentials should be expected.

Studies of Laitenen and Wawzonek were extended by Hoijtink and his co-workers (46,48) and independently by Bergman (49). The latter worker determined halfwave reduction potentials of 78 aromatic hydrocarbons—a most comprehensive investigation of this subject. Some of his results are compared in Table VIII with those reported by

TABLE VIII

Polarographic Half-wave Potentials, $\epsilon_{1/2}$, for Aromatic Hydrocarbons and Some Related Hydrocarbons (in volts in reference to Saturated Calomel Electrode)[a]

Compound	Laitinen and Wawzonek[d]		Bergman[e]		Hoijtink[f]	
	$-\epsilon_{1/2}$	$-\Delta_{1/2}$	$-\epsilon_{1/2}$[b]	$-\Delta\epsilon_{1/2}$	$-\epsilon_{1/2}$	$-\Delta\epsilon_{1/2}$
Biphenyl	2.70	+0.20	2.07	+0.09	2.70	+0.20
Naphthalene	2.50	0.00	1.98(1.99)[c]	0.00	2.50	0.00
Phenanthrene	1.96(?)	—	1.93(1.93)[c]	−0.05	2.45	−0.05
Triphenylene	—	—	1.97	−0.01	2.50	0.00
Chrysene	—	—	1.80	−0.18	—	—
Pyrene	2.11	−0.39	1.61(1.56)[c]	−0.37	2.13	−0.37
Anthracene	1.94	−0.56	1.46(1.41)[c]	−0.52	1.96(2.42)	−0.54
Naphthacene	—	—	1.13	−0.85	1.58(1.84)	−0.92
Perylene	—	—	1.25	−0.73	1.67	−0.83
Styrene	2.35	−0.15	—	—	—	—
1,1-Diphenyl-ethylene	2.26	−0.24	—	—	2.25	−0.25
Stilbene	2.14	−0.36	—	—	2.50	0.00
Tetraphenyl-ethylene	2.05	−0.45	—	—	2.06	−0.44
α-Methyl-styrene	2.54	+0.04	—	—	—	—

[a] All the data of Laitinen and Wawzonek and of Hoijtink were obtained in 75% dioxane + 25% water. Bergman's data were obtained in monomethylether of glycol. The data of Given (in brackets) were observed in dimethylformamide. Extensive comparison of calculated and observed data in $CH_3O \cdot C_2H_4OH$, 96% dioxane and 75% dioxane are given in Table I of Hoijtink's paper, Rec. Trav., 74, 1952 (1955).

[b] Expressed in reference to the mercury pool anode potential. To convert these values to those referred to the calomel electrode 0.52 V should be added to $-\epsilon_{1/2}$.

[c] The data reported by Given (59b) in dimethylformamide. These are referred to the mercury pool anode potential.

[d] H. A. Laitinen and S. Wawzonek, J. Am. Chem. Soc., 64, 1765 (1942); S. Wawzonek and H. A. Laitinen, ibid., 64, 2365 (1942); S. Wawzonek and J. W. Fan, ibid., 68, 2541 (1946).

[e] I. Bergman, Trans. Faraday Soc., 50, 829 (1954).

[f] G. J. Hoijtink and J. van Schooten, Rec. Trav., 71, 1089 (1952); 72, 691 and 903 (1953); G. J. Hoijtink, J. van Schooten, E. de Boer, and W. I. Aalbersberg, ibid., 73, 355 (1954); G. J. Hoijtink, ibid., 73, 895 (1954); 74, 1525 (1955).

the other two groups of investigators. Inspection of the table demonstrates a remarkable degree of agreement between all the reported findings. Even in solvents as different as acetonitrile (59a) or dimethylformamide (59b) the differences in measured potentials were not greater than 0.01 V. These observations indicate that the differences between the free energies of solvation of various aromatic radical-ions are small—certainly less than ± 0.1 eV or ± 2.5 kcal/mole.

The first polarographic reduction takes place extremely rapidly, whereas the protonation is relatively slow.* The measurements provide, therefore, the standard reduction potential of the investigated hydrocarbon. This conclusion was confirmed through ac polarographic studies (55) performed in dimethylformamide, a solvent used previously in dc polarography (59b). The correction term, viz., $(RT/\mathscr{F}) \ln (D_A/D_{A.-})$ is negligible, because the diffusion constants, D_A and $D_{A.-}$, of the parent molecule and of its radical-ion are nearly identical (37). However, the second wave, associated with the formation of A_{2-}, probably represents an irreversible process. Owing to the enhanced reactivity of the dinegative ions toward protons (54,55), A^{2-} is rapidly removed, and hence the measured potential does not provide a thermodynamic value. Indeed, its magnitude is affected by the nature of solvent, by the type of the cation present in the system, and by the concentrations of the reagents.

The polarographic data were used by Matsen (51) to calculate the absolute values of electron affinities in solution. The following sequence of reactions was considered:

$$A(g) + e^- \rightleftarrows A^{.-}(g), \qquad (\epsilon_0, \text{ electron affinity})$$
$$A^{.-}(g) + A(sol) \rightleftarrows A^{.-}(sol) + A(g), \qquad \text{overall solvation energy, } \Delta G'_{solv}$$
$$e^-(\text{in Hg}) \rightleftarrows e^-(g), \text{ work function of mercury, } \phi(\text{Hg}) = -4.54 \text{ eV}$$
$$\left. \begin{array}{l} Hg(l) + Cl^- (1N,aq) \rightleftarrows \frac{1}{2}Hg_2Cl_2 + e^-(\text{in Hg}) \\ Cl^-(\text{sat. } Hg_2Cl_2) \rightleftarrows Cl^- (1N,aq) \end{array} \right\} \begin{array}{l} \text{the absolute potential of a} \\ \text{saturated calomel electrode.} \end{array}$$

The absolute potential of a saturated calomel electrode was calculated by Latimer, Pitzer, and Slansky (52), who arrived at a value of -0.53 V. Thus, $\epsilon_{1/2} = \epsilon_0 + \Delta G'_{solv} - 5.07$ eV.

Alternatively, the above constant (-5.07 eV) correlating the half-wave reduction potential with the electron affinity in solution may be derived from the measured halfwave potential of triphenyl-methyl radical and its electron affinity calculated by Swift (50).

Polarographic studies were extended to heteroaromatics such as pyridine, quinoline, etc. The subject has been reviewed by Kolthoff and Lingane (53), who pointed out that a catalytic hydrogen wave may be

---

* It has been shown recently (101) that proton transfer from alcohols to aromatic radical-ions is indeed slow ($k \sim 10^4 \ M^{-1} \ sec^{-1}$).

observed in addition to the normal type of reduction wave. The catalytic wave arises from reactions (1)–(3).

$$(1)$$

$$(2)$$

$$H + H \longrightarrow H_2 \tag{3}$$

The net result is therefore $2H_2O + 2e^- \rightarrow H_2 + 2OH^-$. At pH < 6 the normal reduction wave is observed.

The catalytic wave should be avoided in aprotic solvents. Indeed, Given (59b) reported an apparently unperturbed polarographic reduction of pyridine and quinoline in dimethylformamide, viz., $\epsilon_{1/2}$ was found to be 2.01 and 1.53, respectively, in reference to a mercury pool anode. These values seem, however, to be too low. Indeed, if the difference between the reported halfwave potentials of naphthalene and quinoline (+0.46 V) are used for quantum mechanical calculation of L_N, the result is much too high—it is 1.1 instead of the usual value of about 0.5. It is possible that the dimerization of the heteroaromatic radical-ions vitiated the results (268).

Polarographic data inspired much theoretical work. In a series of papers (46,48,56) Hoijtink correlated the halfwave potentials with the energy of the lowest unoccupied orbitals E. Thus the relation

$$\epsilon_{1/2} = \gamma \cdot E + \text{const.}$$

was derived and Bergman's data, as well as his own, showed a good linear relation with E. Such a relation applies to alternant as well as to nonalternant hydrocarbons. Various approaches were used in the calculations, e.g., the original Hückel method gives $\gamma = -2.23$; Wheland's approximation (57) with overlap integral of 0.25 gives $\gamma = -1.97$; Longuet-Higgins' method (58) leads to $\gamma = -1.81$; etc. Correlation with the p bands of the absorption spectra of the respective hydrocarbons were reported by Bergman (49), with methyl affinities of aromatic hydrocarbons (275) by Matsen (51), etc.

XI. COMPARISON OF THE ELECTRON AFFINITIES DATA

A correlation between the polarographic data of Bergman (49) and the potentiometric electron affinities, determined in this writer's laboratory and corrected for the ion pairing in THF solution, is shown in Figure 6. The agreement is excellent, adding to the credulity of both sets of data. The slope of the line deviates by 16% from unity, indicating that solvation of the radical-ions by 2-methoxyethanol is more sensitive to the ion's size than the interactions taking place in THF. Since $CH_3OCH_2CH_2OH$ may hydrogen bond to the negative ion, while THF cannot, this conclusion seems to be plausible. The potentiometric data of Hoijtink (33), when used for such a correlation, show some scatter and lead to an even greater deviation of the slope from unity (about 30%).

Comparison of the electron affinities determined in the gas phase with those derived from potentiometric studies is also instructive. This is shown in Figure 7, and the experimental points fit, within the experimental errors, to a 45° line, indicating that the solvation free energies of the investigated radical-ions appear to be virtually constant in THF solution. The only serious deviation is shown by anthracene. Since the

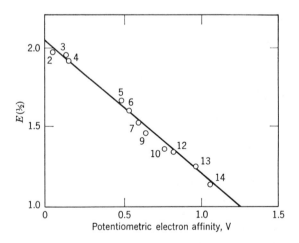

Fig. 6. Polarographic half-wave reduction potentials of aromatic hydrocarbons determined in 2-methoxyethanol-1 compared with the respective potentiometric reduction potentials determined in tetrahydrofuran. Consult Table IV for the meaning of the numbers.

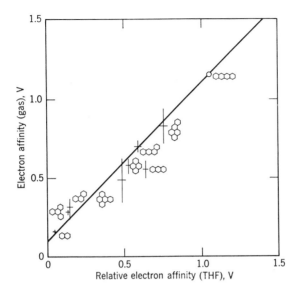

Fig. 7. Comparison of absolute electron affinities of aromatic hydrocarbons determined in the gas phase with the respective potentiometric relative reduction potentials determined in tetrahydrofuran. The line is drawn as the "best" line through experimental points having a fixed slope of unity. The length of the arms of each cross gives the experimental error of the respective determination.

potentiometric data for anthracene were confirmed through studies of equilibria (34,37), it appears that the value determined in the gas phase is in error. Indeed, this value does not fit other correlations reported by Becker and Chen (10). For example, the point representing anthracene deviates from the line correlating the gaseous electron affinities with ionization potential or with methyl affinity of aromatic hydrocarbons (see Fig. 4 of ref. 10). Correlation of gaseous electron affinities with Hoijtink's potentiometric data would lead to an improbable slope of the resulting line.

Electron affinity data obtained by the different techniques are collected and compared in Table IX. Its inspection shows a remarkable agreement between all the reported data, the potentiometric values of Hoijtink being slightly too high. The reason for this deviation is intriguing. In Hoijtink's procedure a sodium biphenyl solution was continually flowed through a narrow and long capillary as the potential was measured. This leads to an electrocapillary effect which increases

TABLE IX
Comparison of Electron Affinities Determined by Different Techniques

| | $\Delta\epsilon$ | | |
System	Perylene–naphthacene	Pyrene–anthracene	Naphthalene–anthracene
Polarographic (Bergman)	0.12	0.15	0.52
Polarographic (Given)	—	0.15	0.58
Polarographic (Hoijtink)	0.09	0.17	0.54
Potentiometric (Hoijtink)	0.19	0.17	0.69
Potentiometric (Szwarc)	0.09	0.11	0.60
Equilibrium (Szwarc)	0.09	0.11	—

the observed potential if the glass wall is negatively charged. Moreover, the effect becomes more pronounced as the resistance of the solution in the reactor decreases, i.e., the added potential would be greater for, say, perylene than for naphthalene. Such an effect was avoided in studies performed in this writer's laboratory. The capillary was short and the flow interrupted when the potential was recorded.

Finally, the effect of liquid junction calls for some discussion. The concentrations of the reagents across the boundary is represented by Figure 8. Diffusion of $B\cdot^-$ between I and II leads to a potential which is, at least, partially balanced by the opposite potential caused by the diffusion of $A\cdot^-$ from III to II. The decrease in the concentration of $B\cdot^-$ taking place between II and III contributes negligibly to the potential of the liquid junction, because in this region the charge is transported mainly by $A\cdot^-$. For the same reason the effect of the decreasing concentration of $A\cdot^-$ between II and I is also insignificant. A thorough discussion of liquid junction potentials is given in *Principles of Electrochemistry* by McInnes (61).

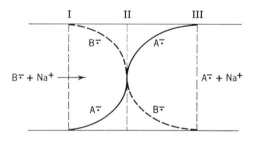

Fig. 8. Schematic representation of the concentration gradients in a liquid junction. Solid line concentration of $A\cdot^-$, dashed line concentration of $B\cdot^-$.

XII. EQUILIBRIA BETWEEN ALKALI METALS AND AROMATIC HYDROCARBONS

The reduction by alkali metals of aromatic hydrocarbons and some related compounds to the respective radical-anions was already discussed. Similar reduction may be achieved by alkaline earth metals (62,63) and their amalgams. It was reported (63) that amalgams of Ca or Mg, but not the pure metals, may reduce aromatic hydrocarbons. This observation was accounted for by the lower photoelectric work function of the amalgam when compared to the pure metal (64). Undoubtedly, the work function of different metals is correlated with their reductive capacity, e.g., sodium or potassium (work functions 2.3 and 2.2 V, respectively) easily reduce anthracene under conditions at which barium and magnesium (work functions 2.7 and 3.7 V, respectively) are ineffective. Nevertheless, the proposed explanation for the amalgam's activity seems unreasonable. Apparently, the lowering of the work function arises from higher heat of reaction $Ca^{2+}(g)$ + mercury(l) → Ca^{2+}(in mercury solution) when compared with the process $Ca^{2+}(g)$ + calcium metal → Ca^{2+} in calcium lattice. Formation of calcium anthracene in solution requires not only emission of an electron but also removal of Ca^{2+} ions; the latter is unfavorable for the amalgam. It is probable that the ease of the amalgam reduction and the inertness of the pure metal result from *kinetic* factors associated with this heterogeneous reaction, and the structure of the reacting surface apparently plays a decisive role in determining the rate of the process.

The equilibrium between an alkali metal, Alk, and an electron acceptor is represented by eq. (4).

$$Alk + Acceptor \rightleftharpoons Alk^+ + Acceptor^{\cdot-} \qquad (4)$$

The heat of this reaction, referred to gaseous ions, solid metallic alkali, and a crystallinic acceptor is given by

$$\Delta H_{(4)} = \Delta H_{sub}(Alk) + I_p(Alk) + \Delta H_{sub}(Acceptor) - \epsilon(Acceptor)$$

where ΔH_{sub} is the relevant heat of sublimation, $I_p(Alk)$ the ionization potential of the gaseous alkali atom, and $\epsilon(Acceptor)$ the electron affinity of the investigated species. The change in free energy is given by a similar equation, viz.,

$$\Delta G_{(4)} = \Delta G_{sub}(Alk) + I_p(Alk) + \Delta G_{sub}(Acceptor) - \epsilon(Acceptor)$$

because the entropy change of the gaseous process

$$Alk(g) + Acceptor(g) \rightleftharpoons Alk^+(g) + Acceptor^{\cdot-}(g)$$

is probably insignificant. However, if the reaction takes place in solution, two additional processes must be considered, namely, the solvation
of the gaseous Alk^+ ions and of the negative $Acceptor^{\cdot-}$ radical-ions.
Hence, the total heat of the reaction is decreased because $\Delta H_S(Alk^+)$
and $\Delta H_S(Acceptor^{\cdot-})$ are negative. In discussing the equilibrium, the
terms $\Delta G_S(Alk^+)$ and $\Delta G_S(Acceptor^{\cdot-})$ have to be added to the
equation giving the free energy change of the solution process. The
problem becomes even more complex when ion pairs, and not free
dissociated ions, are formed. Then, additional enthalpy or free energy
terms referring to the ion-pair association have to be introduced.

The solution equilibria between alkali metals and many acceptors
often lie too far to the right to permit a reliable study of the equilibrium
constant. This is particularly the case when the acceptor is a radical, e.g.,
triphenylmethyl. The difficulty may be avoided if mercury amalgam,
instead of pure alkali metal, is used because the chemical potential of
the alkali is then reduced. The necessary data giving the chemical
potential of sodium in amalgam are available (30), and therefore the
shift in the equilibrium may be calculated. Indeed, it is known that
some metalloorganic compounds may be decomposed into their components if their solutions are stirred with mercury.

It has been known that the reaction of biphenyl with metallic
sodium in ethereal solvents leads only to its partial conversion to
sodium biphenyl (33) and the position of equilibrium is determined
by the nature of solvent. The degree of conversion increases at lower
temperatures proving that the overall process is exothermic. In ethereal
solvents and in not too dilute solutions the reaction yields ion pairs,
the proportion of free ions being negligible. Hence, the free energy of
ion-pair solvation is of paramount importance in determining the
position of such an equilibrium. Under these conditions the reaction is
adequately represented by the equation,

$$\text{sodium (metal)} + \text{biphenyl (solution)} \xrightleftharpoons{} Na^+, biphenyl^{\cdot-} \text{ (solution)}$$

and therefore, the spectrophotometrically determined ratio,

$$[Na^+, biphenyl^{\cdot-}]/[biphenyl] = K$$

is independent of the initial concentration of biphenyl.

The dependence of the equilibrium constant K on the nature of the
solvent and on temperature was investigated by Shatenstein and his
co-workers (65,66). His results are shown in Table X, which also

TABLE X

Equilibrium Constant for the Reaction:

Metallic Sodium + Biphenyl (Solution) $\rightleftarrows$ Biphenyl$\cdot^-$,Na$^+$ (Solution)[a]

$$K = [B\cdot^-,Na^+]/[B]$$

T,°K	MEE	1,2-DMPr	THF	MeTHF[b]	DEE	THP	1,3-DMPr
318	0.07	0.07	0.08	—	—	—	—
313	0.12	0.09	0.10	—	—	—	—
303	0.28	0.20	0.20	—	—	—	—
293	0.75	0.49	0.36	0.02	0.07	—	—
283	2.55	1.40	0.66	0.036	0.11	—	—
273	7.0	5.0	1.50	0.055	0.19	0.06	—
263	—	—	2.90	0.11	0.39	0.10	0.12
258	—	—	—	0.14	0.61	—	0.20
253	—	—	—	0.20	1.25	0.17	0.34
248	—	—	—	0.29	2.25	—	0.61
243	—	—	—	0.45	8.7	0.29	1.20
238	—	—	—	0.64	—	—	2.75
233	—	—	—	1.18	—	0.48	—
228	—	—	—	—	—	0.75	—

[a] MEE = 1,2-methoxy-ethoxy-ethane; 1,2-DMPr = 1,2-dimethoxy-propane; THF = tetrahydrofuran; MeTHF = 2-methyltetrahydrofuran; DEE = 1,2-diethoxyethane; THP = tetrahydropyrane; 1,3-DMPr = 1,3-dimethoxypropane.

[b] Ref. 67.

includes similar data obtained in this writer's laboratory. Plots of log K versus $1/T$ are shown in Figure 9, and the respective heats and entropies of reaction are listed in Table XI. It is obvious that solvation provides much of the driving force for the reaction and that the solvating power is strongly affected by steric factors. For example, at 25°C the equilibrium appears to be almost completely shifted to the right in 1,2-dimethoxyethane (DME), only 7% is converted in 1,2-methoxyethoxyethane, and the conversion is imperceptible in diethoxyethane. Interestingly, no conversion is observed in 1,1-dimethoxyethane; conversion reaches 100% in 1,2-dimethoxyethane and only 6% in 1,3-dimethoxypropane. Also, it is significant to note that 1,2-dimethoxypropane is a much better solvating agent for Na$^+$ ion pairs than 1,3-dimethoxypropane. While the equilibrium constant $K = [Na^+,B\cdot^-]/[B]$ has a value of 5.0 in the former solvent at 0°C, its value in the latter is only 0.04. Obviously, the cooperative coordination of both oxygens with the cation requires some specific spacial configuration which may be attained with the 1,2- but not with the 1,3-derivative.

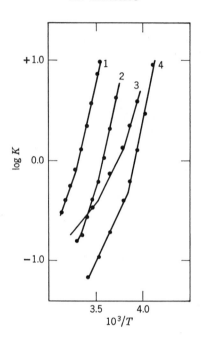

Fig. 9. Equilibrium constants of the reaction. Sodium metal + biphenyl (solution) ⇌ sodium biphenyl (solution) as functions of temperature for various solvents: *1*, dimethoxyethane–heptane; *2*, methoxyethoxyethane–heptane (MEE–Hp); *3*, tetrahydrofuran–heptane; *4*, diethoxyethane.

TABLE XI
Heat and Entropy Change in the Reaction of Biphenyl with Metallic Sodium

Solvent[a]	$K(273°K)$	ΔH[b], kcal/mole	ΔS, eu
MEE	7.2 ± 0.8	-17.4 ± 0.5	-60 ± 2
1,2-DMPr	4.6 ± 0.2	-16.5 ± 0.7	-58 ± 3
THF	1.4 ± 0.1	-11.2 ± 0.5	-40 ± 2
DEE	0.2 ± 0.05	-9.6; at the lowest temp. $\Delta H = -22$	-38 ± 2 at the lowest $\Delta S = -86$
1,3-DMPr	0.04	-15.5 ± 0.3	-63 ± 1
THP	0.06	-6.8 ± 0.2	-31 ± 1

[a] The meaning of the symbols is explained in the footnote of Table X.

[b] These values should be taken with some reservation because the van't Hoff lines are slightly curved. The exothermicity becomes larger at lower temperatures. A striking example is provided by DEE where $-\Delta H = +9.6$ kcal/mole at 20°C and increases to $+22$ kcal/mole at $-30°C$. Furthermore, small experimental errors in determination of K may lead to relatively large errors in ΔH.

Comparison of the equilibrium established between metallic lithium and biphenyl with that involving metallic sodium is most instructive. In diethoxyethane 45% of biphenyl is reduced by lithium, but only 15% by sodium, and the difference is even larger in tetra-hydropyrane, the conversions being 80 and 10%, respectively, for lithium and sodium (65). The heats of sublimation and the ionization potentials are greater for lithium ($\Delta H_{sub} = 37$ kcal/mole, $I = 5.36$ V) than those of sodium ($\Delta H_{sub} = 26$ kcal/mole, $I = 5.12$ V). Therefore, the heat of solvation of lithium ion pairs must be substantially larger than that of the sodium in order to account for the above observations. The solvation effects are, however, most specific because Shatenstein found 4% of the biphenyl to be reduced by sodium in dioxane but none by lithium. Further examples of the specificity of interaction between solvent and cation are provided by some findings concerned with the reducing power of sodium and potassium. Sodium is a more powerful reducing agent than potassium in tetrahydrofuran or tetrahydropyrane, but the reverse order is found in 1,3-dimethoxypropane (66).

A word of caution is needed. Radical-anions become readily associated with other ion pairs in ethereal solvents. For example, shifts in the absorption maximum of the lithium diphenylketyl in tetrahydro-furan were observed when even small amounts of LiBr (less than $10^{-3}M$) were present (75). These findings are illustrated by Figure 10,

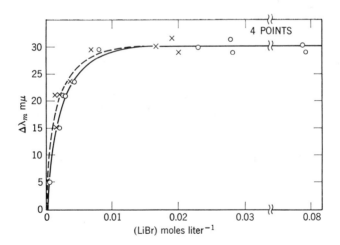

Fig. 10. Shift in the absorption spectrum of lithium diphenylketyl in tetrahydro-furan arising from the addition of lithium bromide.

taken from a paper by Powell and Warhurst, who interpreted their observations in terms of ion-pair agglomerations. It means that a new equilibrium may be established in the presence of a suitable impurity such as lithium or sodium alcoholate or hydroxide, viz.,

$$\text{biphenyl}^{\cdot-},\text{Na}^+ + \text{impurity} \rightleftharpoons \text{biphenyl}^{\cdot-},\text{Na}^+, \text{impurity}$$

This association shifts the main equilibrium to the right,

$$\text{sodium metal} + \text{biphenyl} \rightleftharpoons \text{biphenyl}^{\cdot-},\text{Na}^+ \text{ (complexed or not)}$$

and consequently the concentration of the spectrophotometrically determined sodium biphenyl increases in the presence of impurities. Such artifacts were observed in this writer's laboratory. They led to erroneous equilibrium constants, and to serious errors in the experimental ΔH of the reaction, because the slope of a Van't Hoff line is most susceptible to small but systematic and temperature dependent errors in the respective equilibrium constants. A more detailed discussion of this problem is given in ref. 117.

The complicating effect of impurities discussed in the preceding section may be advantageously exploited in studies of interesting equilibria of ion pairs with various complexing agents (E), e.g.,

$$\text{biphenyl}^{\cdot-},\text{Na}^+ \text{ (solution)} + n\text{E} \rightleftharpoons \text{biphenyl}^{\cdot-},\text{Na}^+ (\text{E})_n, \qquad K_E$$

The results of such studies were reported (67) for the systems sodium metal, biphenyl, and glyme-3 $(CH_3OCH_2CH_2OCH_2CH_2OCH_3)$ or glyme-4 $(CH_3OCH_2CH_2OCH_2CH_2OCH_2CH_2OCH_3)$ in tetrahydropyrane and 2-methyltetrahydrofuran. Denoting the equilibrium constant of the above described complex formation by K_E, one may easily verify that

$$K_E[E]^n = (K_{ap} - K)/K$$

where K_{ap} is the apparent equilibrium constant defined as the ratio

$$K_{ap} = \frac{\text{total concentration of B}^{\cdot-} \text{ in the presence of a complexing agent}}{\text{concentration of the unconverted B}}$$

and K is the respective equilibrium constant found in the absence of the complexing agent, i.e.,

$$K = [B^{\cdot-}{}_t]/[B] \text{ in the absence of E}$$

Here, B and B·⁻ denote biphenyl and any salt or complexed salt of biphenyl·⁻ radical-ion as determined by spectrophotometric method. The temperature dependence of K_{ap} and K for the system metallic

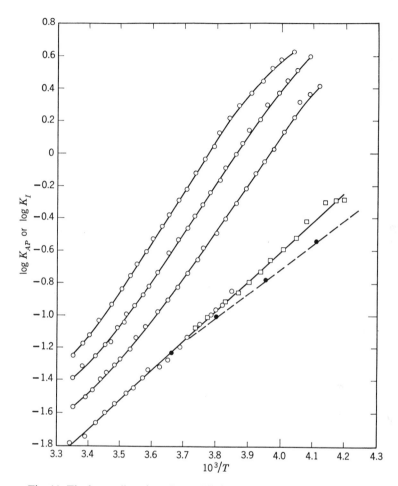

Fig. 11. The lowest line gives the equilibrium constant of the reaction sodium metal + biphenyl (solution) $\rightleftarrows$ sodium$^+$,biphenyl$\cdot^-$ (solution), in tetrahydropyran at temperatures varying from 25 down to $-40°$C. ($\bullet$), Shatenstein data; ($\bigcirc$), Slates and Szwarc data; ($\square$), from a separate experiment. Note the degree of agreement between Shatenstein's data and those of Slates and Szwarc. The three higher lines represent the effect of added glyme-4. The increase in total concentration of B$\cdot^-$ arises from formation of complex between B$\cdot^-$,Na$^+$ and glyme.

sodium, biphenyl in tetrahydropyran solution, and glyme-4 is shown in Figure 11 for various concentrations of glyme. From such graphs plots of log $(K_{ap} - K)/K$ versus log [glyme] were constructed (see ref. 67). They were found to be linear and have a slope of unity, proving that the complexing of glyme-4 (as well as glyme-3) with the sodium biphenyl ion pair corresponds to a well defined 1:1 stoichiometry, i.e., we deal with a *chemical* coordination and not with a physical association.

Ion pairs may exist in some solutions in two forms—contact and solvent separated, the absorption maximum of the latter being shifted to longer wavelength when compared with the former (116). A similar phenomenon is observed when a small amount of powerful solvating agent is added to a solution of ion pairs in a poorly solvating medium (118). The spectrum of sodium biphenyl and of its 1:1 complex with glyme-3 absorbs at $\lambda_{max} = 400$ mμ within the investigated range of temperatures. The λ_{max} for the complex with glyme-4 appears at 407 mμ at $-40°$C, although at 25°C the absorption shows a maximum at 400 mμ. In fact, for every concentration of glyme-4 it was possible to find a temperature at which two peaks of equal optical density were seen, one at 400 mμ, the other at 407 mμ. It could be concluded that one of them corresponds to a noncomplexed pair and the other to a pair separated by the glyme. However, it was shown that the concentration of the former is lower than that calculated from the spectrophotometrically determined overall conversion and that of the latter is higher. Hence, the complex exists in two forms, one possessing the glyme on the outside of the ion pair (like the complex with glyme-3), the other being separated by the glyme. This is the first example of isomerism involving different positions of solvating agent in a solvated ion pair.

XIII. DIRECT CAPTURE OF ELECTRONS IN IRRADIATED SOLUTIONS AND GLASSES

It is well known that the primary action of high-energy γ-radiation, or x-rays, fast electrons, etc., causes ionization or electronic excitation of the irradiated molecules. Two theories of radiation chemistry inspired most of the work in this field. The earlier theory, developed by many workers (see, e.g., ref. 276), proposes a heterogeneous dissipation of the radiation energy and creation of regions of high concentration of fragments—ions, electrons, radicals, and excited molecules. In these so-called spurs or hot spots most of the fragments combine or interact and, thus, they yield the observed products, while only a small fraction diffuses out and eventually achieves a homogeneous distribution throughout the irradiated material.

The second theory was put forward by Platzmann (277). He suggested that the electrons formed in the primary ionization process become solvated and, therefore, "protected." Thus, many thermal electrons are eventually formed in such a process. Their capture by acceptors in the gas phase was discussed in Sec. III. The same event may lead to electron capture in a liquid or glass.

Two methods permit us to study such phenomena: (1) The irradiation may be performed at very low temperatures in an extremely viscous medium, essentially in a glass. This prevents or slows down the recombination of the positively and negatively charged species, and under such conditions their concentration may be increased to a level at which detection and investigation of their chemistry becomes possible. (2) The irradiation is performed by a pulse of a very high intensity but extremely short duration. The events taking place after cessation of the pulse may be observed by suitable electronic techniques in periods as short as a few microseconds.

Radiolysis of organic glasses was extensively investigated by Hamill. His first observations (68) dealt with systems involving frozen solvents, or their mixtures, such as 2-methyltetrahydrofuran, 3-methylpentane, isopentane, methylcyclohexane, diethylether, and ethanol. Samples were irradiated at $-196°C$, the typical dose rate being $\sim 10^{18}$ eV/liter-min, and the dose varied from 1×10^{21} to 5×10^{21} eV/liter. A broad absorption band was developed in irradiated polar glasses, its intensity increasing monotonically from 4000 Å up to the limits of applicability of his spectrophotometer, viz., $\sim 13,000$ Å. This band was attributed to solvated electrons. Indeed, similar absorption spectra, reported in the past (69), had been assigned to solvated electrons. Modern techniques of pulse radiolysis (see Sec. XV) allow us to produce solvated electrons in a variety of media and to determine unequivocally their absorption spectra. For the sake of illustration the spectra of solvated electrons formed in aliphatic alcohols (144) are shown in Figure 12. The absorption observed in frozen glasses is usually broader because, due to the rigidity of the matrix, a variety of irregular "solvation shells" are formed, and slightly different spectra are produced by electrons located in different shells (69).

Addition of naphthalene or biphenyl to the glass dramatically changed the absorption spectrum of the irradiated sample (68). Instead of the broad band, the characteristic absorption spectrum of naphthalene·⁻ or biphenyl·⁻ radical-anions appeared. It has been concluded, therefore, that the electrons formed in the matrix were trapped by the aromatic hydrocarbons, and thus the respective radical anions were produced. Extension of this work showed that the technique may yield

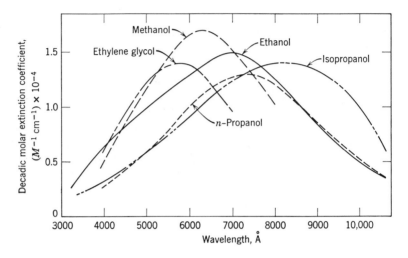

Fig. 12. Spectra of solvated electrons formed in aliphatic alcohols. Determined by pulse radiolysis (144).

radical-anions of benzophenone, nitrobenzene, and tetracyanoethylene (71), and in a later study (72) it was demonstrated that radical-anions may also be formed from phenanthrene, benzanthracene, triphenylene, *ortho*-terphenyl, pyrene, tri- and tetraphenylethylene, cycloheptatriene, and hexamethylbenzene. On the basis of the known extinction coefficients of naphthalene$\cdot^-$ or biphenyl$\cdot^-$ radical-anions, the G value for the captured electrons was determined. In a glass produced by freezing a 0.01 mole % of naphthalene in 2-methyltetrahydrofuran the G value was found to be 1.4. Its value reached the maximum of 4.0 for 1 mole % solution. The presence of carbon tetrachloride in the matrix suppressed the formation of negative radical-anions, e.g., the G value was reduced from 4 to 1.5 and eventually to 0 when 0.16 and 1.6 mole %, respectively, of carbon tetrachloride was mixed with 1 mole % solution of naphthalene in methyltetrahydrofuran. This suggests that carbon tetrachloride competes with the aromatic hydrocarbons for the electrons, whose capture by the former molecules causes their dissociation, viz., $CCl_4 + e^- \rightarrow CCl_3\cdot + Cl^-$.

If two aromatic hydrocarbons are present in the glass, they are expected to compete for the electrons. Therefore, the relative intensities of spectra of the respective radical ions should depend on the mole ratio of the parent hydrocarbons. This, indeed, is the case as shown by

Figure 13. The recorded spectra refer to irradiated samples in which the concentration of one aromatic hydrocarbon was kept constant, whereas the proportion of the other increased. The decrease in the intensity of absorption due to the radical-anion of the former hydrocarbon and simultaneous increase in the absorption intensity of the other is seen clearly. Moreover, the isosbestic point, revealed by Figure 13, proves that the total number of captured electrons remained constant. From such data the relative electron-capture efficiencies of aromatic hydrocarbons were determined, and the relevant results are collected in Table XII. It was expected that the relative efficiencies would increase with the size of the molecule or with its polarizibility. This was not the case. However, Hamill reported a monotonic relation of the efficiency with electron affinities of the investigated aromatic hydrocarbon. This observation raises the question of whether his results refer to an equilibrium established in the glass or whether they are determined by the kinetics of the electron capture. It seems that the latter is the governing factor. The efficiencies determined by Hamill are only slightly affected by the nature of the aromatic substrate; the

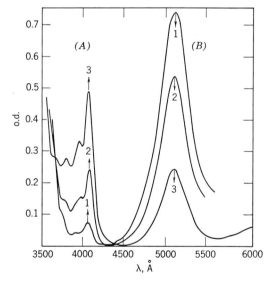

Fig. 13. Competition of two aromatic hydrocarbons for electrons formed by γ irradiation in a frozen glass. A = triphenylene; mole % = 0.20%, 0.10%, and 0.04% for curves 1, 2, and 3, respectively. B = biphenyl; 0.158 mole %. Note that the appearance of the isosbestic points proves that $[A]^{\cdot-} + [B]^{\cdot-}$ = const.

TABLE XII

Relative Electron Capture Efficiency of Aromatic Hydrocarbons in Organic Glasses (2-MeTHF) (72)

Hydrocarbon	Relative electron capture efficiency	Electron affinity in THF solution relative to biphenyl	Molar refraction
Biphenyl	1.00	0.0	52
Triphenylene	1.04	0.132	71
Phenanthrene	1.22	0.142	62
Ortho-terphenyl	1.40	—	74
Cycloheptatriene	1.94	—	30
Pyrene	1.95	0.529	67
Triphenylethylene	2.70	—	87
1,2-Benzanthracene	2.60	0.590	66
Tetraphenylethylene	2.76	—	118
Anthracene	2.76	0.642	65
Hexamethylbenzene	2.80	—	46

observed maximum change is less than a factor of 3. On the other hand, a very large variation by many powers of 10 would be expected had the equilibrium governed the outcome of the experiments. The relative electron capture efficiencies determined in the glass are somewhat similar to those observed in the gas phase (see Sec. III and ref. 9). The gaseous electron capture coefficients are also not much affected by the nature of the acceptor and their values vary only within a factor of 10. However, inspection of the latter data does not reveal any regularities or correlations with electron affinities.

The relative electron-capture cross sections of aromatic hydrocarbons were reinvestigated by Hamill and his co-workers (73,74) by means of an alternative technique, namely, by studying the competition between an aromatic hydrocarbon and organic halides for the electrons. This also permitted the determination of the relative reactivities of halides towards electrons.

The matrix may also compete with the solute for the electrons. At a sufficiently low concentration of an electron acceptor the γ irradiation of the glass produces two spectra, that of the radical-ion and a broad one characteristic of the solvated electron. Subsequent irradiation with the selected visible light bleaches the broad spectrum with concomitant increase in the spectrum of radical-anion. The absorption of the visible light ejects the solvated electrons from their shallow traps and permits their capture by the aromatic hydrocarbon. Light absorbed by the radical-anions eventually bleaches their spectrum. The ionization or excitation of radical-anions facilitates the release of the captured electrons into the glass and leads to their ultimate capture by the positive "holes." Of course, the latter had been formed during γ irradiation of a glass as a consequence of electron ejection from the matrix.

In hydrocarbon glasses the G value for an electron capture is lower than that found in polar glasses. For example, in 3-methylpentane containing naphthalene the G value is only 0.15, but it increases to 0.4 on the addition of 1 mole % of triethylamine. Apparently, the stability of the radical-anion and of the positive hole increases with their solvation, or alternatively the presence of amine facilitates electron transfer to the aromatic acceptor.

Finally, we may stress that the spectra of radical-anions formed in a frozen glass through γ irradiation differ only slightly from those formed in the same medium by chemical reduction with sodium or potassium. On the whole, the agreement between both spectra is remarkably good. There is, however, a slight shift to the red in the spectrum developed in the glass. The fact that the γ irradiation forms free ions, whereas chemical reduction produces ion pairs, accounts for this observation (see also ref. 116). In some systems the spectrum of a γ-irradiated glass shows a marked vibrational structure. Such a spectrum was observed when glass containing anthracene was irradiated. This may be seen from inspection of Figure 14 where, for the sake of comparison, the spectrum of anthracene·⁻ in solution is also depicted. The lack of structure in solution may be due to the higher temperature of the sample, or to the presence of counterion, the soft vibration of which may broaden the spectrum.

An attempt was made to produce the radical-ion of styrene. Radiolysis of a glass containing styrene produced spectra with maxima at 4100 and 6000 Å (71) which were attributed to the styrene radical-anion. However, pulse radiolysis of styrene solution investigated by Dainton and his associates (76) failed to reveal their existence. Only a single peak at 340 mμ, apparently that of living polystyrene, was observed in carefully purified organic solvents such as cyclohexane, dioxane, tetrahydrofuran, and benzene. The problem needs further studies because Dainton's observations differ from those reported by Katayama (77) and by Keene, Land, and Swallow (78).

More conclusive evidence was obtained for the formation of radical-anions of butadiene and its derivatives. Radiolysis of a frozen

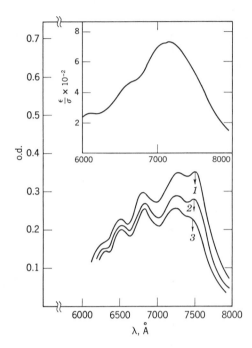

Fig. 14. Spectrum of anthracene·⁻ radical-ion in a frozen glass at 70°K (lower curve) and in solution at room temperature (formed by alkali-metal reduction). Note the vibrational structure in the lower spectrum and its absence in the upper one.

solution of butadiene in methyltetrahydrofuran yields species that absorb at 388 and 570 mμ (79). Similar spectra were recorded on examining glasses containing butadiene derivatives. This is shown by the data collected in Table XIII. However, irradiation of glasses in which nonconjugated dienes, such as 1,4-pentadiene or 1,5-hexadiene, were dissolved resulted in a typical broad spectrum of solvent-trapped electrons.

Two methods of identification of the species produced from butadiene and its derivatives were available. The optical spectrum of 1,3-butadiene radical-anion was calculated on the basis of the simple HMO approach, and the results led to λ_{max} of 382 and 473 mμ if $\beta = -2.62$ eV. The agreement with observation is reasonable, since it is known that this approach is much too crude to give more reliable values. Interestingly, the spectrum is only slightly affected by the

TABLE XIII

Absorption Spectrum of Radical-Anions Derived from Butadiene and Its
Derivatives in γ-Irradiated MeTHF Glass

Hydrocarbon	λ_{max} (higher peak), mμ	λ_{max} (lower peak), mμ
1,3-Butadiene	388	570
Isoprene	390	565
cis-1,3-Pentadiene	388	563
2,4-Hexadiene	390	575
2,3-Dimethyl-1,3-butadiene	390	530
1,3-Cyclohexadiene	395	~ 575
1,3-Cyclooctadiene	380	~ 500

presence of methyl substituents, as may be seen from Table XIII, indicating an insignificant influence of hyperconjugation or induction on the spectra of the *negative* radical-ions. Alternatively, the ESR spectra of the glasses could be examined. That observed in a sample containing 1,3-butadiene revealed a quintuplet resembling the one reported by Levy and Meyers (80) for the butadiene·⁻ radical-anion generated electrolytically. However, these workers obtained a better resolution and could, therefore, determine the hyperfine splitting due to the four terminal hydrogens as well as the triplet due to the internal protons.

The irradiation performed in methanol glass leads to obvious complications. Methanol glass stabilizes the positive "holes" by the reaction

$$CH_3OH^+ + CH_3OH \longrightarrow CH_3OH_2^+ + CH_3O\cdot$$

The released electrons are mobile and may be trapped by aromatic hydrocarbons or by some related compounds, e.g., by styrene. However, the resulting radical-anions eventually remove protons from neighboring methanol molecules forming the respective radicals (86). Thus, the cyclohexadienyl radical resulting from the reaction $C_6H_6^- + CH_3OH \rightarrow C_6H_7\cdot + CH_3O^-$ was identified by its optical spectrum (74) as well as by its ESR spectrum (86,87). Interestingly, in the presence of styrene a very sharp peak appeared at about 320 mμ. This spectrum agrees well with that obtained in methyltetrahydrofuran glass in the presence of PhCHBrCH$_3$ (74). In the latter case, the following reaction took place:

$$PhCHBrCH_3 + e^- \longrightarrow Ph\dot{C}HCH_3 + Br^-$$

Hence, it appears that on irradiation of styrene in methanol an electron is captured and the respective radical-ion is initially formed. However, this species abstracts a proton from a neighboring methanol molecule and produces the $PhCHCH_3$ radical. The absolute rate constants of such proton-transfer reactions are discussed in Sec. XV.

Radical-anions may be formed in the course of γ irradiation by an entirely different route. For example, irradiation of formate salts dissolved in potassium halide pellets (KCl, KBr, or KI) results in the formation of $CO_2{}^{\cdot-}$ through rupture of the H—COO^- bond (81). The radical-anion was identified through its infrared and ESR spectra. This interesting technique was developed and explored by Hartman and Hisatsune (82).

$CO_2{}^{\cdot-}$ radical-anions were also formed by an electron capture process (86). The γ-irradiation of methanol glass saturated with carbon dioxide produced a species identified through its ESR spectrum. This agreed with that reported previously (88) for the $CO_2{}^{\cdot-}$. The $CO_2{}^{\cdot-}$ spectrum is observed even if methanol contains benzene. Apparently, carbon dioxide competes efficiently with benzene for the electrons. Hence, the proportion of products derived from benzene decreases significantly on the CO_2 addition (86).

XIV. FORMATION OF POSITIVE RADICAL-IONS BY IRRADIATION

Let us discuss more thoroughly the actual process of electron formation and its trapping in organic glasses. The interaction of an ionizing agent with a molecule results in ejection of an electron (which may ionize adjacent molecules by virtue of its high kinetic energy) and leaves a positive residue in the glass. In frozen methyltetrahydrofuran, or more generally, in glasses composed of ethers or some related compounds, the positive residue becomes stabilized by proton transfer. For example, $C_5H_{10}O^{\cdot+} + C_5H_{10}O \rightarrow C_5H_{10}\overset{+}{O}H + \dot{C}_5H_9O$ (radical). Such a stabilization immobilizes (traps) the positive hole and makes it relatively unreactive, whereas the electron may move through a crystal by tunneling, its velocity being calculated as 1Å in 10^{-14} sec (83). Eventually, the electron is trapped by an acceptor or by a suitably distorted site in a glass. The latter represents, therefore, a shallow "trap" of low potential energy.

In some glasses, or microcrystalline solids, the electrons become stabilized by a process leading to immobile and inert negative centers.

Carbon tetrachloride or other organic halides are good examples of these microcrystalline solids (84). Thus, $CCl_4 + e^- \rightarrow CCl_3 \cdot + Cl^-$, or more generally, $RCl + e^- \rightarrow R \cdot + Cl^-$. What happens to the positive center? Such a center, e.g., $CCl_4 \cdot ^+$, moves through the matrix by resonance involving the neighboring CCl_4 molecules, viz., $CCl_4 \cdot ^+ + CCl_4 \rightarrow CCl_4 + CCl_4 \cdot ^+$, and this leads to its stabilization. We may remark in passing that no CCl_4^+ ion is observed in the mass spectrum of CCl_4, even at relatively high pressures when the deenergization of an excited ion should be fast (85). Apparently either the gaseous CCl_4^+ ion is in a repulsive state, or its potential energy minimum is very shallow.

Addition of substrates of sufficiently low ionization potential to glasses in which mobile positive holes are created leads to the formation of positive radical-cations. For example, addition of tetramethyl-phenylenediamine (TMPD) to frozen carbon tetrachloride dramatically changed the spectrum of this polycrystalline solid when it was irradiated by γ-rays. Instead of the broad absorption at about 400 mμ attributed to the positive holes, a characteristic spectrum of Würster blue (TMPD$\cdot^+$) appeared (84). This is shown in Figure 15 where curve *4* refers to the irradiated pure CCl_4 and curve *1* to the solid in which

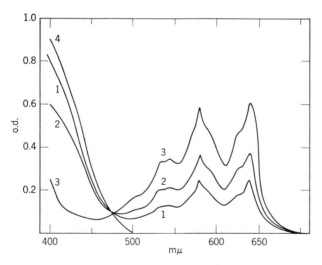

Fig. 15. Absorption spectra of γ-irradiated TMPD (7 mM liter^{-1}) in CCl_4 at 77°K: *1*, irradiated to 3.6×10^{18} eV ml^{-1}; *2*, same as *1*, then photobleached; *3*, same as *1*, after warming to $\sim$140°K and then chilling to 77°K; *4*, pure CCl_4, same dose.

TMPD was dissolved. Photobleaching of the irradiated sample decreased the intensity of the 400 mμ absorption and increased that due to the TMPD$\cdot^+$. This is shown by curve 2 of Figure 14. The UV light increased the mobility of the residual CCl_4^+ holes by enhancing their exchange with CCl_4 and, thus, some "diffused" to the amine and subsequently reacted with it, viz., $CCl_4\cdot^+ + TMPD \rightarrow CCl_4 + TMPD\cdot^+$. Such a "diffusion" becomes fast when the sample is heated to 140°K and then chilled to 77°K. In an actual experiment all the $CCl_4\cdot^+$ holes disappeared and the intensity of the spectrum of Würster salt greatly increased (curve 3 of Fig. 14).

The above technique was applied to the generation of many radical-cations, e.g., $Ph_3N\cdot^+$, $Ph_3P\cdot^+$, the radical-cations of tetramethyl benzidine, etc. (84). All these species were produced by an electron transfer from a donor to a positive hole, and not by a direct ejection of an electron from the donor. The results shown in Figure 16 confirm this conclusion. The intensity of the spectrum of $Ph_3N\cdot^+$ radical-cation increased with concentration of the parent molecule, but eventually its value became constant and independent of amine's concentration. Such a situation arises when virtually all the positive holes are scavenged by the solute.

Irradiation of solutions of aromatic hydrocarbons in frozen carbon tetrachloride produced monopositive radical-ions of benzene, toluene, biphenyl, naphthalene, phenanthrene, anthracene, etc. (89). Of the 18 hydrocarbons investigated by this technique, anthracene was

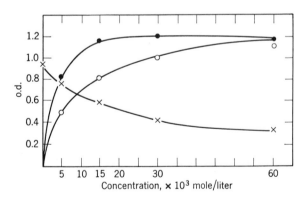

Fig. 16. Optical densities of ϕ_3N^+ at 640 mμ ($\bigcirc$) and of CCl_4^+ at 400 mμ ($\times$) vs. concentration of ϕ_3N. Sample irradiated by γ-rays at 77°K, dose 3.6 × 10^{18} eV ml^{-1} ($\bullet$). After warming samples to 142°K, then returning to 77°K.

the only one for which the spectrum of the positive radical-ion had been reported previously (90). However, spectra of other aromatic, alternant radical-cations, not investigated by Hamill's technique, are known (90,91), viz., those of triphenylene, pyrene, tetracene, perylene, coronene, 3,4-benzpyrene, and quaterphenyl. These species were obtained either by photoionization of the parent hydrocarbons in rigid glasses or by using concentrated sulfuric acid at room temperature as an oxidizing agent.

Finally, the γ-irradiation technique led to the formation of radical-cations derived from nonconjugated olefins (92). Isobutene, cis- and trans-butene-2, tetramethylethylene and many other olefins were investigated. The spectra of the resulting radical-cations showed absorption maxima between 600 and 750 mμ.

How are hydrocarbon glasses affected by γ irradiation? This problem was investigated by Gallivan and Hamill (93) who showed that positive holes are mobile even in hydrocarbon matrices. In fact, their results indicate that such holes may migrate as far as the electrons, i.e., within a range of about 2×10^4 Å. Gallivan and Hamill presented also some evidence indicating that both positive and negative radical-ions of biphenyl were formed when its solution in 3-methylpentane glass was irradiated. Simultaneous formation of positive and negative radical-ions was demonstrated for glasses formed from ketones (94), and their recombination led to chemiluminescence. For example, when pure frozen acetone was irradiated with γ-rays and then removed from liquid nitrogen it began to emit a bluish light which continued for more than a minute.

XV. PULSE RADIOLYSIS

Pulse radiolysis, as has been stated before, involves studies of transient species produced in an investigated sample by a pulse of ionizing radiation delivered in about a microsecond or less. Each pulse delivers energy of about 5–50×10^{18} eV per cubic centimeter of the sample. This energy is provided in the form of fast electrons accelerated to 1–15 million volts, or in the form of x-rays. The initial reaction involves solvent molecules which become ionized or electronically excited, and the fate of the resulting fragments may be followed, e.g., by a spectrophotometric technique. Addition of a chosen solute does not affect the primary process, although it alters the fate of the initial fragments. First studies of pulse radiolysis were reported in 1960 (95–97), and an excellent review of this subject was published in 1965 (98). The reader may find there many technical details and a complete list of the experimental results available at that time. Therefore, we shall limit our discussion to some general remarks and then concentrate our attention on some specific topics pertinent to the subject of this review.

Much of the work in the field of pulse radiolysis is concerned with the generation of hydrated electrons in pure water and in aqueous solutions. The absorption spectrum of hydrated electron has a maximum at 7200 ± 100 Å corresponding to the extinction coefficient of about 1.6×10^4. The most reliable extinction coefficient was determined by Rabani, Mulac, and Matheson (121), who converted the solvated electron into $C(NO_2)_3^-$ by reacting it with tetranitromethane. The extinction coefficient of the latter anion is known. With equipment available at the present time, the products of pulse radiolysis may be detected at concentrations as low as $10^{-8}M$, i.e., optical densities as low as 0.002 are commonly measured (119) and sensitivity down to 0.00002 was reported (120).

Although spectrophotometric methods are most frequently used for identification of the transients and for tracing their formation and decay, other techniques have been also utilized, for example, studies of electrical conduction (99) and of ESR spectra. The difficulty in conductance studies arises from the fact that it is not easy to maintain a constant voltage across the electrodes of the cell when its resistance changes rapidly. Moreover, the excess charge due to the absorbed primary electrons, as well as the noise generated in the pulse, pose serious difficulties. To eliminate much of the noise, two oscilloscope tracings are always taken with reverse polarities of the cell electrodes. The conductivity portions of the two signals are then reversed while all the noise components remain unaffected.

Studies of pulse radiolysis render valuable kinetic data. It was possible to determine the absolute rate constants of capture of hydrated electrons by aromatic hydrocarbons, amines, quinones, etc. The results obtained by Hart, Gordon, and Thomas (100) for hydrated electrons are collected in Table XIV, and in the same table the results obtained in ethanol (101) are also listed. Their comparison shows that the rate is insignificantly affected by the solvent. The last four results given in Table XIV show again only a small change in the rate constant of the electron capture process with the size of the aromatic system. These findings may be compared with those of Hamill listed in Table XI. In

TABLE XIV

The Absolute Rate Constant for Capture of a Solvated Electron

Substrate	Solvent	$k, M^{-1}sec^{-1}$
Tetracyanoethylene	Water (pH $\sim$ 7)	1.5×10^{10}
1,3-Butadiene	Water (pH $\sim$ 7)	8×10^9
Benzene	Water (pH $\sim$ 7)	7×10^6
Naphthalene	Water (pH $\sim$ 7)	3.1×10^8
Nitrobenzene	Water (pH $\sim$ 7)	3×10^{10}
Pyridine	Water (pH $\sim$ 7)	1×10^9
Styrene	Water (pH $\sim$ 7)	1.5×10^{10}
Styrene	Water (pH $\sim$ 13)	1.1×10^{10}
Quinone	Water (pH $\sim$ 6.6)	1.2×10^9
Biphenyl	Ethanol	4.3×10^9
Naphthalene	Ethanol	5.4×10^9
para-Terphenyl	Ethanol	7.2×10^9
Naphthacene	Ethanol	10.2×10^9

both investigations the rate of capture seems to increase as one compares biphenyl and naphthalene with anthracene and naphthacene, but by not more than a factor of 2 or 3. The identity of the resulting radical-anions was established beyond any doubt by their absorption spectra (101–104). In alcoholic solvents the aromatic radical-anions decay due to proton transfer from adjacent molecules. The rate constant of the proton transfer from alcohols such as methanol, ethanol, n-propanol, and isopropanol to various aromatic radical-ions was investigated by Arai and Dorfmann (101), and subsequent work led to the respective activation energies (105). They found a good correlation between the observed rates and the acidity of the alcohols. The pertinent results are listed in Table XV.

TABLE XV

Rate Constants of Proton Transfer from Aliphatic Alcohols to
Aromatic Radical Anions at 25°C

Radical-anion	Alcohol	$k, M^{-1}sec^{-1} \times 10^{-4}$
Biphenyl$\cdot^-$	CH_3OH	7
Biphenyl$\cdot^-$	C_2H_5OH	2.6
Biphenyl$\cdot^-$	n-C_3H_7OH	3.2
Biphenyl$\cdot^-$	iso-C_3H_7OH	0.5
Anthracene$\cdot^-$	CH_3OH	8
Anthracene$\cdot^-$	C_2H_5OH	2.3
Anthracene$\cdot^-$	n-C_3H_7OH	2.4
Anthracene$\cdot^-$	iso-C_3H_7OH	0.4

Recent improvements of electronic techniques permitted the extension of pulse radiolysis to the nanosecond region (278). The results demonstrated directly, for the first time, the reality of spurs and made their quantitative studies feasible. This is shown in Figure 17 which gives the decay of the hydrated electron after a 12-nsec pulse. The rapid decay taking place within the time period AB is independent of the radiation intensity, whereas the slower decay BC does depend on the dose. The slower decay arises from the homogeneous bimolecular reaction of hydrated electrons in the bulk of the liquid, the initial rapid one is the heterogeneous process in the spur. This reaction is over in just over 100 nsec (10^{-7} sec).

The spur effect is quite small; in deaerated water the G value for the solvated electrons destroyed in spurs is 0.6, while the yield of those which diffused out is about 2.8. The hydrated electrons are destroyed in the spur by two reactions,

$$e_{aq}^- + OH \longrightarrow OH^-$$
$$e_{aq}^- + H^+ \longrightarrow H$$

The OH radicals and protons are formed in the spur, viz.,

$$H_2O \underset{\longleftarrow}{\overset{h\nu}{\longrightarrow}} e_{aq}^- + H^+ + OH$$

The evidence for the proposed scheme is provided by two observations: (a) The addition of sodium hydroxide in excess of $10^{-2}M$ reduces the

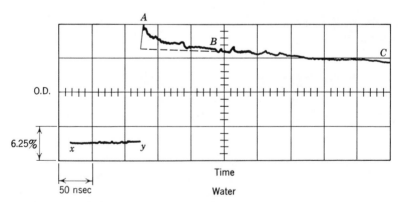

Fig. 17. Oscilloscope tracing of decay of solvated electrons in water. A-B (100 nsec), the rapid decay in spurs. B-C, the slow, homogeneous decay of electrons which diffused into the bulk of the liquid.

spur effect by about 50%. Alkali removes H^+ and prevents the destruction of electrons through proton trapping. (b) The addition of $1M$ ethanol and $10^{-2}M$ alkali removes the spur effect completely, because interaction with alcohol removes OH radical. This interpretation, if correct, gives an estimate for the rate constant of the reaction $C_2H_5OH + OH \rightarrow$ products and $H^+ + OH^- \rightarrow H_2O$. The rate constant of the latter reaction was measured by two methods and its value is $1.4 \times 10^{11} \ M^{-1} \ sec^{-1}$ (279).

Similar studies were performed in ethanol and isopropyl alcohol. The yield of electrons lost in spurs is given by $G = 0.5$, those reacting homogeneously correspond to $G = 1.0$.

XVI. SOLVATED ELECTRONS

Although our knowledge of solvated electrons was extended by pulse radiolysis, this subject is not really new. The blue solutions of alkali metals in liquid ammonia have been known for more than a century (145) although their exact nature is still in dispute. Extensive studies of their conductance, initiated by the pioneering investigations of Kraus (124), left no doubt that the dissolved metal dissociates into ions. At sufficiently low concentration the equivalent conductance increases with dilution and eventually reaches its limiting value, e.g., $\Lambda_0 = 1022$ for the sodium solution at $-33°C$. Since the limiting conductance of Na^+ ions in liquid ammonia is only about 130, the anion must be extremely mobile. These were the first observations which led to the idea of solvated electrons, i.e., it was postulated that

$$Na + solvent \rightleftarrows Na^+ \ (solvated) + e^- \ (solvated)$$

The system alkali metal–liquid ammonia must be more complex than it appears, because at high concentration of metal the equivalent conductance *increases* with concentration (125). The problems associated with this increased conductance cannot be discussed here, and the reader who wishes to explore this subject is referred to the original literature (126,127,129). It appears that a transition to a metallic state takes place in such systems; the valence electrons may move independently of the liquid since their orbitals may overlap and form conductance bands. Several alternative explanations have also been proposed, e.g., those involving a "microscopic Wien effect" caused by the field of the ions (128).

Let us return now to the conducting behavior of dilute alkali metal solutions in liquid ammonia and in amines. The deviations of the equivalent conductance from the Debye-Hückel-Onsager law, which is applicable to dilute, ideal (i.e., completely dissociated) electrolyte solutions, may be explained in terms of ion pairing. Indeed, Kraus interpreted his data in this way and visualized the pair in a vague sense as {M^+ (solvated), e^- (solvated)} bound through coulombic interaction. Detailed analysis of the dependence of Λ/Λ_0 on concentration showed, however, that one equilibrium constant, K_1, is not sufficient to account for the observations,

$$M^+ \text{ (solvated)} + e^- \text{ (solvated)} \xrightleftharpoons{} M^+ \text{ (solvated)}, e^- \text{ (solvated)} \cdots K_1$$

but an agreement could be obtained if an additional dimerization was postulated, e.g.,

$$2(M^+, e^-) \xrightleftharpoons{} M_2$$

Even a better fit may be secured if three equilibria are postulated (130), e.g.,

$$M \xrightleftharpoons{} M^+ + e^-$$
$$M^- \xrightleftharpoons{} M + e^-$$
$$2M \xrightleftharpoons{} M_2$$

This is not surprising because three adjustable coefficients are employed in calculations which account for the experimental data.

The existence of solvated M^+ and e^- ions in equilibrium with their pair is confirmed by several observations:

1. Electrolysis of alkali metal solution leads to the accumulation of the alkali ions (and of the blue colored species) in the cathode compartment, but nothing is evolved or deposited on the anode.

2. The Walden rule applies to alkali solutions in liquid ammonia and in various amines.

3. The salts having a common cation with the dissolved metal affect the conductance in a way accounted for by the mass law (132).

4. The calculated equilibrium constant K_1 for alkali metal and for alkali salt solutions are of the same magnitude.

The M^+ ion formed from the dissolved metal does not differ from that produced on dissociation of other M^+, X^- salts; however, the nature of the solvated electron, of the pair M^+, e^- (often referred to as the monomer), and of the dimer need further elaboration (147).

There are several plausible models proposed for solvated electrons (127). In nonpolar solvents, e.g., in liquid rare gases, the electron seems to be located in a cavity (133–135). Electron–atom repulsion, if sufficiently strong, may stabilize the state of local fluid dilation and thus create the cavity within which the electron moves. Indeed, it is known that the electron–helium atom repulsion is large (136), and therefore this model seems to be applicable to this system. However, in liquid argon the electron–argon interaction is attractive and therefore a different model is needed to describe excess electron in this liquid. Preliminary data of Schnyders, Meyer, and Rice (137), who determined the magnitude of electron mobility (400 cm^2/V-sec at 90°K) and its temperature dependence, favor the free electron model in which the electron is treated as a quasi-free particle scattered by the atoms of the dense fluid.

In polar solvents the electron produces a polarization field which leads to its self-trapping. The binding energy is large, e.g., the heat of solution of an electron is liquid ammonia or in water is of the order 30–40 kcal/mole (138). The short-range electron solvent repulsion leads, however, to the formation of a cavity in which the electron becomes localized; e.g., in liquid ammonia the radius of the cavity is calculated to be about 3 Å. The long-range polarization effect makes such a cavity smaller than that formed in a nonpolar medium (e.g., in liquid helium). Another factor influencing the radius of the cavity arises from the surface energy term which is related to the surface tension of the liquid. The polarization of the liquid is also responsible for the leakage of electron charge density outside the cavity. For example, in liquid ammonia 30–40% of its charge leaks out.

The localization of the excess electron on a solvent molecule, such as ammonia, amines, or ethers is unikely because no low energy antibonding orbitals are available in these species. If the resulting orbital for the electron is spread over many solvent molecules, the resulting system is referred to as a polaron and the solvent may then be treated as continuous dielectric. A concise review treating the excess electrons in liquids has been published recently (327).

A substantial decrease in the density of alkali-metal solution relative to that of the components is a clear manifestation of cavity formation. The volume expansion accompanying dissolution of sodium in liquid ammonia was reported first by Kraus and Lucasse (139). The magnitude of the dilation is conveniently measured by $\Delta V = [V_{\text{solution}} - (V_{\text{NH}_3} + V_{\text{metal}})]/$(g-atoms of metal), ΔV being given as a function of metal concentration. The largest expansion was observed for the lithium solutions; in fact, a saturated solution of lithium in ammonia is the lightest liquid known at room temperature, having a density of only 0.477 g/cc. This indicates creation of cavities with a volume of about 100 Å^3 each. Such a volume is normally occupied by three ammonia molecules.

Much of our knowledge of the structure of solvated electrons was derived from studies of their absorption spectra. The spectrum of a hydrated electron, which was determined by the pulse-radiolysis technique, has been mentioned previously. The technique of pulse-radiolysis also permitted determination of the spectra of solvated electrons in other solvents (141–144). The reported spectra resemble those found for the hydrated electron and for the "blue" ammonia solution, being broad and showing lack of structure even at −78°C. The oscillator strength varies from 0.6 to 0.9 and implies that no intense bands exist at energies higher than 4 eV.

The mobile cavities which accommodate solvated electrons formed in solutions may be compared with localized electron traps formed in crystals doped with alkali metals. These traps represent vacancies in the crystal lattice which should be occupied by anions but which contain free electrons instead. ESR studies show that, although a trapped electron leaks out of the cavity and interacts with the neighboring cations, and to a lesser extent with the more remote ions, its highest density is in the trap. Such electrons are often referred to as F centers.

Let us consider now the neutral but paramagnetic species referred to as the M^+,e^- ion pair or as the monomer. Several workers attempted to describe its structure. The model proposed by Becker, Lindquist, and Alder (148) represents the associate as a pseudoatom. The electron is supposed to circulate in an expanded orbital among the solvent molecules oriented by the field of the cation. The "monomer" is a distinct species different from an alkali atom. This is clearly demonstrated by the ESR spectra: the spin density on the alkali nucleus in a monomer is only 0.1% of that observed in gaseous alkali atoms.

At higher concentrations of alkali the paramagnetism of the solution decreases (164) indicating that some diamagnetic species are formed. This observation, as well as the detailed conductance studies, led to the assumption of dimer formation. Becker, Lindquist, and Alder described the dimer as a diamagnetic species similar to the hydrogen molecule resulting from the association of two "expanded atom" alkali monomers. Such a species should have a very different spectrum from that of a solvated electron or a "monomer."

The above model has been subsequently modified by Gold, Jolly, and Pitzer (146). They pointed out that the ammonia–alkali metal solutions obey Beer's law over a concentration range within which the solvated electrons associate into paramagnetic, nonconducting pairs, even when the latter become dimerized into diamagnetic species. It is unlikely that the spectra of all these species are identical, and therefore it was concluded that the association yielding the "monomer" must be a "loose" one, like ion pairing, and that the dimer is formed by further electrostatic aggregation leading to quadrupolar species held together by coulombic forces. A similar suggestion was made by Symons and his co-workers (140).

The subject is, however, further confused by the ESR and NMR data. The Knight shifts in ammonia solution were measured by Hughes (149) and by Acrivos and Pitzer (150), and each group drew different conclusions from their own data. It appears that some other species may be present in such solutions. For example, Douthit and Dye (165) suggested that two monomers coexist in equilibrium with each other, the cation centered species represented by an expanding atom model, and an ion pair. A similar situation may be encountered when dimers are considered. Indeed, Becker, Lindquist, and Alder (148) pointed out that it is necessary to consider the existence of two different diamagnetic species, the "dimer" and a diamagnetic M^- which was treated by Pitzer as a triple ion. The structure of the latter species has been reinterpreted recently by Golden, Guttman, and Tuttle (131) as a solvated (ammoniated) alkali anion, and a reasonable estimate of its solvation energy makes this suggestion plausible (154).

There is no doubt that various species exist in various amine solutions. In these media Beer's law does not hold, and spectrophotometric studies show that the "monomer" and "dimer" differ in their spectra (166). Moreover, ESR studies concerned with the structure of the spectrum, the shape of the line, and its dependence on temperature and solvent (151–153) demand new models. The large hyperfine splitting due to the cation indicates a much higher electron density on the metal nucleus in amines than in liquid ammonia and supports the idea of expanded atom, at least for these solutions. In fact, no single model can yet account for all the data presently available.

Potassium and cesium dissolve in tetrahydrofuran and in dimethoxyethane forming characteristic blue solutions (155). However, the dissolved species are diamagnetic (156,157) although still very reactive; e.g., they initiate rapid anionic polymerization of styrene (158). It was suggested that the active entity is composed of two coupled electrons, $(e^-)_2$, but it is more probable that such solutions contain negative alkali ions produced by the reaction, $2K \rightarrow K^+ + K^-$ (154).

A powerful aprotic solvent capable of dissolving alkali and alkali earth metals has been reported recently, namely, hexamethylphosphoramide. The formation of a blue solution on contacting alkali metals with this liquid was observed in 1961, but reports describing this system were not published until 1965 (159–162). Hexamethylphosphoramide dissolves lithium, sodium, and potassium. The blue solution is paramagnetic and stable for about an hour at room temperature. It is extremely reactive and bursts into flame on contact with air. On standing, it turns red and becomes diamagnetic.

Solvated electrons may also be produced by photoionization. The pioneering work of Lewis and his students (169,170) demonstrated that electrons may be ejected from some phenols, amines, and dyes by irradiating their solutions in rigid glasses at low temperatures with UV or visible light. Observation of the irradiated glass demonstrated the presence of positive radical-ions; however, the solvated electrons were not detected at that time. They were detected later in the course of studies of photolysis of lithium metal and of N-lithiumcarbazol solutions in frozen methylamine glass (167). The characteristic bands of solvated electrons appeared in the near infrared, and these disappeared on softening the glass—an obvious indication of the electron–positive-ion recombination. It is also interesting to note that the spectrum of the solvated electron observed at $70°K$ differed somewhat from that recorded after the temperature of the glass was raised slightly. Linschitz explained this phenomenon by postulating the formation of a "poor" trap at $70°K$, because the low mobility of molecules of solvent, hindered by the rigidity of the medium, prevented them from attaining the most profitable orientation around the electron. At higher temperature the mobility of solvent molecules became sufficiently high to permit their orientation, and then a "proper" trap was formed.

Photoejection of electrons from radical-anions of aromatic hydrocarbons dissolved in frozen 2-methyltetrahydrofuran was investigated by Zandstra and Hoijtink (171). Although the spectrum of the solvated electron was not observed in this initial study, the subsequent investigation (168) led to identification of absorption at $\nu_{max} = 7.7$ kK. This spectrum was independent of the nature of the generator (Perylene$\cdot^+$,Li$^+$ or Tetracene^{2-},2Li$^+$). Formation of the solvated electron was also confirmed by ESR spectra of the irradiated solution. The question was raised whether the electron is trapped in the bulk of solvent or around the cations. The answer was provided by investigating the competitive trapping of the ejected electrons by dissolved pyrene and the "traps" (whether in the bulk of the solvent or in the vicinity of cations). Such competitive studies showed that "traps" are more numerous than cations and, therefore, the ejected electrons had to be located in solvent cavities and not around the cations (Li^+).

Solvated or hydrated electrons may be produced by photolysis of other negative ions (176,177). Recent studies involving photoionization of $K^+NH_2^-$ led to better understanding of the kinetics of elementary processes which occur in this reaction (178).

An interesting method for production of solvated electrons was suggested by Baxendale and Hughes (173), namely,

$$H + OH^- \rightleftarrows e^- \text{ (solvated)} + H_2O$$

and studies of Jortner and Rabani (174) confirmed its feasibility. A similar reaction has been investigated in liquid ammonia, i.e.,

$$NH_2^- + \tfrac{1}{2}H_2 \rightleftarrows e^- \text{ (ammoniated)} + NH_3$$

Using ESR for determining the concentration of solvated electrons, the equilibrium constant of the above reaction was found to be about $10^{-6}M$ (175).

The reactions of solvated electrons were often confused with reactions of H atoms. Such mistakes were particularly common in studies of radiolysis of aqueous solutions. Discrimination between the reactions of solvated electrons and H atoms are possible if N_2O is used as the reagent. This oxide reacts extremely rapidly with solvated electrons producing $N_2 + O^-$ but does not react with hydrogen atoms.

XVII. DISPROPORTIONATION OF RADICAL-ANIONS

Many electron acceptors may acquire a second electron and form dianions, e.g., anthracene^{2-}, benzophenone^{2-}, etc. In such systems the radical-anions may disproportionate. Denoting by $A^{\cdot-}$ the mononegative radical and by A^{2-} its dianion, one describes the disproportionation by

$$2A^{\cdot-} \rightleftarrows A + A^{2-} \qquad K_d$$

Electron affinity of the parent molecule, A, is usually higher than that of $A^{\cdot-}$, because the coulombic repulsion hinders the addition of a second electron. In terms of a simple Hückel MO theory, the repulsion energy, ΔE_{rep}, is given by the integral

$$\Delta E_{rep} = \int \int \bar{\phi}_{N+1}(1)\bar{\phi}_{N+1}(2)[e^2/r_{1,2}]\phi_{N+1}(1)\phi_{N+1}(2) \, d\tau_1 \, d\tau_2$$

where ϕ_{N+1} is the wave function of the lowest antibonding orbital. Calculations performed for aromatic alternant hydrocarbons (27) show that the magnitude of this repulsion is sufficiently large to make the disproportionation insignificant in the gas phase. However, a different situation is encountered in solution where ionic species enjoy an extra stability arising from their solvation. Accepting the original Born approach and assuming that both $A^{\cdot-}$ and A^{2-} may be represented by identical spheres of radius r, we find that the total free energy of solvation increases on disproportionation. Its gain, ΔS, is given by $(e^2/r)(1 - 1/D)$, i.e., it is twice as large as the free energy of solvation of a single $A^{\cdot-}$ ion. Hence, the free energy of disproportionation of free radical-ions in solution is given by

$$\Delta G_{dispr} = \Delta E_{rep} - \Delta S$$

if the change in entropy of disproportionation of gaseous ions is neglected. Most probably this term is small. The above equation served to calculate ΔG_{dispr} of aromatic radical-ions (27), and the numerical results of such computations are collected in Table XVI. Comparison with experiments shows again that the theory needs much refinement,

TABLE XVI

Calculated Free Energy of Disproportionation of Radical Anions in Solution in Electron Volts

Parent molecule	ΔE_{rep}	$2\Delta G_{solv.A.^-}$	ΔG_{dispr}	
			Calculated[a]	Observed
Benzene	7.03	3.6	−3.4(3.0)	—
Naphthalene	5.64	2.2	−3.4(1.9)	—
trans-Stilbene	5.51	1.7	−3.8(2.4)	−0.23[b]
Phenanthrene	5.17	1.6	−3.6 —	—
Anthracene	5.10	1.6	−3.5(1.7)	−0.58[c]
Tetraphenylethylene	4.76	1.2	−3.6 —	—

[a] The values given in brackets in the fourth column arise from Lyon's calculation of solvation energy (see Ref. 42).

[b] Ref. 106.

[c] Ref. 33.

although it correctly predicts that the repulsion and solvation energies simultaneously decrease with increasing size of the hydrocarbon. Such a compensating effect may account for the nearly constant difference between the first and the second reduction potential, as had been observed in the polarographic studies of aromatic hydrocarbons (33) and polyphenyls (106).

Let us stress again that the treatment discussed here applies only to *free* radical-ions in solution. However, in most solvents the radical-ions, and certainly the dianions, are paired with their respective counterions. In such a case, the disproportionation is represented by the equation,

$$2A^{\cdot-},M^+ \rightleftarrows A + A_2^-,2M^+$$

where M^+ is the counterion. As may be easily proved, the free energy of disproportionation of ion pairs is given by the difference of the first and second reduction potentials directly observed in potentiometric titrations performed in solvents favoring a substantial pairing, e.g., in tetrahydrofuran. The ion pairing is expected to favor disproportionation, because binding of two counterions by the negative dianions is more than twice as powerful as that responsible for the linkage of $A^{\cdot-}$ with its partner. The situation is, however, more complex, as shown by the temperature effects revealed in the spectra of radical-ions in methyltetrahydrofuran (107), and even more strikingly from the findings of de Boer (108). De Boer observed that lithium naphthalene in 2-methyltetrahydrofuran exists as the radical-ion at $-120°C$. Optical and ESR spectra conclusively identified this species. However, at higher temperatures the spectrum was drastically changed and the paramagnetism disappeared. Evidently, the radical-ions disproportionated, and indeed the spectrum of the resulting solution became identical with that reported for the naphthalene dianion (65,109). Since the disproportionation was favored by higher temperature, the reaction must be endothermic. Apparently the gain in energies of solvation and association is still not sufficient to overcome the energy loss caused by the electron repulsion and, therefore, the disproportionation process is driven by entropy and not by energy.

The question arises as to why disproportionation increased the entropy of the system. Of course, conversion of a free $A^{\cdot-}$ ion into A^{2-} dianion must *decrease* the entropy of the system, but the situation may be different for the ion pairs. The naphthalene$^{\cdot-}$,Li$^+$ ion pair probably forms a strong dipole which interacts powerfully with the solvent, orients its molecules and, therefore, decreases the entropy of the solution. Disproportionation produces a tight quadrupole which weakly interacts with the solvent. Hence, the solvent molecules become free, and the entropy of the system increases. This accounts for the findings. Let us add, however, that the association is exothermic, and thus it partially compensates for the increasing repulsion energy of the electron and for the *loss* of solvation energy.

The most thoroughly investigated disproportionation is that of the radical-ions of tetraphenylethylene, species which will be denoted here by $T^{\cdot-}$. It was known, since the early studies of Schlenk et al. (2), that the reaction of this hydrocarbon (T) with metallic sodium in ether or dioxane yields only the disodium salt $(T^{2-}, 2Na^+)$. No radical-ions seemed to be formed even when an excess of the parent hydrocarbon was available. This observation was most puzzling, especially when the sensitive technique of ESR spectroscopy still failed to detect the radical-ions (110). However, recent investigations of Evans and his students (111a) proved that the ESR signal *may* be obtained if the process is performed in tetrahydrofuran, and this finding was soon confirmed by independent studies of Garst and Cole (112). His results emphasized the important role of solvent; although the ESR spectrum was observed in tetrahydrofuran and dimethoxyethane, no signal was discerned in dioxane. Further studies of this reaction revealed that the intensity of the signal depends not only on the solvent but also on the nature of counterion and temperature (111–113). Obviously, we deal here with a complex system and, therefore, the disproportionation required further studies. The results were most illuminating and deserve detailed discussion.

It was remarked earlier that the disproportionation in ethereal solvents is represented by an equation such as

$$2T^{\cdot-}, M^+ \rightleftharpoons T + T^{2-}, 2M^+$$

and, therefore, the degree of dissociation should not be affected by dilution. The disproportionation could be investigated by a spectrophotometric technique because the spectra of radical-ions and of dianions are different. Careful studies of this equilibrium in tetrahydrofuran revealed, however, that the apparent equilibrium constant, K_{ap}, defined by the ratio

$$[T^{2-} \text{ total}][T]/[T^{\cdot-} \text{ total}]^2$$

decreases on dilution (113b,114). Here $[T^{2-}$ total] denotes the concentration of *all* the dianions, in whatever form they are present, while

[$T^{\cdot-}$ total] refers to the concentration of all the radical-ions. The effect of dilution on K_{ap} may be accounted for if the system is governed by three independent equilibria, each determined by its own equilibrium constant (114). For example,

$$2T^{\cdot-},Na^+ \rightleftharpoons T + T^{2-},2Na^+ \qquad K_1$$
$$T^{\cdot-},Na^+ + T^{\cdot-} \rightleftharpoons T + T^{2-},Na^+ \qquad K_2$$
$$2T^{\cdot-} \rightleftharpoons T + T^{2-} \qquad K_3$$

The last equilibrium may be neglected because the concentration of the dianions, T^{2-}, is espected to be vanishingly small. The equilibrium constants K_1 and K_2 are correlated through a simple relation, viz.,

$$K_1/K_2 = K_{diss,T^{\cdot-},Na^+}/K_{diss,T^{2-},2Na^+}$$

The latter two constants refer to the dissociation processes

$$T^{\cdot-},Na^+ \rightleftharpoons T^{\cdot-} + Na^+ \qquad K_{diss,T^{\cdot-},Na^+}$$

and

$$T^{2-},2Na^+ \rightleftharpoons T^{2-},Na^+ + Na^+ \qquad K_{diss,T^{2-},2Na^+}$$

The apparent disproportionation constant, K_{ap}, is then given by

$$K_{ap} = K_1(1 + K_{diss,T^{2-},2Na^+}/[Na^+])/(1 + K_{diss,T^{\cdot-},Na^+}/[Na^+])^2$$

and, in accord with experimental findings, its value decreases on dilution, if $K_{diss,T^{2-},2Na^+} < K_{diss,T^{\cdot-},Na^+}$.

Dissociation of both salts was investigated (114) and the results are shown in Figure 18. Its inspection reveals two facts: (1) The dissociation constant of the $T^{\cdot-},Na^+$ ion pair is relatively high, although the exothermicity of the process is low. Such relation is characteristic for solvent separated pairs. (2) The dissociation constant of $T^{2-},2Na^+$ is low, but this process is strongly exothermic. This behavior characterizes a contact pair which dissociates into solvated ions. It is possible now to estimate the term $(K_{diss,T^{2-},2Na^+})[Na^+]^{-1}$ and show that its value is much smaller than unity. Therefore, the approximation

$$K_{ap} = K_1/\{1 + (K_{diss,T^{\cdot-},Na^+})[Na^+]^{-1}\}^2$$

is valid, and this leads to the equation

$$1/K_{ap}^{\frac{1}{2}} = 1/K_1^{\frac{1}{2}} + (K_{diss,T^{\cdot-},Na^+}^{\frac{1}{2}}/K_1^{\frac{1}{4}})\{[T]\cdot[T^{2-} \text{ total}]\}^{-\frac{1}{4}}$$

Plots of $1/K_{ap}^{\frac{1}{2}}$ versus $\{[T][T^{2-} \text{ total}]\}^{-\frac{1}{4}}$ are shown in Figure 19, and from their intercepts and slopes the values for K_1 and $K_{diss,T^{\cdot-},Na^+}$ were

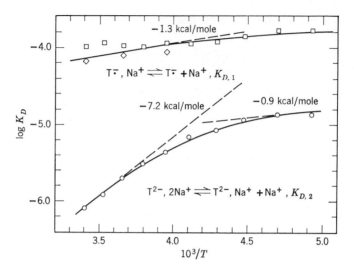

Fig. 18. Dissociation constant of monosodium salt of radical-anion ($T^{\cdot-}$,Na^+) derived from tetraphenylethylene (T) and of the disodium salt of its dianion (T^{2-},$2Na^+$). Both spectrophotometric and conductance studies were used in determining the dissociation constants of the respective ionic agglomerates. ($\diamondsuit$), spectrophotometry; ($\square$ and $\bigcirc$), conductance.

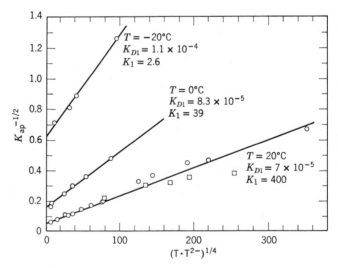

Fig. 19. Linear plots of $1/K_{ap}^{1/2}$ versus $(T \cdot T^{2-})^{-1/4}$ for the system tetraphenylethylene and its radical-ion and dianion. ($\bigcirc$), Roberts and Szwarc, ref 114; ($\square$) Garst, ref. 113.

derived. Thus, K_1 and K_2 were computed and the final results, given in Table XVII, are shown graphically in Figure 20.

TABLE XVII

$$T^{\cdot-},Na^+ + T^{\cdot-},Na^+ \rightleftharpoons T^{2-},2Na^+ + T \qquad K_1$$
$$T^{\cdot-},Na^+ + T^{\cdot-} \rightleftharpoons T^{2-},Na^+ + T \qquad K_2$$

$T,°C$	K_1	K_2
20[a]	400	3.3
0	39	0.72
−20	2.6	0.10
−37	0.44	0.038

[a] At 20°C $\Delta H_1 = 19 \pm 2$ kcal/mole, $\Delta S_1 = 75$ eu; $\Delta H_2 = 13 \pm 2$ kcal/mole, $\Delta S_2 = 45$ eu.

Let us assess now the factors governing these disproportionations. Both reactions 1 and 2 are endothermic. Electron repulsion and *loss* of solvation energy of the cation contribute to this effect, which is partially balanced by the greater binding energy of Na^+ to the dianion. On the other hand, the gain of entropy, due to the desolvation of the ions or ion pairs provides the driving force of this process. Note that two $T^{\cdot-},Na^+$ (dipoles) orient solvent molecules more strongly than does one $T^{2-},2Na^+$ (quadrupole), and a similar situation is established in disproportionation 2.

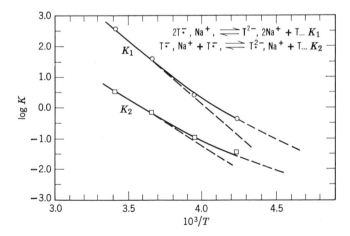

Fig. 20. Plot of log K_1 and log K_2 versus $1/T$ for the disproportionation of radical-ions of tetraphenylethylene.

The exothermicity of solvation of $T^{\cdot -}$,Na^+ (solvent separated ion pair) is indeed substantial. Consequently, disproportionation 1 is *more* endothermic than disproportionation 2 (disproportionation causes desolvation), i.e., $\Delta H_1 > \Delta H_2$. In fact, this experimental result indicates that solvation by tetrahydrofuran of the $T^{\cdot -}$,Na^+ ion pair is even stronger than that of the free negative $T^{\cdot -}$ ion.

Let us pause and compare our present interpretation of the disproportionation with that outlined at the beginning of this section. The early treatment was concerned solely with the solvation of anions. It appears, however, that the solvation of *cations* is more important, at least for some ion pairs. The early approach stressed the contribution of the solvation energy (of anions) to the driving force of the disproportionation. Our present position is that the driving force arises from the gain in entropy of disproportionation caused by the *desolvation* of solvent separated ion pairs. On the basis of the early approach one could expect *more* disproportionation in a more powerfully solvating medium (implicitly in a better solvent for anions—a point not always appreciated). The findings of Evans (111) and of Garst et al. (112,113) demonstrate the contrary; disproportionation is hindered in tetrahydrofuran, and even more so in dimethoxyethane or glyme, i.e., in media which powerfully solvate the cations. The process is favored, however, in ether or dioxane, i.e., in a solvent of low dielectric constant which interacts poorly with cations. All of these observations are now rationalized.

In hexamethylphosphoramide no disproportionation takes place (268,280). In this solvent $T^{\cdot -}$ and $T^=$ exist as free ions and, therefore, the driving force (entropy gain) of disproportionation is lost.

Formation of $T^{\cdot -}$,M^+ is favored by strong solvation of M^+ ions if $T^{\cdot -}$,M^+ is a solvent separated pair. Such an interaction is lost in the tight T^{2-},$2M^+$ ion agglomerate. Therefore, the concentration of $T^{\cdot -}$,M^+ radical-ions should be higher for $M^+ = Li^+$ and negligible for the salts of cesium. This again is confirmed by experiments.

Finally, let us ask why T^{2-},$2Na^+$ forms contact ion pairs (at least at temperatures not too low) while $T^{\cdot -}$,Na^+ forms solvent separated pairs. It seems that the geometry of the T^{2-} ion is different from that of the $T^{\cdot -}$ radical-ion. The latter resembles the parent hydrocarbon tetraphenylethylene. Steric hindrance caused by placing phenyl groups in the *cis* position prevents the $=CPh_2$ groups from attaining a coplanar conformation (181), and the shape of the molecule is shown by Figure 21(*a*). The same factors are responsible for the lack of coplanarity of the

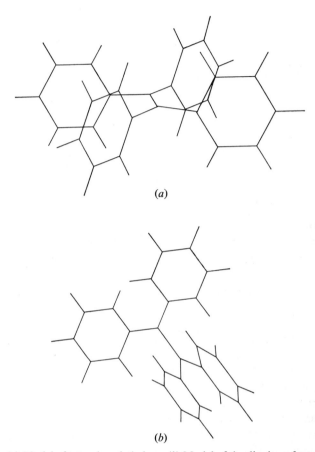

(a)

(b)

Fig. 21. (a) Model of tetraphenylethylene. (b) Model of the dianion of tetraphenylethylene.

radical-ion $T^{.-}$, and consequently the counterion cannot approach
closely the negative charge which mainly resides around the
$>C=C<$ bond.

The $C=C$ double bond is destroyed when the dianion, T^{2-}, is
formed. Consequently, the rotation around the $>C-C<$ axis becomes
free and the dianion attains a skewed conformation shown in Figure 21(b).
This conformation is favored by resonance, by the reduced repulsion
between the extra electrons, and by the reduced steric hindrance; and it
permits a close approach of the counterions to the coplanar negative

centers of the $-C\underset{Ph}{\overset{Ph}{\diagup}}$ groups. It is obvious that the expulsion of the

solvation shell surrounding a free cation, which takes place in the course of formation of a contact pair, is possible if the system gains a sufficiently large energy through the cation–anion interaction. Let us denote the thickness of the shell by Δx. The gain in Coulombic energy is given by the approximate expression

$$\text{Const.} \{(r_1 + r_2)^{-1} - (r_1 + r_2 + \Delta x)^{-1}\}$$

where $r_1 + r_2$ is the distance separating the ions of a contact pair. For a constant Δx the gain in energy increases as the distance $r_1 + r_2$ decreases. Hence, the approach to a bulky $T^{\cdot-}$ radical-ion ($r_1 + r_2$ large) does not provide a sufficient gain to favor the expulsion of solvent from around the cation. Consequently, the resulting product forms predominantly solvent-separated $T^{\cdot-},Na^+$ ion pairs. The gain is large, however, in the pairing with T^{2-}, because the cation may approach closely the planar fragments of the anion. Thus, $T^{2-},2Na^+$ forms a contact-type agglomerate.

The different behavior of contact and solvent-separated pairs accounts for the apparent anomaly stressed by Zabolotny and Garst (228). They noted that the degree of disproportionation of radical-ions derived from stilbene *increases* as the size of the cation decreases, i.e., the respective equilibrium constant for the lithium salt is greater than that for the potassium. This trend is the reverse of that observed in tetraphenylethylene systems. It is probable that the radical-anion of stilbene, as well as its dianion, forms a contact pair even with Li^+ as the counterion, whereas a solvent-separated pair is formed when Li^+ or Na^+ becomes associated with a radical-anion of tetraphenylethylene. Thus, the gain arising from solvation contributes to the stability of radical-anion in the latter system but not in the former.

The above discussion clarifies the mechanisms which are responsible for the trends in the disproportionation constants caused by temperature, solvent, and the nature of cation. At ambient temperatures the disproportionation usually favors radical-anions, if an excess of parent hydrocarbon is available. Tetraphenylethylene is an exception. The stability of $T^{2-},2Na^+$ results from its geometry being different from that of the radical-ion. Various factors then contribute to the shift of the pertinent equilibrium to the right. The idea of different geometry for radical-ion and dianion is supported by the interesting findings of Garst et al. (210). They noted that the sodium salt of α-methylstilbene radical-anion disproportionates in 2-methyltetrahydrofuran at 25°C to a much greater extent that the salt of stilbene radical-anion, the equilibrium constants of disproportionation being $\geqslant 1000$ and 0.09, respectively. The extent of disproportionation of α-methylstilbene radical-anion is, under these conditions, comparable to that found for the tetraphenyl-ethylene system ($K_1 \geqslant 1000$) and greater than observed in the triphenylethylene system ($K_1 = 36$). As pointed out by Garst, the alternative explanation invoking the reduced electron–electron repulsion in the tetraphenyl- or triphenylethylene dianions, as compared with dianions of stilbene, cannot account for the behavior of α-methylstilbene radical-anions.

Further indication for the skewed geometry of the dianions of stilbene, tri- and tetraphenylethylene and their derivatives comes from studies of triplets of stilbene. The geometry of such triplets may be similar to that of the corresponding dianions (210). Photochemical studies of Hammond and Saltiel (211) led them to postulate the existence of three triplets, the *cis*, *trans*, and the "phantom." It is probable that the last corresponds to the most stable skewed form.

The importance of geometrical factors is seen in some other systems, e.g., cyclooctatetraene. Cyclooctatetraene exists in a nonplanar tub conformation (D_{2d} symmetry group) and has a pronounced olefinic character. In tetrahydrofuran solution it reacts readily with alkali metal giving a dianion which yields on hydrolysis cyclooctatriene and on carboxylation the corresponding acid (182). Potentiometric and polaro-graphic studies (183) showed that the parent hydrocarbon is reversibly reduced in a two-electron process, i.e., the first reduction potential is *higher* than the second—a situation similar to that found for the tetraphenylethylene (34). ESR studies showed, nevertheless, that the respective radical-ion is present in the system, and its spectrum was analyzed (184). It consists of nine equally spaced lines having the intensities predicted for an octuplet. The hyperfine splitting constant was 3.21 G, the g value 2.0036. The spectrum was sharp when the reduction was performed in tetrahydrofuran with lithium but broadened when potassium was used. Since the broadening was large for high conversion but insignificant for low conversion (low K/hydrocarbon ratio), it was concluded that the exchange is rapid between radical-ion and dianion, but slow between radical-ion and the parent hydrocarbon.

The study of ^{13}C splitting (185), amounting to 1.28 G, demonstrated that the three valences radiating from each carbon atom of the ring have to be coplanar. This fact, in conjunction with other observations, some of which we shall discuss later, indicates that the radical-ion is planar and forms a regular octagon. The symmetry implies charge density of $1/8$ on each carbon atom, however, the width of the spectrum Q is -25.7 G. According to the theory of π–σ interaction, originally developed by McConnell (186) and refined by Fraenkel et al. (187), the value of Q depends on the hybridization of the carbon atoms, and it should be -23.4 G for three equivalent sp_2 bonds ($120°$ apart). This indeed was confirmed by examining the spectra of the benzene radical-ion and the CH_3 free radical. The same treatment leads, however, to the conclusion that $Q = -10$ G for a regular octagon (the bond angle $135°$). Thus, the observed value is very different from the predicted one. It was proposed (188) that this contradiction may be resolved by assuming "bent" bonds (189) still having $120°$ bond angles. Similar conclusions would apply to the cycloheptatrienyl radical for which Q was found to be between 25.6 and 27.6 G (190,191).

The planar, regular octagonal conformation of the cyclooctatetraene radical-anion would require a Jahn-Teller distortion which should increase the g value over and above $g = 2.0023$ characteristic of a free electron. In addition, this effect should lead to a large line width and to a decrease in spin–lattice relaxation time (192a). This is not the case, and indeed the reexamination of the problem (192b) indicates that in the cyclooctatetraene radical-ion, in contradistinction to that derived from benzene, the Jahn-Teller effect may be insignificant.

The behavior of the cyclooctatetraene anion is reflected in the reactions of its dibenzo derivative. The parent hydrocarbon is again nonplanar and olefinic in character. Its radical-anion, formed by lithium reduction in tetrahydrofuran, was studied by Carrington et al. (229), who concluded from the analysis of its ESR spectrum that the odd electron is usually located on the two olefinic C=C bonds, the conjugation with the benzene rings being negligible (i.e., $\beta' \approx 0$). The spectrum was reexamined by Katz et al. (230), who produced this species by alkali metal reduction (with lithium, sodium, or potassium) as well as by electrolytical reduction in dimethylformamide or dimethyl sulfoxide. Their analysis of the ESR spectrum shows that its structure may be accounted for by any value of β' from 0 to β, i.e., the conclusion $\beta' \approx 0$ is not unique. Additional information about the nature of this species was obtained from polarographic studies. As in the cyclooctatetraene system, the exchange $A^{·-} + A$ is slow, whereas $A^{·-} + A^{2-}$ is fast.

Still another system resembling cyclooctatetraene was investigated by Dauben and Rifi (231). They showed that tropylmethylether reacts with potassium in tetrahydrofuran giving a deep blue anion $C_7H_7^-,K^+$. The existence of this anion was questioned in the past. Simple Hückel molecular orbital treatment indicates that the system should have two degenerate antibonding orbitals occupied by one electron in the $C_7H_7\cdot$ radical and by two in the anion. Thus, energetically the anion should be unstable. However, as predicted by the Jahn-Teller theorem, such a symmetrical cyclic π system with a degenerate ground state should be stabilized by some perturbation which would remove its degeneracy. Indeed, a stabilization by 6 kcal/mole was anticipated by distortion of the bond length.

The blue species is diamagnetic; however, if the reduction is continued, a deep green radical-anion is formed (232). Its ESR spectrum was recorded at $-100°C$; it is an octet ($\alpha = 3.86$ G), further split into a heptet ($\alpha = 1.74$ G) by the two magnetically equivalent Na^{23} ions. The latter splitting provides excellent evidence for its dianion character.

This system furnishes, therefore, an interesting example of a diamagnetic monoanion and a paramagnetic dianion. The dianion should be subject to Jahn-Teller stabilization, but no evidence is available on this point.

Disproportionation of other radical-ions was less extensively studied. Radical-anions derived from aromatic ketones (ketyls) were discovered by Beckman and Paul (227) and recognized as radicals by Schlenk et al. (28). An excellent review of the earlier literature of this subject may be found in the monograph by Hückel (212). Ketyls may perhaps dimerize to pinocone derivatives as well as disproportionate to dianions. The evidence for dimerization comes from synthetic studies, e.g., those of Bachman (213), who isolated quantitative amounts of the pinacone from products of hydrolysis of sodium benzophenone by dilute acetic acid. However, other preparative investigations led to different conclusions, e.g., hydrolysis by water or alcohol yielded the monomeric alcohols (214).

Recently the equilibria established in the ketyl system were investigated by physical methods. For example, variation of optical density with concentration was studied by Warhurst and his students (215). The validity of Beer's law was established for a number of ketyls in dioxane in the concentration range 10^{-4} to $10^{-3}M$. It was concluded, therefore, that under these conditions association does not take place.* However,

* Alternatively this may indicate a complete association and negligible dissociation.

deviations from Beer's law were observed in ether (216). Extensive spectrophotometric and ESR studies of ketyl systems have been recently reported by Hirota and Weissman (217,218). The ESR spectra revealed the existence of two types of paramagnetic species, a monomeric and a dimeric one, the latter being present, e.g., in dioxane solution of sodium benzophenone or sodium hexamethylacetone (218). The monomeric species undergo rapid electron exchange with the respective ketones (217) and hence their ESR spectra collapse into four lines (due to Na^{23}) at sufficiently high concentrations of ketone. Under these conditions the ESR spectrum of the dimer may be conveniently observed, since the electron transfer from the dimer to ketone is extremely slow (see Table XX on p. 414).

The paramagnetic dimers are probably bonded through dipole–dipole interaction. Both sodium ions are magnetically identical suggesting the following structure of the dimer:

$$\begin{array}{ccc} Ar & & Ar \\ \diagdown & Na^+ & \diagup \\ \overset{\cdot}{C}\!-\!\bar{O} & \quad & \bar{O}\!-\!\overset{\cdot}{C} \\ \diagup & Na^+ & \diagdown \\ Ar & & Ar \end{array}$$

Closer examination of ESR spectra of the Mg^{2+} salt permitted discrimination between several possible stereochemical structures of such dimers.

In addition to a paramagnetic dimer, a diamagnetic dimer was also observed. Equilibria between these two species were investigated by spectrophotometric techniques. The diamagnetic species become abundant in nonpolar solvents and most probably these are the salts of pinacols. Hence, the equilibrium between a diamagnetic and paramagnetic dimer may be represented by eq. (5):

$$\begin{array}{ccc} Ar & & \\ \diagdown & & \\ & C\!-\!\bar{O},Na^+ & \\ Ar \diagup & | & \\ & | & \\ Ar & | & \\ \diagdown & | & \\ & C\!-\!,\bar{O}Na^+ & \\ Ar \diagup & & \end{array} \rightleftarrows \begin{array}{ccccc} Ar & & & & Ar \\ \diagdown & & Na^+ & & \diagup \\ & \overset{\cdot}{C}\!-\!\bar{O} & & \bar{O}\!-\!\overset{\cdot}{C} & \\ \diagup & & Na^+ & & \diagdown \\ Ar & & & & Ar \\ & & \Updownarrow \text{ at high dilution} & & \end{array} \qquad (5)$$

$$2 \begin{array}{c} Ar \\ \diagdown \\ \overset{\cdot}{C}\!-\!\bar{O},Na^+ \\ \diagup \\ Ar \end{array}$$

Disproportionation of radical-anions derived from diketones (benzil) were reported by Evans et al. (219). The equilibrium constant was determined to be about 0.6–0.7 for the sodium salt in tetrahydrofuran at 20°C. Apparently no dimerization takes place in this system.

XVIII. DIMERIZATION OF RADICAL-IONS AND INITIATION OF ADDITION POLYMERIZATION

Radical anions derived from aromatic hydrocarbons, such as naphthalene, anthracene, etc., do not dimerize. A substantial loss of resonance stability of the aromatic systems, which arises from the formation of a new covalent C—C bond, prevents such a process. The quantum mechanical treatment of this obvious conclusion was presented by McClelland (220). On the other hand, the radical-ions formed from hetero-aromatics, e.g., pyridine, quinoline, etc., dimerize if present in the form of ion pairs. Dimerization of some of them is prevented if they are present as free ions. A rapid dimerization is expected when radical-ions are formed from vinyl or vinylidene monomers, and this step was proposed to explain the initiation of polymerization by an electron-transfer process (193).

When a molecule of a vinyl or vinylidene monomer acquires an extra electron from a suitable electron donor the resulting radical-anion has only a fleeting existence. In solution it couples with another, similar radical-ion and forms a dimeric dicarbanion, or it may react with an ordinary molecule of the parent monomer and then form a dimeric radical-anion (193). These processes are illustrated by the following equations, styrene being used as an example:

$$2 \; CH_2:CH(Ph)^{\cdot-} \; \rightleftharpoons \; {}^{-}CH(Ph)CH_2CH_2CH(Ph)^{-}; \text{ dimeric dicarbanion}$$
$$CH(Ph):CH_2^{\cdot-} + CH_2:CH(Ph) \; \rightleftharpoons \; {}^{-}CH(Ph)CH_2CH_2CH(Ph)^{\cdot}; \text{ dimeric radical-anion}$$

Although the ionic species are denoted here by symbols representing free ions, in actual processes the reaction involves mainly ion pairs.*

In the presence of an excess of monomer the dicarbanions initiate anionic polymerization producing linear macromolecules possessing two

* Reactions of free ions may be studied in hexamethylphosphoramide because in this solvent the association of radical-ions with counterions is negligible (220). It was shown that the free ions dimerize much more slowly than the respective ion pairs of radical ions (T. Staples, J. Jagur-Grodzinski, and M. Szwarc, to be published).

growing ends. The dimeric radical-anions may also initiate polymerization—anionic growth ensues then from the carbanion end, whereas the radical end initiates a radical-type propagation. The anionic growth is fast and it proceeds usually without termination giving rise to living polymers, whereas the radical propagation is rapidly terminated because free radicals destroy each other by dimerization or disproportionation. Thus, the contribution of the radical propagation to the overall polymerization is usually insignificant.

There are three distinct ways in which radical-anions derived from vinyl or vinylidene monomers may couple, viz.,

$$CXY\overset{\ominus}{:}CH_2 + CH_2\overset{\ominus}{:}CXY \rightleftharpoons \bar{C}XYCH_2CH_2\bar{C}XY,$$

$$CXY\overset{\ominus}{:}CH_2 + CX\overset{\ominus}{Y}:CH_2 \rightleftharpoons \bar{C}XYCH_2CXY\bar{C}H_2,$$

and

$$CH_2\overset{\ominus}{:}CXY + CXY\overset{\ominus}{:}CH_2 \rightleftharpoons \bar{C}H_2CXYCXY\bar{C}H_2.$$

The first mode of dimerization is greatly favored on thermodynamic grounds when X or Y (or both) is an unsaturated group, e.g., —COOR, —CN, —CH=CH$_2$, Ph, or other aromatic or heteroaromatic moiety. At equilibrium, if it is established, the first dimer should predominate to the virtual exclusion of the remaining two. One may question, however, whether the products of dimerization are determined by the equilibrium or by the kinetics of the coupling. Although a general answer to this problem is not known, it was found that in all those cases in which dimers were isolated and identified, only the thermodynamically stable species were produced. For example, only the 1,4-dicarboxylic acids, derived from the first type of dimers, were quantitatively isolated from the products of carboxylation of dicarbanions derived from styrene, α-methylstyrene (194) or 1,1-diphenyl-ethylene (2).

The dimerization is, in principle, reversible. Not much is known, however, about the equilibrium constant of such a reaction. ESR study showed that the equilibrium concentration of radical-ions which coexist with their respective dimers is vanishingly small and hence they could not be detected. For example, the concentration of 1,1-diphenyl ethylene radical-ions in equilibrium with the relevant dimeric dianions is less than $10^{-7}M$ when the dianions are at about $0.1M$ concentration (195). Similar results were obtained for the dimeric dianions of α-methylstyrene.

The rate of dissociation of the dimeric dianions of α-methylstyrene into radical-ions was investigated by Asami and Szwarc (196). The dianion was prepared by quantitative reaction of α-methylstyrene with metallic potassium in tetrahydrofuran. An identical solution was prepared using, however, deuterated α-methylstyrene, viz., $CD_2:C(Ph)CD_3$ instead of the ordinary monomer. Both solutions were stable. The dianions could be converted into the corresponding hydrocarbons by adding a drop of water, i.e., $CH(Ph)(CH_3)CH_2CH_2CH(Ph)(CH_3)$ (mass 238) is then formed from the ordinary monomer and $CH(Ph)(CD_3)CD_2CD_2CH(Ph)(CD_3)$ (mass 248) is produced on protonation of the dianion resulting from the deuterated monomer.

To measure the rate of dissociation of the dimer, both solutions were mixed and the mixture was kept at constant temperature in a thermostat. Aliquots were withdrawn at various times and the dianions protonated, isolated, and analyzed on a suitable mass spectrograph operating with 8 V acceleration potential. Under these conditions only the parent peaks appear in the mass spectrum. The aliquot withdrawn immediately after the mixing of the solution showed only the 238 and 248 mass peaks; no peak was observed at 243 mass. However, after a day a peak appeared in the analyzed mixture at the mass 243 and it grew more intense with time. This peak is due to the mixed dimer, viz.,

$$CH(CH_3)(Ph)CH_2CD_2CH(Ph)CD_3 \text{ (mass 243)}$$

and its formation results from reaction (6)–(8):

$$^-\alpha MeS_H\alpha MeS_H{}^- \underset{k_r}{\overset{k_d}{\rightleftarrows}} 2\ \alpha MeS_H{}^{\bar{\cdot}} \tag{6}$$

$$+ \rightleftarrows {}^-\alpha MeS_H.\alpha MeS_D{}^- \tag{7}$$

$$^-\alpha MeS_D.\alpha MeS_D{}^- \underset{k_r'}{\overset{k_d'}{\rightleftarrows}} 2\ \alpha MeS_D{}^{\bar{\cdot}} \tag{8}$$

where αMeS denotes α-methylstyrene and the subscripts H and D refer to the ordinary and deuterated compound. The isotope effect is probably negligible, i.e., $k_r = k_r'$ and $k_d = k_d'$. Therefore, for a 50:50 mixture the initial rate of formation of the mixed dimer is one-half the rate of dissociation. Thus, the rate constant of dissociation, $^-\alpha MeS\alpha MeS^- \rightarrow 2\alpha MeS^{\cdot-}$ was found to be $6 \times 10^{-8}\ sec^{-1}$ at 25°C. Similar experiments were performed with the dimer of 1,1-diphenylethylene, but some experimental difficulties prevented reliable determination of the respective rate constant. It was shown, however, that its value is even lower than that found for the α-methylstyrene system.

Spach et al. (197) developed an alternative method of studying the dissociation of the dimeric dicarbanion derived from 1,1-diphenylethylene. This hydrocarbon is a most reactive vinylidene "monomer" which, however, does not homopolymerize. The enormous steric strain, caused by the presence of two bulky phenyl groups on every second carbon atom of a hypothetical polymer, prevents the normal head-to-tail polymerization. Let us remark in passing that steric hindrance is larger when the bulky substituents are located on carbon atoms separated by another C than if they are attached to two adjacent atoms, a point easily appreciated from inspection of models.

The reasons outlined above explain why 1,1-diphenylethylene, D, does not react with its dimeric dianion, $^-DD^-$, viz., no reaction

$$^-DD^- + D \rightleftharpoons {}^-DDD^-$$

is observed. However, if $^-DD^-$ is in equilibrium with its radical-ion, $D^{\cdot-}$, an exchange should be observed between $^-DD^-$ and D. Such an exchange was investigated in tetrahydrofuran (197), radioactive D and nonradioactive $^-DD^-$, or vice-versa, being used for labeling the reagents. Their separation, needed for counting, could be easily accomplished by carboxylation of the dianion. $^-DD^-$ yields then a substituted succinic acid which may be easily freed from the hydrocarbon. The following mechanism was envisaged:

$$^-DD^- \xrightarrow{k_1} 2D^{\cdot-} \quad \text{slow, rate-determining step}$$
$$D^{\cdot-} + D^{14} \rightleftharpoons D^{14\cdot-} + D \quad \text{very fast}$$
$$D^{14\cdot-} + D^{\cdot-} \longrightarrow \text{radioactive } {}^-DD^- \quad \text{again fast}$$

Hence, the rate of exchange, R_{ex}, should be first order in $^-DD^-$ and independent of D. However, experimental results showed a more complex kinetics, namely, in addition to the expected term $k_1[^-DD^-]$, another one, proportional to $[D][^-DD^-]^{1/2}$, appeared, i.e.,

$$R_{ex} = k_1[^-DD^-] + \text{const. } [D][^-DD^-]^{1/2}$$

or $R_{ex}/[^-DD^-]$ was found to be a linear function of $[D]/[^-DD^-]^{1/2}$ as shown in Figure 21. Such a relation implies that the following route contributes to the exchange:

$$^-DD^- + D \rightleftharpoons {}^-DD^{\cdot} + D^{\cdot-}$$
$$^-DD^{\cdot} \underset{k_2}{\overset{k_2}{\rightleftharpoons}} D^{\cdot-} + D$$

A rapid electron transfer between $D^{\cdot-}$ and D^{14} as well as between the radioactive $^-DD^{\cdot}$ and the nonactive $^-DD^-$ leads then to the radioactive $^-DD^-$. Note that the radioactive $^-DD^{\cdot}$ can be formed only by the combination of $D^{\cdot-}$ and D, either of which acquires radioactivity.

The formation of $^-DD\cdot$ *cannot* affect the equilibrium concentration of $D\cdot^-$ which is determined by dissociation (1). This process provides, however, an alternative route for the *production* and *consumption* of the radical-ions and therefore it contributes to the rate of the ultimate exchange. The experimental results indicate that the formation of $^-DD\cdot$ from $^-DD^-$ and D is much faster than its decomposition into $D\cdot^- + D$ and, therefore, the rate of exchange arising from this route is governed by reaction (2). Since the system is in equilibrium, apart from the exchange, the investigated rate is given by

$$k_{-2}[D][D\cdot^-] = k_{-2}K_1^{1/2}[D][^-DD^-]^{1/2}$$

i.e., the slope of the line shown in Figure 22 gives $k_{-2}K_1^{1/2}$.

In the original paper of Spach et al., k_1 was reported to be 10^{-7} sec^{-1} at 25°C. However, it appears that the intercept is too small to be determined reliably, and the results should be interpreted by the inequality, $k_1 \leqslant 10^{-7}$ sec^{-1}. In fact, other evidence (195) suggests that $k_1 = 10^{-9}$ sec^{-1}. The rate of combination of $D\cdot^-$ into $^-DD^-$ was investigated directly by a flow technique (198) and its value seems to be 1–2×10^6 M^{-1} sec^{-1}. Hence, the equilibrium constant, K_1, of dissociation of $^-DD^-$ into radical ions is about $10^{-15}M$. On this basis the value of k_{-2} was determined as about 800 M^{-1} sec^{-1} (195).

Two comments are appropriate at this point. Judging from the results obtained in the 1,1-diphenylethylene system, the recombination of labile radical-ions, derived from monomers, is relatively slow when

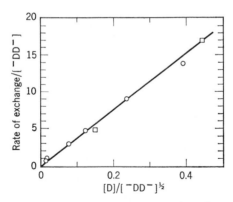

Fig. 22. Rate of exchange between dimeric dianion of 1,1-diphenylethylene ($^-DD^-$) and its radioactive monomer D. The rate of exchange per $^-DD^-$ is plotted versus $[D]/[^-DD^-]^{1/2}$: (◯), radioactive D; (☐), radioactive $^-DD^-$.

compared with that of recombination of ordinary radicals. The latter reaction is diffusion controlled, the respective rate constants being about 10^{10} M^{-1} sec^{-1}. Apparently the delocalization of the odd electron and the need of rehybridization of the bonds is responsible for the inertia in radical-ions combination. The latter factor is manifested in the system, anthracene$\cdot^-$ + styrene$\cdot^-$. Recent study indicates that this mixed combination is even slower than combination of two $D\cdot^-$, its rate constant being 10^4 M^{-1} sec^{-1} (199). This is not surprising. The rehybridization of bonds in anthracene$\cdot^-$, i.e., in a *stable* radical-ion, and the resulting loss of its resonance energy increases substantially the chemical inertia of combination. A similar value for the combination constant was obtained for the system anthracene$\cdot^-$ + 1,1-diphenylethylene$\cdot^-$ $\rightleftarrows$ $^-DA^-$ (195).

How much does a radical reaction contribute to polymerization initiated by radical-ions? The quantitative results obtained for the system $D\cdot^- + D\cdot^- \overset{k_{-1}}{\rightleftarrows} {}^-DD^-$ and $D\cdot^- + D \overset{k_{-2}}{\rightleftarrows} {}^-DD\cdot$ may serve as a starting point of our discussion. The rate constant k_{-2} appears to be about 800 M^{-1} sec^{-1}, whereas $k_1 = 1\text{--}2 \times 10^6$ M^{-1} sec^{-1}, i.e., the latter is at least 1000 times greater than the former. If a similar relation is valid for other monomer systems, the formation of dimeric radical-ions should be probable whenever the concentration of monomer exceeds that of its monomeric radical-ion by a factor of 1000 or more. The stationary concentration of radical-ions in polymerizing systems is difficult to assess, but its value must be lower at least by a factor of 1000 than that of the monomer if the degree of polymerization of the resulting product exceeds 1000. Hence, the dimeric radical-ions may play some role in systems yielding high molecular weight polymers if the polymerization is initiated by electron-transfer reaction.

Participation of a radical-growth in polymerization reactions initiated by electron transfer might be revealed in a copolymerizing system composed of two monomers—one, say A, growing by anionic mechanism but unable to participate in a radical propagation, whereas the other, say B, being reactive in radical polymerizations but inert in anionic processes. If an electron-transfer initiation of such a system yields a block polymer, A.A...A.B.B...B, or A.A...A.B.B... B.A.A...A, this would provide a clear manifestation of the participation of both the anionic and the radical ends in the growth process.

Such a copolymerizing system is not known yet. However, the system styrene–methyl methacrylate initiated by metallic lithium, investigated by Tobolsky and his students (200), could serve as a diagnostic tool in studies of these problems. A typical reaction mixture investigated by Tobolsky contained 20 ml of solvent (tetrahydrofuran or heptane) and 2 ml of each monomer. Fifty milligrams of lithium dispersion were used to initiate the reaction, and the polymerization was quenched at the desired time by adding methanol. A reaction carried out at 20°C in the absence of solvent yielded a product containing up to 60% of styrene if the process was interrupted at 1% of conversion. As the polymerization proceeded, the proportion of styrene in the polymer rapidly decreased, being only 20% at 10% conversion. The resulting product had an intrinsic viscosity in benzene of about 0.2 at 25°C, indicating its high molecular-weight nature. In tetrahydrofuran the reaction was much faster and it was not feasible to stop the polymerization before 15% of the mixture had polymerized. The content of styrene was less than 10%; its proportion remained constant up to about 40% conversion and then it gradually increased. This is a trivial result since, as methyl methacrylate becomes consumed, styrene had to be incorporated into the product because lithium emulsion is still available for initiation.

Fractionation and extraction with suitable solvents indicated that the products were not mixtures of homopolymers; neither were they composed of ordinary radical type, random copolymers. It was suggested, therefore, that the reaction yielded a block polymer (201). The electron-transfer initiation produced monomeric radical-ions, and eventually the dimeric radical ions which initiate two types of growth. The anionic growth led to a block of homopoly(methyl methacrylate), virtually free of styrene, whereas the chain grown from the radical end should form a block of 50:50 random copolymer of styrene and methyl methacrylate, as expected on the basis of the well-known study of Walling and Mayo (202). In fact, under otherwise identical conditions the polymerization initiated by butyllithium produced pure homopoly(methyl methacrylate), indicating that the anionic polymerization with lithium counterion behaves conventionally. These findings were confirmed by other workers (261).

Other results reported by Tobolsky et al. (200) seemed to fit this scheme well. In mixed solvents (tetrahydrofuran–heptane) the proportion of styrene in the polymer decreased with increasing concentration of tetrahydrofuran in the polymerizing medium. The ether accelerates anionic polymerization and has a negligible effect on the radical growth. Hence, the observed result seems to be accounted for. However, the intrinsic viscosity of the product is also decreased at higher ether concentration—a fact not considered by Tobolsky and contradicting his mechanism. Indeed, the scheme discussed by him and elaborated mathematically in a paper with Hartley (203), as well as in an earlier paper by Stretch and Allen (204), assumes a conventional termination of radical polymerization and lack of termination in anionic growth. Consequently, the higher the concentration of radical-ions the less significant the contribution of radical process. For example, the extremely rapid initiation by sodium naphthalene yields only poly-(methyl methacrylate) (203). Strangely enough, polymerization initiated by metallic sodium yielded again only poly(methyl methacrylate) (205), and it was postulated therefore that the anionic growth or initiation is many times faster with sodium than with the lithium counterion. This is questionable.

Some inconsistency becomes apparent when the data of Tobolsky are scrutinized. The anionic growth is faster than the radical growth, and therefore it should not be possible to incorporate more than 25% of styrene into the polymer, if the hypothesis of the simultaneous anionic growth and propagation applies. Nevertheless, styrene content as high as 60% was observed. A mathematical treatment developed by Tobolsky and Hartley (203) claimed to account for this difficulty. Let's denote the rate of simultaneous formation of the radical and anion ends through electron transfer by R_i and the respective rate constants of their propagation by $k_{p,r}$ and $k_{p,a}$. The polymerizing system was assumed to be in a *stationary* state in respect to radical growth, and therefore, $R_i = k_t[R]^2$. Hence, the amount of the radical-grown polymer formed in time t is $P_r = k_{p,r}(R_i/k_t)^{1/2}M_0 t$ (we are interested in a low conversion, i.e., $[M] \sim [M]_0$). The amount of the polymer formed by the anionic propagation is $P_a = k_{p,a}(1/2)R_i t^2 M_0$, because the anionic ends are assumed to be continually formed but not destroyed. Thus, the fraction of the radical-formed material $P_r/(P_r + P_a) = \{1 + \frac{1}{2}(k_{p,a}k_t/k_{p,r}^2 M_0)P_r\}^{-1}$. Its value therefore approaches unity when P_r and P_a tend to zero, i.e., the polymer produced in the very early stage of the process should result from the radical growth only. Unfortunately, this conclusion is erroneous because the extrapolation to $P_r = 0$ invalidates the stationary-state assumption.

The attractive hypothesis of Tobolsky was challenged by other investigators. Overberger and Yamoto (206), who repeated Tobolsky's experiments, examined the resulting products by the NMR technique. The high content of styrene in the initially formed material was confirmed but, contrary to the expectation, the resulting polymers were shown to involve blocks of *homopolystyrene* instead of the predicted 50:50 random copolymer blocks of styrene and methyl methacrylate.

An alternative hypothesis was suggested by Overberger et al. (207,206). Styrene is assumed to be preferentially adsorbed on the surface of lithium particles and the initial polymerization is visualized to occur in the adsorbed layer. Therefore, a block of living homopolystyrene is produced initially, and only after this species is desorbed from the lithium particles and enters the liquid phase does it react with methyl methacrylate. The latter reaction is irreversible because living polymers with methyl methacrylate end groups cannot react with styrene (209). Hence, further polymerization produces blocks of homopoly(methyl methacrylate). Such a model circumvents the objection raised by George and Tobolsky (208).

In conclusion, the participation of radical ends in polymerization initiated by an electron-transfer process is not ruled out, but no convincing evidence for this phenomenon is as yet available. The behavior of the system styrene–methyl methacrylate initiated by lithium metal emulsion cannot be explained by such a hypothesis and, therefore, no evidence for simultaneous radical and anionic polymerization is provided by its study. Some electron donors may initiate polymerization by mechanisms other than electron transfer, see, e.g., refs. 271–273.

We remarked at the beginning of this section that radical-anions derived from aromatic hydrocarbons do not dimerize—the formation of a new C—C bond is energetically unprofitable. The same reasons suggest that analogous radical-cations should not form dimers, and even less probable would be a dimerization in which a covalent bond is formed between a radical-ion and its parent aromatic hydrocarbon. Indeed, all these processes have not been observed. However, in the presence of an excess of aromatic hydrocarbon its radical-cation undergoes an interesting association. The ESR studies of such systems revealed the formation of new species (221,222). Their ESR spectra are similar to those of the respective radical-cations, but each line is doubled and the splitting is halved. It was concluded that association of $A^{\cdot+}$ and A (A being an aromatic hydrocarbon) yields a sandwich compound in which the two partners are arrayed parallel, one on top of the other, separated perhaps by a counterion and bonded by virtue of delocalization of the "positive hole." Obvious quantum mechanical reasons make such an association improbable for the system $A^{\cdot-} + A$ and, indeed, this phenomenon was not observed in the anionic systems.

The quantum mechanical treatment of bonding of $A \cdot^+$ and A is similar to that of charge-transfer complexes. Another type of dimer, resembling the $A \cdot^+ A$, is found in photochemical studies when an excited molecule associates with the nonexcited one (223). The formation of such an associate, known as excimer (224) was manifested by a change in the fluorescence of the irradiated hydrocarbon, e.g., pyrene, the spectrum being affected by the concentration of the hydrocarbon. At low concentration the fluorescence of pyrene is violet and shows fine structure, but at concentrations about $10^{-4}M$ a broad, structureless band appears in the blue region, its intensity being proportional to the square of concentration. The quantum mechanical treatment of the excimer was discussed by Hoijtink (225), who represents its function by $E^- = A_1^* A_2 - A_1 A_2^*$, A being the ground state and A^* the second excited singlet wave function of pyrene. An interesting example of an intramolecular excimer, formed from cyclophane, was described by Ron and Schnepp (226).

Excimers are also formed in processes involving interaction between the negative, $A \cdot^-$, and positive, A^+, radical-ions. This topic is discussed in Section XXI.

Finally, dimerization is observed between dianions of aromatic hydrocarbons and vinyl monomers [for example, eq. (9)].

$$\text{Anthracene}^{2-} + \text{styrene} \longrightarrow \quad \text{CH}_2.\text{CH(Ph)}^-, \quad (^-\text{AS}^-) \tag{9}$$

Reaction (9) takes place directly (199) and not through the sequence of two processes:

$$\text{Anthracene}^{2-} + \text{Styrene} \rightleftarrows \text{Anthracene} \cdot^- + \text{Styrene} \cdot^-$$
$$\text{Anthracene} \cdot^- + \text{Styrene} \cdot^- \longrightarrow {}^-\text{AS}^-$$

A similar dimer is directly formed when A^{2-} reacts with 1,1-diphenylethylene (D). However, the resulting product, $^-AD^-$, rapidly decomposes into $A \cdot^- + D \cdot^-$ (195), i.e., at low concentrations the equilibrium favors the formation of the mixed dimer from A and D but not from $A \cdot^-$ and $D \cdot^-$.

XIX. KINETICS OF ELECTRON-TRANSFER REACTIONS BETWEEN RADICAL-IONS AND THEIR PARENT MOLECULES

Electron exchange reactions between radical-ions and their parent molecules are perhaps the simplest electron-transfer processes. Their study was pioneered again by Weissman (233), who determined the bimolecular rate constants of the reaction:

Naphthalene$^{\cdot-}$,M$^+$ + Naphthalene $\rightleftharpoons$ Naphthalene + Naphthalene$^{\cdot-}$,M$^+$

from the broadening of the ESR lines of naphthalene$^{\cdot-}$. The results are collected in Table XVIII and show the influence of counterions and

TABLE XVIII

Naphthalene$^{\cdot-}$,M$^+$ + Naphthalene $\xrightarrow{k}$ Naphthalene + Naphthalene$^{\cdot-}$,M$^+$ at 25°C

Counterion	Solvent[a]	k, M^{-1} sec^{-1}
K$^+$	DME	$(7.6 \pm 3) \times 10^7$
K$^+$	THF	$(5.7 + 1) \times 10^7$
Na$^+$	DME	$\sim 10^9$
Na$^+$	THF	$\sim 10^7$
Li$^+$	THF	$(4.6 \pm 3) \times 10^8$

[a] DME, dimethoxyethane; THF, tetrahydrofuran.

solvents on the rate of transfer. The rate was found to be especially high for the sodium salt in dimethoxyethane (DME) and, to a lesser extent, for the lithium salt in tetrahydrofuran (THF). These observations have been accounted for by assuming participation of free naphthalene$^{\cdot-}$ radical-ions in these two reactions, whereas ion pairs were involved in the other processes. However, recent conductance studies indicate that the sodium salt in DME and the lithium salt in THF form solvent-separated ion pairs, the other solutions being composed of contact pairs. Hence, the results might be interpreted as evidence for rapid electron transfer from solvent-separated pairs which seem to be about 20 times as reactive as the contact pairs. Both exchanges proceed with low activation energy, probably not more than 2–3 kcal/mole. Recent evidence for the participation of two, or more, types of ion pairs in electron transfer processes is provided by Hirota (70).

The sodium salt in DME produces sharp ESR lines, the splitting caused by the Na^{23} being negligible because the cation is relatively far away from the radical-ion. Addition of naphthalene broadens the signal (the consequence of exchange), but the lines sharpen again on addition of potassium iodide. Apparently the counterions exchange, i.e.,

$$N^{\cdot-},Na^+ + KI \rightleftharpoons N^{\cdot-},K^+ + NaI$$

and the electron transfer from the potassium salt is too slow to cause an appreciable broadening of the lines.

The technique described above was refined by Zandstra and Weissman (238). Studies of Atherton and Weissman (239) demonstrated that ESR spectroscopy enables one to discriminate between sodium naphthalene ion pairs and the free naphthalene$^{\cdot-}$ radical-ions (or perhaps between the contact and solvent separated pairs of sodium naphthalene; see ref. 70). If both are present, the ESR spectrum of the investigated solution contains two sets of lines, the intensity of each being proportional to the concentration of the respective species.

On addition of naphthalene the lines of each set became broader (238), the broadening being again proportional to the concentration of naphthalene and independent of the concentration of naphthalene$^{\cdot-}$. The degree of broadening was, however, different for each set. Using the method described in detail by Weissman (235), it was possible to determine from the broadening the rate constants of electron transfer from each species. Fortunately, the dissociation of pairs into free ions (or transformation of contact pairs into solvent separated pairs, if we accept the alternative interpretation of the ESR signals), was found to be sufficiently slow not to cause any perturbation of the respective spectra.

The pertinent results are summarized in Table XIX. The following

TABLE XIX

Electron Transfer between Naphthalene$^{\cdot-}$ Ion Pair (Na$^+$), or Its Free Ion and Naphthalene[a]

Solvent	Species	T, °C	$k \times 10^{-6}$, $M^{-1} \sec^{-1}$	E, kcal/mole
Tetrahydropyran (THP)	Ion pair	21	4.4	18
Tetrahydropyran	Free ion	−23	9.7	19
2-Methyltetrahydrofuran (MeTHF)	Ion pair	21	4.5	12
2-Methyltetrahydrofuran	Free ion	−23	2.8	13
Tetrahydrofuran (THF)	Ion pair	50	34	5
Tetrahydrofuran	Free ion	13	38	13

[a] Most of these activation energies seem to be too high.

points deserve comments. Rate constants of exchange are substantially lower in THP and MeTHF than in THF, and the respective "activation energies" are higher. However, reactivities of the free ions are not much different from those of ion pairs, the THF system being an exception. In this solvent the activation energy for the free naphthalene·⁻ ion was reported to be much higher than that for the ion pair (13 kcal/mole as compared with 5 kcal/mole)—a strange observation. In fact, the behavior of the sodium ion pair in THF is most bizarre. As may be seen from Figure 23, the rate constant decreases with temperature but increases again at still lower temperatures. Several suggestions were made to account for such a temperature dependence, e.g., association of ion pairs with naphthalene which could become appreciable at lower temperatures. Although this explanation was not rejected, it was not considered satisfactory either.

Alternatively, one may visualize several vibrational subspecies, each contributing to the process, and postulate the highest rate constant for the lowest state. Denoting the *temperature independent* rate constants of each subspecies by k_0, k_1, etc., the ratios of concentrations [subspecies$_i$]/[subspecies$_0$] by K_1, K_2, etc., and finally the respective statistical

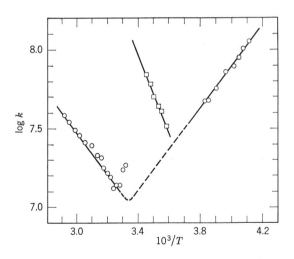

Fig. 23. Plot of exchange constant N·⁻ + N $\rightleftarrows$ N + N·⁻ (N = naphthalene) versus $1/T$. The exchange is measured by ESR broadening. ($\bigcirc$) ion pairs; ($\square$) free ions. Units M^{-1} sec^{-1}.

weights and the separation energies by g_i and E_i, we find the observed k to be given by the equation

$$k = (k_0 + \sum_i k_i K_i)/(1 + \sum_j K_j)$$

with $K_i = g_i \exp(-E_i/RT)$. Obviously, k *increases* with *decreasing* temperature of $k_0 > k_i$, but the function $k(T)$ has no minimum.* Hence, this treatment, presented in ref. 238 for 2 subspecies only, cannot account for the observations. Furthermore, the change seen in Figure 23 is much too abrupt for a physical phenomenon of this type. It seems, therefore, that the problem needs additional investigation.

Further information on this system comes from the interesting work of Chang and Johnson (240), who utilized the yet little exploited "fast exchange limit" technique. In the presence of a large excess of naphthalene the spectrum collapses into one very sharp line (for the free ion reaction) or into four lines (for the exchange involving sodium ion pairs). The collapsed spectrum is shown in Figure 24. The line is

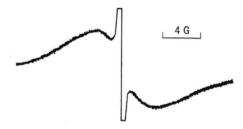

Fig. 24. ESR first derivative spectrum obtained when naphthalene (1.4M) was reduced with potassium in tetrahydrofuran. Only a small fraction of naphthalene is reduced and the presence of a large excess of naphthalene leads to a rapid exchange. The clipped center peak indicates an off-scale reading.

rigorously Lorentzian, thus confirming the fast exchange limit. The second-order rate constant of exchange can then be calculated from the equation derived by Piette and Anderson (241), namely, $k = 2.04 \times 10^7 \times \nabla/\Delta H[A]$. Here ∇ is the second moment, in gauss, of the spectrum of the investigated species in the absence of exchange, ΔH is the line width of the collapsed spectrum (separation between the extremes in the first derivative spectrum) corrected for modulation, natural line width, etc., and [A] is the concentration of the hydrocarbon. The method is useful if the two investigated species react with sufficiently

* $dk/dT = -\{\sum_i (k_0 - k_i) E_i Ki\}/RT^2 (1 + \sum_j K_j)^2 < 0$.

different rates. Only one line was observed at high naphthalene concentration for $N^{\cdot-},Na^+$ and $N^{\cdot-},K^+$ in DME at temperatures ranging from -30 to $40°C$. However, two lines of appreciably different width were found for $N^{\cdot-},K^+$ in THF (see Fig. 24), and a single line superimposed upon a broad uncollapsed spectrum was observed for $N^{\cdot-},Na^+$ in THF.

The faster rate was assigned to free ions. The smaller rate constants, attributed to ion pairs, agreed well with those reported by Zandstra and Weissman (238) from the results of "slow exchange limit."

Studies of exchange between $N^{\cdot-}$ and N were followed by similar investigations of electron transfer between ketyl of benzophenone and its parent ketone (234). The ESR spectrum of the sodium ketyl in DME consists of more than 80 lines and it was conclusively shown that a splitting by the nuclear moment of Na^{23} does take place. The magnitude of this splitting is about one gauss, indicating that each molecule of ketyl retains its sodium counterion for at least 3×10^{-7} sec.

On addition of ketone, an electron transfer takes place. The question was raised whether the transfer takes place to benzophenone and then the newly formed free radical-ion becomes paired with *any* free cation, or whether the electron transfer is coupled with the transfer of Na^+ ion. These two alternatives may be represented schematically by the equations (a) and (b)

$$Na^+,OC\phi_2^{\cdot-} + \phi_2CO \longrightarrow Na^+ + OC\phi_2 + \phi_2CO^{\cdot-}$$
$$\text{any free } Na^+ + \phi_2CO^{\cdot-} \longrightarrow \phi_2CO^{\cdot-},Na^+ \tag{a}$$

and

$$\phi_2CO^{\cdot-}Na^+ + OC\phi_2 \rightleftarrows \phi_2CO + Na^+,OC\phi_2 \tag{b}$$

For a rapid transfer (large concentration of benzophenone) the ESR spectrum should collapse into *one* line if the process takes place through route a. On the other hand, four lines, each corresponding to one possible orientation of the nuclear spin, should be formed if the transfer proceeds by mechanism b. The latter spectrum was observed, indicating that we deal here with an atom transfer rather than electron transfer. It is significant that the splitting due to Na^{23} is larger (1.1 gauss) in the collapsed spectrum than in the isolated ketyl (1.0 gauss). Hence, the electron density on the sodium nucleus *increases* in the course of transfer process.

The second-order rate constant was determined to be about $5 \times 10^7 \ M^{-1} \sec^{-1}$, i.e., for a $2M$ concentration of benzophenone the mean lifetime of each ketyl molecule is less than 10^{-8} sec, whereas a sodium ion retains its association with radical-ion for more than 3×10^{-7} sec.

More comprehensive kinetic studies of electron exchange between ketyls and ketones were reported by Hirota and Weissman (217). Their results, summarized in Table XX, show that the transfer is faster in

TABLE XX

Rate Constants, k, and Activation Energies, E, for Electron Transfer Reaction Ketyl + Ketone $\rightarrow$ Exchange

Ketyl	Solvent[a]	T, °C	k, M^{-1} sec^{-1} $\times 10^{-8}$	E, kcal/mole
Na$^+$,Xanthone$\cdot^-$	DME	37	4.6	5.1
Na$^+$,Xanthone$\cdot^-$	THF	25	4.5	4.3
Na$^+$,Xanthone$\cdot^-$	THP	7	4.7	—
Na$^+$,Xanthone$\cdot^-$	MeTHF	0	4.3	—
Rb$^+$,Xanthone$\cdot^-$	DME	25	2.5	—
Na$^+$,Banzophenone$\cdot^-$	DME	25	1.1	6.3
Na$^+$,Benzophenone$\cdot^-$	THF	12	1.1	6
K$^+$,Benzophenone$\cdot^-$	THF	25	2.5	—
Rb$^+$,Benzophenone$\cdot^-$	DME	25	1.6	—
Mg^{2+}(Benzophenone$\cdot^-$)$_2$	DME	24	$< 10^{-2}$	—
Ca^{2+}(Benzophenone$\cdot^-$)$_2$	DME	24	$< 10^{-2}$	—
(Na$^+$,Benzophenone$\cdot^-$)$_2$	MeTHF	24	$< 10^{-2}$	—
(Na$^+$,Benzophenone$\cdot^-$)$_2$	Dioxane	24	$< 10^{-2}$	—
Free Xanthone$\cdot^-$ ion	Acetonitrile	25	10	—

[a] DME = dimethoxyethane; THF = tetrahydrofuran; THP = tetrahydropyrane; MeTHF = 2-methyltetrahydrofuran.

tetrahydrofuran than in tetrahydropyrane, and even faster in dimethoxyethane. The transfer is extremely slow for the dimers. Apparently the transition state has a sandwich-type structure, as shown by eq. (b), and for steric reasons such a structure is impossible for a dimer. It may be that the electron transfer requires dissociation of the dimer and the dissociation process is then the rate determining step.

The activation energies are relatively high, i.e., the A factors are of the order $10^{12}-10^{13}$ M^{-1} sec^{-1}. Apparently some desolvation of ion pair takes place in the transition state.

Electron exchange processes between parent molecules and their radical-anions formed electrolytically in dimethyl formamide were investigated by Adams and his co-workers (282). Under these conditions the radical ions are *not* associated with counterions. The results obtained for the systems anthracene, *para*-benzoquinone, duroquinone, and 1,4-naphthaquinone led to the following bimolecular rate constants of exchange: 5×10^8, 4×10^8, 0.6×10^8, and 4×10^8, respectively (in units M^{-1} sec^{-1}). These studies were extended to substituted nitrobenzenes and led to similar results. It was concluded that the exchange occurs via molecular orbital overlap, i.e., an electron is not transferred over long distance in the solution. This statement is consistent with the current opinion on processes involving radical-ions not associated with counterions.

Exchange of an electron between potassium 2,2′ bipyridyl radical-anion and its parent molecule was investigated by Reynold (252). His intention was to determine whether "complex" formation takes place in the electron transfer when the investigated species is a potent complexing agent. No peculiarity was shown in this system. The bimolecular rate constant of exchange was found to be $\sim 4 \times 10^6$ M^{-1} sec^{-1} at 20°C, the activation energy of transfer, 10 kcal/mole, and the entropy of activation, about $+5$ eu.

Electron exchange between a radical and its carbanion was investigated by Jones and Weissman (236), namely,

$$(p\text{-}NO_2 \cdot C_6H_4)_3C\cdot + Me^+, \bar{C}(C_6H_4\text{-}p\text{-}NO_2)_3 \longrightarrow \text{exchange}$$

Again, broadening of the ESR lines of the neutral radical permitted them to determine the respective bimolecular rate constants. Their values were in the range of 10^8–10^9 M^{-1} sec^{-1} at 25°C, like in other systems, but the activation energies were only of the order 1–2 kcal/mole. The rate constants were affected by the solvent and counterion, indicating the participation of ion pairs. The exchange was found to be 2–4 times faster in acetonitrile than in ethereal solvents and, if the dissociation of pairs is complete in the former medium, the free ions appear to be only slightly more reactive than ion pairs.

A very interesting problem was posed by Zandstra and Weissman (237). Consider a system composed of radical-ions and their parent molecules in which electron transfer takes place. Denote by t the average time spent by the odd electron on a molecule. The molecules exist in N nuclear-spin states, of which g_k states corresponding to the same energy level and give rise to the same line k in the ESR spectrum. Electron transfer leads, therefore, to broadening of this line only if the transfer takes place from a molecule belonging to group g_k, to another

that does not belong to this group. Hence, the characteristic time
determining the broadening is given by t_k where

$$1/t_k = (1/t)(N - g_k)/N$$

Obviously, t_k is always greater than t and it becomes infinity if there is
only one spin state (in such a case the molecular magnetic field within
which the electron moves is not changed by the electron transfer). The
above equation is based on the assumption that the probability of
transfer is independent of the spin state of the acceptor. Is this assumption valid?

The treatment given above indicates that the addition of parent
molecule broadens different lines in the ESR spectrum to a different
extent, because g_k is, on the whole, different for different lines. Moreover, this approach permits one to calculate the broadening ratios of,
say, line k and k'. The result of this calculation may be verified experimentally and the agreement would then confirm the basic assumption
of constant probability of transfer. Studies of Zandstra and Weissman
(237) verified this relation for several lines of the ESR spectrum of
potassium naphthalene in the presence of naphthalene (237) and
showed, therefore, that the probability of electron transfer is indeed
independent of the spin state of the acceptor.

Although the ESR method is versatile and conveniently adaptable
for electron transfer studies, other methods should not be neglected.
For example, a recent investigation of Doran and Waack (270) of *cis–
trans* isomerization induced by radical-anion formation provides an
interesting method for studying the kinetics of some electron-transfer
processes. These workers demonstrated that *cis*-stilbene is isomerized
into the *trans* compound when a small fraction of olefine is converted
into stilbene·⁻. The reaction was followed spectrophotometrically and
it was shown that it does not take place in the absence of radical-ion.
Presumably, the process is governed by the following reaction:

cis-stilbene + stilbene·⁻ $\rightleftarrows$ stilbene·⁻ + *trans*-stilbene

and therefore the rate of the electron transfer determines the rate of
isomerization. It was not elucidated whether, and to what extent if any,
the dianion participates in the reaction. A similar behavior was observed
in the 1,2,3,4-tetraphenyl butadiene system in which three geometrical
isomers exist.

Kinetic studies by ESR technique of electron transfer reactions of
the type $A\cdot^- + B \rightarrow B\cdot^- + A$ were attempted (242), but only qualita-

tive results were obtained. These reactions were successfully investigated by means of pulse radiolysis (243). The following principle is applied. Solvated electrons, formed by a pulse in isopropyl alcohol, are scavenged by aromatic hydrocarbons. Since the rate of electron capture is nearly independent of the electron affinity of the scavenger (see Table XII), the hydrocarbon A, which is present in large excess, initially captures all the electrons. The scavenging is completed in about 0.5 μsec. The system contains, also, another hydrocarbon B, of high electron affinity but present at low concentration. Thus, the reaction

$$\text{A·}^- + \text{B} \xrightarrow{k_{AB}} \text{B·}^- + \text{A}$$

follows the initial electron capture, and its progress is monitored by the usual spectrophotometric method. Biphenyl and *para-*, *meta-*, and *ortho*-terphenyls were used as the scavengers which produced the respective electron donors, whereas naphthalene, phenanthrene, *para*-terphenyl, pyrene, and anthracene were the electron acceptors. The kinetics is slightly complicated by protonation, and hence,

$$-d[\text{A·}^-]/dt = k_{AB}[\text{A·}^-][\text{B}] + k_{2A}[\text{A·}^-][\text{PrOH}] + k_{3A}[\text{A·}^-][\text{PrOH}_2{}^+]$$

and

$$d[\text{B·}^-]/dt = k_{AB}[\text{A·}^-][\text{B}] - k_{2B}[\text{B·}^-][\text{PrOH}]$$

The rate constants k_{2A} and k_{2B} refer to the protonation of A·$^-$ and B·$^-$, respectively, by isopropyl alcohol, and these were determined in an independent study (see Table XIV); k_{3A} is the rate constant of protonation by the $\text{PrOH}_2{}^+$ acid but the term $k_{3A}[\text{A·}^-][\text{PrOH}_2{}^+]$ turned out to be very small. Solution of these differential equations led to the results given in Table XXII.

It is obvious that the investigated processes involve free radical-ions. The $\text{PrOH}_2{}^+$ are the positive counterions, and their interaction with radical-anions leads to protonation and not to ion pair formation. It is also interesting to note that these reactions, although very fast and favorable energetically, are not all diffusion controlled. This is certainly the case for the slower process, e.g., diphenyl with naphthalene or phenanthrene, i.e., in those reactions not every encounter leads to electron transfer.

Let us compare the electron transfer reactions

$$\text{A·}^- + \text{A} \rightleftharpoons \text{A} + \text{A·}^-$$
$$\text{B·}^- + \text{B} \rightleftharpoons \text{B} + \text{B·}^-$$

and

$$\text{A·}^- + \text{B} \rightleftharpoons \text{A} + \text{B·}^-$$

TABLE XXII

Electron-Transfer Rate Constants for Aromatic Molecules in
Isopropanol at 25°C

Donor anion radical	Acceptor molecule	Absolute rate constant, $M^{-1}\,sec^{-1} \times 10^{-9}$
Diphenylide	Naphthalene	0.26 ± 0.08
Diphenylide	Phenanthrene	0.6 ± 0.3
Diphenylide	para-Terphenyl	3.2 ± 0.7
Diphenylide	Pyrene	5.0 ± 1.8
Diphenylide	Anthracene	6.4 ± 2.0
para-Terphenylide	Pyrene	3.6 ± 1.1
para-Terphenylide	Anthracene	5.5 ± 0.9
meta-Terphenylide	Pyrene	3.5 ± 1.2
ortho-Terphenylide	Pyrene	4.0 ± 1.8

and denote by K_{2A}, k_{2B}, and k_{AB} the respective rate constants and by K_{AB} the equilibrium constant of the last reaction. Marcus has shown (295) that

$$k_{AB} = (k_{2A}, k_{2B} \cdot K_{AB} \cdot f)^{\frac{1}{2}}$$

where f is a unique function of K_{AB}, such that $f \equiv 1$ when $K_{AB} = 1$.

In an ingenious experiment Bruning and Weissman (294) tested this theory. They investigated, by ESR technique, the exchange between the optically pure radical-ion of $(\alpha\text{-naphthyl})(\text{Ph})\text{CHCH}_3$ and its parent hydrocarbon, as well as the exchange in the racemic mixture. The first system provides the rate constant $k_{2d} = k_{2l}$, i.e.,

$$A(d)^{\cdot -} + A(d) \rightleftarrows A(d) + A(d)^{\cdot -}$$

whereas from the rate constant of exchange in the racemic system one may calculate the rate constant k_{dl},

$$A(d)^{\cdot -} + A(l) \rightleftarrows A(d) + A(l)^{\cdot -}$$

It may be easily verified that $k_{2d} = k_{2l} = k_{dl}$ if Marcus's equation applies. The experimental results obtained for the potassium salts at 25°C were as follows: in DME $k_{2d} = (0.68 \pm 0.07) \times 10^8\ M^{-1}\,sec^{-1}$; $k_{dl} = (1.1 \pm 0.3) \times 10^8\ M^{-1}\,sec^{-1}$; in THF $k_{2d} = (0.19 \pm 0.04) \times 10^8\ M^{-1}\,sec^{-1}$; $k_{dl} = (0.2 \pm 0.1) \times 10^8\ M^{-1}\,sec^{-1}$. It seems, therefore, that Marcus's rule does apply, although Weissman felt that a small, but distinct, deviation is observed in dimethoxyethane.

The kinetics of electron transfer to vinyl monomers was studied, and it was shown that the rate of the overall reaction is determined by the dimerization of the resulting monomeric radical-anion and not by the transfer (195,198,244). For example, polymerization of styrene may be initiated by sodium naphthalene and the initial rate of disappearance of the radical-ion seems to be given by the equation:

$$(-dN\cdot^-/dt)_0 = \text{const} \, (N\cdot^-/N)^2$$

Hence, the equilibrium $N\cdot^- + \text{styrene} \rightleftarrows N + \text{styrene}\cdot^-$ is rapidly established.

Relatively slow electron transfer processes were reported by Evans et al. (245), who studied the reactions of sodium chrysene and sodium picene with 1,1,3,3-tetraphenylbutene-1. The reaction is claimed to be first order in the olefin and in the radical-ion (presumably the ion pair), and the respective rate constants were 0.013 M^{-1} sec^{-1} for chrysene$\cdot^-$,Na$^+$ and 10^{-5} M^{-1} sec^{-1} for picene$\cdot^-$,Na$^+$, both values being obtained at 40°C in tetrahydrofuran. It was shown previously (248) that butyllithium does not react with 1,1,3,3-tetraphenylbutene-1, because steric hindrance prevents the approach of the Bu$^-$ moiety to the C$=$C bond. The reaction with electron donors presumably involves the transfer of electron through the phenyl groups of the olefin and, consequently, it becomes sterically possible in spite of the shielding. The reliability of these results is questionable. The rates are much too slow for genuine electron-transfer processes. Judging from the data given by Evans, the relevant equilibrium constants cannot be smaller than 0.01 and, hence, the endothermicity of the reaction cannot account for the results. It seems that the observed process is favored by some other reaction, the nature of which is not understood by the authors.

Donors derived from aromatic hydrocarbons of lower electron affinity, e.g., naphthalene or phenanthrene, react too fast to permit a kinetic study (a conventional mixing batch technique was used). Those derived from hydrocarbons of higher electron affinity, e.g., pyrene or anthracene, do not react at all—the equilibrium is obviously unfavorable.

An analogous study was performed with tetraphenylethylene (246). Here, we encounter another hindered hydrocarbon which does not react with butyllithium. Nevertheless, those electron-transfer reactions which take place with tetraphenylethylene were too fast for a kinetic investigation. Also, the reverse reactions, such as the transfer from the dianion of tetraphenylethylene to anthracene, were found to be very fast.

Finally, we should consider some intramolecular electron-transfer processes. This problem was treated by McConnell (249), who discussed radical-ions having the general structure

$$\left\{ \bigcirc\hspace{-0.5em}-(CH_2)_n-\hspace{-0.5em}\bigcirc \right\}^{\overline{}}$$

ESR spectra of such molecules were observed by Weissman (250) and by Voevodskii et al. (251) for $n = 1$ or 2. They found the rate of electron transfer from one ring to another to be fast as compared to the σ-proton hyperfine splitting ($\sim 10^7$ sec^{-1}).

When an electron moves through a molecule or a solid, there is a tendency for the nuclear motions to become correlated with the electronic motion. This correlation represents a breakdown of the Born-Oppenheimer approximation, and such a tendency of the nuclear motion to follow the electronic motion leads to "self-trapping" of the electron (formation of a polaron). The self-trapping greatly reduces the rate of transfer which decreases exponentially with n. Calculation suggests that the increase by one CH_2 group in the length of the polymethylene chain reduces the rate by roughly a factor of 10. No experimental data are yet available to test these predictions.

XX. ELECTRON-TRANSFER REACTIONS FROM CARBANIONS TO AROMATIC HYDROCARBONS

In the presence of suitable electron acceptor a carbanion may be oxidized by electron-transfer process. For example,

$$Ph_3C^-, Na^+ + A \text{ (anthracene)} \longrightarrow Ph_3C\cdot + A\cdot^-, Na^+$$

Electron transfer from Ph_3C^-, Na^+ to benzophenone was described at the beginning of this century by Schlenk (262), and a similar reaction involving cyclooctatraene was reported by Wittig (263). Powerful electron acceptors, such as aromatic nitrocompounds, may acquire electrons even from a poor donor, e.g., alkoxides (264) or thiolates (265).

Organolithium compounds, e.g., butyllithium, may add to olefins or some aromatic hydrocarbons, a new C—C covalent bond being formed in such a reaction. However, if bulky groups shield the C=C bond, the addition may be prevented by steric hindrance. For example, BuLi does not add to tetraphenylethylene or 1,1,3,3-tetraphenyl-butene-1 (248), and apparently the electron affinities of these olefins are not sufficiently high to permit an electron-transfer process either, although electron transfer is observed with naphthalene or phenanthrene radical anions (266). Electron transfer becomes, however, possible with 1,2,3,4-tetraphenylbutadiene (267). Formation of respective radical-anions was demonstrated by ESR studies as well as by spectrophotometric observations. Structure of the organolithium compound has a pronounced effect upon the rate of the electron-transfer process. The radical ion is formed almost instantaneously with benzyllithium in tetrahydrofuran, somewhat slower with n-butyllithium and very slow with methyl or phenyllithium.

Kinetic studies of electron-transfer reactions involving carbanions are meager. A thorough kinetic study of the system,

$$\text{Ph}_2\bar{\text{C}}\text{CH}_2\text{CH}_2\bar{\text{C}}\text{Ph}_2(^-\text{DD}^-) + \text{A} \rightleftharpoons \text{Ph}_2\bar{\text{C}}\text{CH}_2\text{CH}_2\dot{\text{C}}\text{Ph}_2(^-\text{DD}\cdot) + \text{A}\cdot^- ; K_\text{A}$$

were reported by Szwarc and his associates. This reaction involves ion pairs; however, for the sake of simplicity, the free ion notation is used in the equations. The dimeric radical-ion, $^-\text{DD}\cdot$, decomposes into monomeric radical ion $\text{D}\cdot^-$ and 1,1-diphenylethylene D, viz.,

$$^-\text{DD}\cdot \xrightarrow{k_{-2}} \text{D}\cdot^- + \text{D}$$

and the rapid electron transfer, $\text{D}\cdot^- + \text{A} \rightarrow \text{A}\cdot^- + \text{D}$, completes the process. The system attains its stationary state in a few seconds, and thereafter the rate of the overall reaction, determined by the disappearance of $^-\text{DD}^-$ or by the formation of $\text{A}\cdot^-$, is given by the equation

$$d[\text{A}\cdot^-]/dt = k_{-2}K_\text{A}[^-\text{DD}^-][\text{A}]/[\text{A}\cdot^-]$$

The validity of this scheme was established by testing the integrated form of the kinetic equation or by investigating the retardation of the overall process caused by addition of $\text{A}\cdot^-$.

The kinetics of decomposition of $^-\text{DD}^-$ retains its mathematical form whatever electron acceptor is added, provided that its electron affinity is sufficiently high. This conclusion was established by Gill, Jagur-Grodzinski, and Szwarc (254), who studied the processes involving 9,10-dimethylanthracene or pyrene instead of anthracene. Each of these investigations led to a value for the combined rate constants, i.e., $k_{-2}K_\text{A}$ was calculated from the studies of the anthracene system, while $k_{-2}K_\text{DMA}$ and $k_{-2}K_\pi$ were derived from the results concerned with the 9,10-dimethylanthracene (DMA) and pyrene (π) systems. The constants K_DMA and K_π refer to the equilibria:

$$^-\text{DD}^- + \text{DMA} \rightleftharpoons {}^-\text{DD}\cdot + \text{DMA}\cdot^- ; K_\text{DMA}$$

and
$$-DD^- + \pi \; \underset{\longleftarrow}{\overset{\longrightarrow}{\rightleftharpoons}} \; -DD\cdot + \pi\cdot^-; K_\pi$$

It is easy to verify that the ratios K_{DMA}/K_A and K_π/K_A are the equilibrium constants of electron-transfer processes

$$A\cdot^- + DMA \; \rightleftharpoons \; A + DMA\cdot^-; K_{DMA}/K_A$$
and
$$A\cdot^- + \pi \; \rightleftharpoons \; A + \pi\cdot^-; K_\pi/K_A$$

Hence, such kinetic studies permit us to calculate the differences of electron affinities of acceptors, which were independently determined by other direct methods (see Table VII). The agreement between the kinetic data and the potentiometric one, or those concerned with direct studies of equilibria, is satisfactory and this verifies the proposed kinetic scheme.

Does electron transfer take place directly or through an intermediate adduct? In the $^-DD^-$ system no intermediate could be seen, indicating either direct transfer or extremely short-lived intermediate. However, analogous studies involving a dimeric dianion of α-methylstyrene,

$$\bar{C}(CH_3)(Ph)CH_2CH_2\bar{C}(CH_3)(Ph) = {}^-\alpha\alpha^-$$

conclusively showed that an intermediate adduct is formed (255). The dimer, $^-\alpha\alpha^-$ (potassium counterion, tetrahydrofuran solvent), absorbs at $\lambda_{max} = 341$ mμ. On addition of anthracene this absorption peak disappears and a new one is formed at $\lambda_{max} = 451$ mμ. The addition is very rapid and virtually quantitative, its bimolecular rate constant being about 4000–5000 M^{-1} sec^{-1} at ambient temperature (257), and therefore the reaction is virtually over in 0.2 sec. The optical density of the new absorption peak was determined by a flow technique, and the plot of o.d. (451 mμ) (measured 0.2 sec after mixing the reagents) versus [anthracene]/$[^-\alpha\alpha^-]$ ratio is given in Figure 25 for a constant concentration of the dimer. It is obvious that each carbanion end of the dimer is capable of reacting with anthracene and therefore the plot shown in Figure 25 levels off for a $[A]/[^-\alpha\alpha^-]$ ratio of two. In the adduct the carbanion is bonded through a covalent C—C bond with the 9 carbon atom of the aromatic. This addition resembles the one taking place between anthracene and living polystyrene (258,259) or ethyl- or butyllithium (260). However, while the latter adducts are stable and do not decompose, the adduct formed with $^-\alpha\alpha^-$ decomposes within a few seconds yielding anthracene radical-anion. The initial rate of this decomposition was measured at constant concentration of the dimer

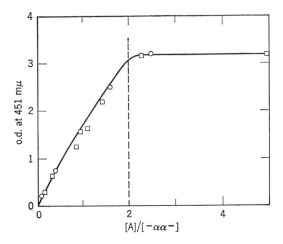

Fig. 25. Change in optical density at 451 mμ on addition of anthracene (A) to dimeric α-methylstyrene dianion ($^-\alpha\alpha^-$). For a constant concentration of the dimer ($^-\alpha\alpha^-$).

and variable concentration of anthracene and the results are presented in Figure 26 as plots of $(d[A^{\cdot-}]/dt)_0$ versus $[A]/[^-\alpha\alpha^-]$. As may be seen from the figure, the initial rate increases with increasing $[A]/[^-\alpha\alpha^-]$ ratio, it reaches maximum for $[A]/[^-\alpha\alpha^-] = 1$ and then decreases, eventually becoming very low for $[A]/[^-\alpha\alpha^-] = 2$. This most unusual behavior was explained by postulating a rapid decomposition of an α-methylstyrene dimeric dianion associated with *one* molecule of anthracene. The dianion possessing *two* molecules of anthracene, one on each of its ends, was found to be relatively stable, measurable decomposition being observed only after about an hour. Let us denote by p the ratio $[A]/2[^-\alpha\alpha^-]$. If the addition of anthracene to one end of the dimer is independent of the fate of the other, then in a mixture of $^-\alpha\alpha^-$ and A a fraction $2p(1 - p)$ of dimers should be associated with one molecule of anthracene, a fraction p^2 with two, and $(1 - p)^2$ with none. This conclusion was confirmed by gas chromatographic analyses of protonated products.

The assumption of rapid decomposition of the monoanthracenated dimer, $^-\alpha\alpha A^-$, leads to proportionality between the initial rate of decomposition and $p(1 - p)$. Hence, the rate increases with p until for $p = 1/2$, i.e., $[A]/[^-\alpha\alpha^-] = 1$, it reaches its maximum value, and then decreases virtually to zero.

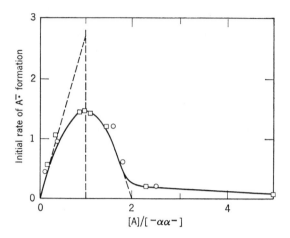

Fig. 26. Initial rate of decomposition of anthracene — $^-\alpha\alpha^-$ adduct at various initial ratios $[A]/[^-\alpha\alpha^-]$. Measured by the rate of formation of $A\cdot^-$ radical-ion. For a constant concentration of the dimer ($^-\alpha\alpha^-$).

The kinetics of the decomposition was thoroughly investigated. It was shown that the following mechanism accounts for the facts:

$$^-\alpha\alpha A^- \; \underset{\longleftarrow}{\overset{\longrightarrow}{\rule{0pt}{0pt}}} \; ^-\alpha\alpha\cdot + A\cdot^- \qquad \text{rapid, equilibrium constant } K_\alpha$$

$$^-\alpha\alpha\cdot \; \overset{k_2'}{\longrightarrow} \; \alpha\cdot^- + \alpha \qquad \text{slow rate determining}$$

$$\alpha\cdot^- + A \; \longrightarrow \; \alpha + A\cdot^- \qquad \text{again rapid}$$

The symbol α denotes a molecule of α-methylstyrene and $\alpha\cdot^-$ its radical anion.

One may wonder now why $^-\alpha\alpha A^-$ decomposes rapidly while the decomposition of $^-A\alpha\alpha A^-$ is slow. It seems that the repulsion between the two negative ends of $^-\alpha\alpha A^-$ contributes to the decomposition and provides enough driving force to shift the equilibrium K_α sufficiently to the right. This repulsion is greatly reduced in $^-A\alpha\alpha A^-$, and consequently the equilibrium concentration of $^-A\alpha\alpha\cdot$ may be much lower than that of $^-\alpha\alpha\cdot$. Moreover, decomposition of $^-\alpha\alpha\cdot$ into $\alpha\cdot^- + \alpha$ may be faster than that of $^-A\alpha\alpha\cdot$ into $^-A\alpha\cdot + \alpha$ or $A\cdot^- + 2\alpha$. It is known that the adduct of anthracene and α-methylstyrene tetramer dianion, $^-\alpha\alpha\alpha\alpha^-$, decomposes only slowly yielding $A\cdot^-$. No maximum is observed in the dependence of the initial rate of this decomposition on the $[A]/[^-\alpha\alpha\alpha\alpha^-]$ ratio.

XXI. ELECTROLYTIC GENERATION OF RADICAL-IONS

In electrolytic process the electrode acts as a donor (cathode) or as an acceptor (anode). The substrate in the vicinity of a cathode becomes reduced, whereas oxidation takes place in the vicinity of an anode. An electrode process involving one-electron transfer may convert a suitable substrate into a radical-ion, and such reactions occur, e.g., in polarographic systems which were discussed in Sec. X. Adaptation of polarographic techniques for ESR studies was first reported by Geske and Maki (179), and since then these methods have become widely used for generation of radical-ions in the cavity of ESR spectrometer.

Electrochemical generation of radical-ions is advantageous and often superior to the conventional chemical reduction (e.g., by alkali metals) or oxidation by electron acceptors. It allows the use of a wide range of solvents, it is more controllable because the potential may be adjusted at will, and it permits the variation of the counterions by judicious choice of the supporting electrolyte. The various handling procedures employed in such investigations are now fairly standard; their details are described, e.g., by Adams (281).

The study of electrode processes permits determination of k_{el}, viz. the rate constant of a heterogeneous electron transfer at an electrode surface. The theoretical treatment of such reactions (283) led to an important correlation between k_{el} and the homogeneous rate constant of the respective electron exchange, k_{ex}, namely, $k_{ex} \approx 10^3 \times k_{el}^2$. Numerous relaxation methods (see, e.g., ref. 284) lead to the determination of k_{el}; however, their utility is somewhat obviated by experimental and theoretical difficulties. The steady-state direct current voltametry has proven to be more reliable (285). For example, this technique, in conjunction with rotated-disk polarography, was adopted by Adams and his students (282) in their investigation of exchanges between radical-anions and their parent molecules. Unfortunately, the data showed a poor agreement with Marcus's theory (283), and a possible explanation of this failure was discussed in their paper.

An interesting electrochemical redox process, leading through electron transfer to chemiluminescence, was reported by Hoijtink (286) and then investigated in greater detail by Hercules (287), by Visco and Chandross (288), and by Santhanam and Bard (289). Positive and negative radical-ions may be formed successively on the *same* electrode if an alternating square-wave potential is applied (287). These species diffuse then into the surrounding liquid where they annihilate each other.

Consider such a reaction which is represented below with the aid of diagrams giving the energy levels of bonding and antibonding molecular orbitals of an alternate hydrocarbon,

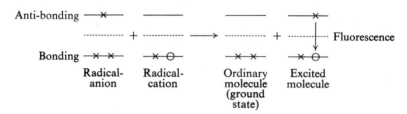

A collision between radical-anion and radical-cation leads to electron transfer; however, since the process is adiabatic the transferred electron initially moves into the antibonding orbital of the acceptor. This yields an electronically excited molecule which eventually returns to its ground state with emission of fluorescence light. The collision of the radical-ions may yield also excited triplets which luminesce then through triplet–triplet annihilation.

Closer examination of this phenomenon shows that the system is rather complex. For example, some radical-cations may be reduced directly on the electrode and yield radical-anions; some of the excited molecules may undergo spontaneous nonradiative decay which reduces the chemiluminescence, and finally some quenching by the parent species may also take place. Furthermore, it was shown by Chandross et al. (290) that the cation and anion may dimerize, i.e., form an excimer which then radiates,

$$A^{\cdot -} + A^{\cdot +} \underset{\longleftarrow}{\overset{\longrightarrow}{\rightleftharpoons}} A_2^* \overset{h\nu}{\longrightarrow} 2A$$

The diffusive character of these processes implies that the intensity of the resulting luminescence is a function of current density and of the distance from the electrode. Mathematical treatment of this problem was developed by Feldberg (291), who showed how the pertinent rate constants may be computed from the experimental data which give the spatial distribution of light intensity for various current densities.

Variation of luminescence intensity with the applied voltage and frequency (sinusoid wave) was measured (288). Intensity increased linearly with the voltage up to a point (~ 15 V) and then leveled off. At frequencies below 100 cps more than 95% of light was modulated, its intensity decreased as the frequency was raised. The luminescence was observed with 9,10-diphenyl anthracene, anthracene, perylene, rubrene, quaterphenyl, and 1,1,4,4-tetraphenyl butadiene. Dimethyl formamide was the most frequently used solvent.

Further studies showed that luminescence may be also observed with donors other than the corresponding radical-anions, e.g., electroluminescence of rubrene was induced by dimethyl formamide, water, n-butylamine, or triethylamine, all of which may serve as donors (292). Recently it was shown that a direct capture of solvated electrons, which were produced by an electrolytic process, also leads to light emission (293).

The electroluminescence phenomena of similar nature were reported by other investigators, e.g., Maricle, Maurer, and their colleagues (296,297). An interesting problem was raised: Is a direct electrode generation of excited molecules possible? Theoretical work of Marcus (303) makes this suggestion unlikely and further arguments against this hypothesis were raised recently by Chandross and Visco (304).

Electrochemically generated radical-ions may initiate anionic polymerization. Such processes were investigated by Funt and his students. The early work (298–300) demonstrated that anionic polymerization may take place under such conditions; however, termination processes were not excluded and consequently only dead polymers were produced. Later Yamazaki et al. (301) showed that under properly chosen conditions living polymers are formed, and their findings were soon confirmed by Funt et al. (302). Similar results were obtained at low temperature in liquid ammonia with 4-vinyl pyridine as the monomer (323). A comprehensive review of electrolytically initiated polymerization was published by Funt (325).

XXII. CHARGE-TRANSFER COMPLEXES AND THEIR SIGNIFICANCE AS INITIATORS OF IONIC POLYMERIZATION

Charge-transfer complexes are well recognized entities, their structure being best explained in terms of Mulliken's theory (305). The association of a donor D with an acceptor A gives a complex described in quantum mechanics language by:

$$\psi = a\psi_1(D,A) + b\psi_2(D^+,A^-)$$

with $|a| > |b|$, whereas the excited state is described by ψ',

$$\psi' = b\psi_1(D,A) - a\psi_2(D^+,A^-)$$

The excitation from state ψ to ψ' leads to the characteristic spectra of charge-transfer complexes. Its maximum absorption is affected by solvent, and qualitatively these changes were explained by assuming that the polar form is stabilized more through solvation than the ground state (ψ). Such stabilization may lead to formation of ion pairs D^+,A^-. For example, the solid charge-transfer complex of tetramethyl-phenylene diamine (TMPD) with chloranile dissociates in acetonitrile into $TMPD\cdot^+$ and chloranile$\cdot^-$, both radical-ions being recognized by their characteristic absorption spectra (306). Such a dissociation does not take place in nonpolar solvents. These types of investigations were extended to other complexes by Foster (e.g., 307) and independently by Briegleb et al. (308).

Similar phenomena may be induced by photoactivation. For example, photoaddition of maleic anhydride to benzene apparently involves photoactivation of a charge-transfer complex (309). The photo-induced ESR signals of charge-transfer complexes provide perhaps the most direct evidence for such a transformation of a nonpolar form into a polar one (310,311).

Studies of ESR spectra of charge-transfer complexes revealed several interesting features. A complete transfer of an electron from a donor to an acceptor produces two independent paramagnetic species, and hence the ESR spectrum should be given by the superposition of those of $D\cdot^+$ and $A\cdot^-$, if the interaction between these species is weak. When the interaction is strong the energy levels become split, the lower corresponding to a diamagnetic singlet and the higher to a paramagnetic triplet; the latter level may then be thermally populated if the energy gap is small. Examples of such phenomena were described in the literature (e.g., 312–316).

Formation of radical-ions from charge-transfer complexes suggests that such systems may initiate ionic polymerization, particularly if the monomer is the component of the complex. For example, polymeriza-tion of N-vinyl carbazole was initiated by electron acceptors (317,318) and it was proposed that this process is caused by an electron transfer. Nevertheless, the details of initiation were not discussed, although many features of the reaction were recorded (319,320). Several other poly-merizations were subsequently attributed to electron-transfer initiation (321,322) but no attempt was made to verify the mechanism.

Extensive kinetic study of vinylcarbazole polymerization initiated by tetranitromethane in nitrobenzene was reported by Pac and Plesch (324). These authors concluded that the reaction involves electron transfer and the formation of radical-cations is responsible for the initiation which leads to conventional cationic polymerization. The radical end was assumed to be inactivated by tetranitromethane, which often acts as a radical trap. Further investigation of this polymerization, and studies of some related systems were completed recently in this writer's laboratory. The results confirmed the cationic character of the propagation; however, they indicated that the initiation does not involve radical-cations but it is due to a direct transfer of NO_2^+ cation leading to the formation of carbonium-ions and the $C(NO_2)_3^-$ counterions (328).

References

1. M. Berthelot, *Ann. Chim.*, *12*, 155 (1867).
2. W. Schlenk, J. Appenrodt, A. Michael, and A. Thal, *Chem. Ber.*, *47*, 473 (1914).
3. R. Willstätter, F. Seitz, and E. Bumm, *Chem. Ber.*, *61*, 871 (1928).
4. W. Schlenk and E. Bergmann, *Ann. Chem.*, *463*, 1 (1928); *464*, 1 (1928).
5. (a) N. D. Scott, J. F. Walker, and V. L. Hansley, *J. Am. Chem. Soc.*, *58*, 2442 (1936); (b) J. F. Walker and N. D. Scott, *J. Am. Chem. Soc.*, *60*, 951 (1938).
6. W. Hückel and H. Bretschneider, *Ann. Chem.*, *540*, 157 (1939).
7. (a) D. Lipkin, D. E. Paul, J. Townsend, and S. I. Weissman, *Science*, *117*, 534 (1953); (b) S. I. Weissman, J. Townsend, D. E. Paul, and G. E. Pake, *J. Chem. Phys.*, *21*, 2227 (1953).
8. (a) J. E. Lovelock, *Nature*, *189*, 729 (1961); (b) J. E. Lovelock, *Anal. Chem.*, *33*, 162 (1961); *35*, 474 (1963); (c) J. E. Lovelock and N. L. Gregory, *Gas Chromatography*, N. Brenner, Ed., Academic Press, New York, 1962, p. 219; (d) J. E. Lovelock, P. G. Simmonds, and W. J. A. Van den Heuvel, *Nature*, *197*, 249 (1963).
9. W. E. Wentworth, E. Chen, and J. E. Lovelock, *J. Phys. Chem.*, *70*, 445 (1966).
10. R. S. Becker and E. Chen, *J. Chem. Phys.*, *45*, 2403 (1966).
11. W. E. Wentworth and R. S. Becker, *J. Am. Chem. Soc.*, *84*, 4263 (1962).
12. G. Briegleb, *Angew. Chem.*, *76*, 326 (1964).
13. N. E. Bradbury, *Phys. Rev.*, *44*, 883 (1933).
14. (a) L. Rolla and G. Piccardi, *Atti. Accad. Lincei*, *2*, 29, 128, 173 (1925); (b) G. Piccardi, *ibid.*, *3*, 413, 566 (1926); (c) G. Piccardi, *Z. Physik*, *43*, 899 (1927); (d) L. Rolla and G. Piccardi, *Atti. Accad. Lincei*, *5*, 546 (1927).
15. P. P. Sutton and J. E. Mayer, *J. Chem. Phys.*, *2*, 145 (1934); *3*, 20 (1935).
16. G. Glockler and M. Calvin, *J. Chem. Phys.*, *3*, 771 (1935).
17. F. M. Page, *Trans. Faraday Soc.*, *56*, 1742 (1960).
18. A. L. Farragher and F. M. Page, *Trans. Faraday Soc.*, *62*, 3072 (1966).
19. A. L. Farragher, Ph.D. Thesis, University of Aston, Birmingham, England, 1966.
20. J. Kay and F. M. Page, *Trans. Faraday Soc.*, *62*, 3081 (1966).

21. J. E. Mayer, Z. Physik., 61, 798 (1930).
22. W. B. Person, J. Chem. Phys., 38, 109 (1963).
23. N. S. Hush and J. A. Pople, Trans. Faraday Soc., 51, 600 (1955).
24. R. M. Hedges and F. A. Matsen, J. Chem. Phys., 28, 950 (1958).
25. J. R. Hoyland and L. Goodman, J. Chem. Phys., 36, 12 (1962).
26. A. Streitwieser, J. Am. Chem. Soc., 82, 4123 (1960).
27. N. S. Hush and J. Blackledge, J. Chem. Phys., 23, 514 (1955).
28. (a) W. Schlenk and T. Weickel, Chem. Ber., 44, 1182 (1911); (b) W. Schlenk and A. Thal, Chem. Ber., 46, 2840 (1913); 47, 473 (1914).
29. A. F. Gaines and F. M. Page, Trans. Faraday Soc., 62, 3086 (1966).
30. J. H. Hildebrand and H. E. Bent, J. Am. Chem. Soc., 49, 3011 (1927).
31. (a) H. E. Bent and N. B. Keevil, J. Am. Chem. Soc., 58, 1228, 1367 (1936); (b) N. B. Keevil, J. Am. Chem. Soc., 59, 2104 (1937).
32. N. B. Keevil and H. E. Bent, J. Am. Chem. Soc., 60, 193 (1938).
33. G. J. Hoijtink, E. de Boer, P. H. van der Meij, and W. P. Weijland, Rec. Trav. Chem., 75, 487 (1956).
34. J. Jagur-Grodzinski, M. Feld, S. L. Yang, and M. Szwarc, J. Phys. Chem., 69, 628 (1965).
35. M. G. Scroggie, Wireless World, 1952, 14.
36. J. Chaudhuri, J. Jagur-Grodzinski, and M. Szwarc, J. Phys. Chem., 71, 3063 (1967).
37. R. V. Slates and M. Szwarc, J. Phys. Chem., 69, 4124 (1965).
38. P. Chang, R. V. Slates, and M. Szwarc, J. Phys. Chem., 70, 3180 (1966).
39. D. Nicholls, C. Sutphen, and M. Szwarc, J. Phys. Chem., 72, 1021 (1968).
40. K. H. J. Buschow, J. Dieleman, and G. J. Hoijtink, J. Chem. Phys., 42, 1993 (1965).
41. D. E. Paul, D. Lipkin, and S. I. Weissman, J. Am. Chem. Soc., 78, 116 (1956).
42. L. E. Lyons, Nature, 166, 193 (1950).
43. H. A. Laitinen and S. Wawzonek, J. Am. Chem. Soc., 64, 1765 (1942).
44. S. Wawzonek and H. A. Laitinen, J. Am. Chem. Soc., 64, 2365 (1942).
45. S. Wawzonek and J. W. Fan, J. Am. Chem. Soc., 68, 2541 (1946).
46. G. J. Hoijtink, J. van Schooten, E. de Boer, and W. I. Aalbersberg, Rec. Trav. Chim., 73, 355 (1954).
47. C. Carvajal, K. J. Tölle, J. Smid, and M. Szwarc, J. Am. Chem. Soc., 87, 5548 (1965).
48. G. J. Hoijtink and J. van Schooten, Rec. Trav. Chim., 71, 1089 (1952); 72, 691, 903 (1953).
49. I. Bergman, Trans. Faraday Soc., 50, 829 (1954); 52, 690 (1956).
50. E. Swift, J. Am. Chem. Soc., 60, 1403 (1938).
51. F. A. Matsen, J. Chem. Phys., 24, 602 (1956).
52. W. M. Latimer, K. S. Pitzer, and C. M. Slansky, J. Chem. Phys., 7, 108 (1939).
53. I. M. Kolthoff and J. L. Lingane, Polarography, Volume 2, 2nd ed., Interscience, New York, 1952, p. 809.
54. A. C. Aten and G. J. Hoijtink, Z. Phys. Chem. (Frankfurt), 21, 192 (1959).
55. A. C. Aten, C. Büthker, and G. J. Hoijtink, Trans. Faraday Soc., 55, 324 (1959).

56. G. J. Hoijtink, *Rec. Trav. Chim.*, *73*, 895 (1954).
57. G. W. Wheland, *J. Am. Chem. Soc.*, *63*, 2025 (1941).
58. H. C. Longuet-Higgins, *J. Chem. Phys.*, *18*, 265 (1950).
59. (a) S. Wawzonek, E. W. Blaha, R. Berkey, and M. E. Runner, *J. Electrochem. Soc.*, *102*, 235 (1955); (b) P. H. Given, *J. Chem. Soc.*, *1958*, 2684.
60. E. de Boer, *Advan. Organometal. Chem.*, *2*, 115 (1964).
61. D. A. McInnes, *The Principles of Electrochemistry*, Reinhold, New York, 1939.
62. W. E. Bachmann, *J. Am. Chem. Soc.*, *55*, 1179 (1933).
63. A. Mathias and E. Warhurst, *Trans. Faraday Soc.*, *58*, 948 (1962).
64. Quoted by E. de Boer in his review on p. 123 of ref. (60).
65. A. I. Shatenstein, E. S. Petrov, and M. I. Belonsova, *Organic Reactivity*, *1*, 191 (1964).
66. A. I. Shatenstein, E. S. Petrov, and E. A. Yakovlava, *J. Polymer Sci.*, *16*, 1729 (1967).
67. R. V. Slates and M. Szwarc, *J. Am. Chem. Soc.*, *89*, 6043 (1967).
68. P. S. Rao, J. R. Nash, J. P. Guarino, M. R. Ronayne, and W. H. Hamill, *J. Am. Chem. Soc.*, *84*, 500 (1962).
69. H. Linschitz, M. G. Berry, and D. Schweitzer, *J. Am. Chem. Soc.*, *76*, 5833 (1954).
70. N. Hirota, *J. Phys. Chem.*, in press.
71. (a) M. R. Ronayne, J. P. Guarino, and W. H. Hamill, *J. Am. Chem. Soc.*, *84*, 4230 (1962); (b) see also J. P. Guarino, M. R. Ronayne, and W. H. Hamill, *Radiation Res.*, *17*, 379 (1962).
72. N. Christodouleas and W. H. Hamill, *J. Am. Chem. Soc.*, *86*, 5413 (1964).
73. E. P. Bertin and W. H. Hamill, *J. Am. Chem. Soc.*, *86*, 1301 (1964).
74. W. H. Hamill, J. P. Guarino, M. R. Ronayane, and J. A. Ward, *Discussions Faraday Soc.*, *36*, 169 (1963).
75. D. G. Powell and E. Warhurst, *Trans. Faraday Soc.*, *58*, 953 (1962).
76. K. Chambers, E. Collinson, F. S. Dainton, and W. Seddon, *Chem. Commun.*, *1966*, 498.
77. M. Katayama, *Bull. Chem. Soc. Japan*, *38*, 2208 (1965).
78. J. P. Keene, E. J. Land, and A. J. Swallow, *J. Am. Chem. Soc.*, *87*, 5284 (1965).
79. T. Shida and W. H. Hamill, *J. Am. Chem. Soc.*, *88*, 5371 (1966).
80. D. H. Levy and R. J. Myers, *J. Chem. Phys.*, *41*, 1062 (1964); *44*, 4177 (1966).
81. K. O. Hartman and I. C. Hisatsune, *J. Chem. Phys.*, *44*, 1913 (1966).
82. K. O. Hartman and I. C. Hisatsune, *J. Phys. Chem.*, *69*, 583 (1965).
83. W. L. McCubbin and I. D. C. Gurney, *J. Chem. Phys.*, *43*, 983 (1965).
84. T. Shida and W. H. Hamill, *J. Chem. Phys.*, *44*, 2369 (1966).
85. L. P. Blanchard and P. LeGoff, *Can. J. Chem.*, *35*, 89 (1957).
86. T. Shida and W. H. Hamill, *J. Am. Chem. Soc.*, *88*, 3689 (1966).
87. J. A. Leone and W. S. Koski, *J. Am. Chem. Soc.*, *88*, 224 (1966).
88. P. M. Johnson and A. C. Albrecht, *J. Chem. Phys.*, *44*, 1845 (1966).
89. T. Shida and W. H. Hamill, *J. Chem. Phys.*, *44*, 2375 (1966).
90. W. I. Aalbersberg, G. J. Hoijtink, E. L. Mackor, and W. P. Weijland, *J. Chem. Soc.*, *1959*, 3049, 3055.

91. P. Bennema, G. J. Hoijtink, J. H. Lupinski, L. J. Oosterhoff, P. Selier, and J. D. W. van Voorst, *Mol. Phys.*, *2*, 431 (1959).

92. T. Shida and W. H. Hamill, *J. Am. Chem. Soc.*, *88*, 5376 (1966).

93. J. B. Gallivan and W. H. Hamill, *J. Chem. Phys.*, *44*, 2378 (1966).

94. T. Shida and W. H. Hamill, *J. Am. Chem. Soc.*, *88*, 3683 (1966).

95. M. S. Matheson and L. M. Dorfman, *J. Chem. Phys.*, *32*, 1870 (1960).

96. R. L. McCarthy and A. McLachlan, *Trans. Faraday Soc.*, *56*, 1187 (1960).

97. J. P. Keene, *Nature*, *188*, 843 (1960).

98. L. M. Dorfman and M. S. Matheson, *Progr. Reaction Kinetics*, *3*, 237 (1965).

99. J. W. Boag, *Am. J. Roentgenology*, *90*, 896 (1963).

100. E. J. Hart, S. Gordon, and J. K. Thomas, *J. Phys. Chem.*, *68*, 1271 (1964).

101. S. Arai and L. M. Dorfman, *J. Chem. Phys.*, *41*, 2190 (1964).

102. G. E. Adams, J. H. Baxendale, and J. W. Boag, *Proc. Roy. Soc. (London), Ser. A*, *277*, 549 (1964).

103. I. A. Taub, M. C. Sauer, and L. M. Dorfman, *Disc. Faraday Soc.*, *36*, 206 (1963).

104. I. A. Taub, D. A. Harter, M. C. Sauer, and L. M. Dorfman, *J. Chem. Phys.*, *41*, 979 (1964).

105. S. Arai, E. L. Tremba, J. R. Brendon, and L. M. Dorfman, *Can. J. Chem.*, *45*, 1119 (1967).

106. G. J. Hoijtink and P. H. van der Meij, *Z. Phys. Chem. (Frankfurt)*, *20*, 1 (1959).

107. J. Dieleman, Thesis, Amsterdam, 1962.

108. E. de Boer, unpublished results.

109. K. H. J. Buschow and G. J. Hoijtink, *J. Chem. Phys.*, *40*, 2501 (1964).

110. D. W. Ovenall and D. H. Whiffen, *Chem. Soc. (London)*, Spec. Publ., *12*, 139 (1958).

111. (a) A. G. Evans, J. C. Evans, E. D. Owen, and B. J. Tabner, *Proc. Chem. Soc.*, *1962*, 226; (b) J. E. Bennett, A. G. Evans, J. C. Evans, E. D. Owen, and B. J. Tabner, *J. Chem. Soc.*, *1963*, 3954; (c) A. G. Evans and B. J. Tabner, *ibid.*, *1963*, 4613.

112. J. F. Garst and R. S. Cole, *J. Am. Chem. Soc.*, *84*, 4352 (1962).

113. (a) J. F. Garst, E. R. Zabolotny, and R. S. Cole, *J. Am. Chem. Soc.*, *86*, 2257 (1964); (b) J. F. Garst and E. R. Zabolotny, *ibid.*, *87*, 495 (1965).

114. R. C. Roberts and M. Szwarc, *J. Am. Chem. Soc.*, *87*, 5542 (1965).

115. D. N. Bhattacharyya, C. L. Lee, J. Smid, and M. Szwarc, *J. Phys. Chem.*, *69*, 608 (1965).

116. T. E. Hogen-Esch and J. Smid, *J. Am. Chem. Soc.*, *88*, 307 (1966).

117. R. V. Slates, Ph.D. Thesis, Syracuse University, Syracuse, New York, 1967.

118. L. L. Chan and J. Smid, *J. Am. Chem. Soc.*, *89*, 4547 (1967).

119. J. P. Keene, *J. Sci. Instr.*, *41*, 493 (1964).

120. I. A. Taub, *Proc. Manchester Symposium on Pulse Radiolysis*, Academic Press, New York, 1965.

121. J. Rabani, W. A. Mulac, and M. S. Matheson, *J. Phys. Chem.*, *69*, 53 (1965).

122. W. E. Wentworth and E. Chen, *J. Phys. Chem.*, *71*, 1929 (1967).

123. W. E. Wentworth, R. S. Becker, and R. Tung, *J. Phys. Chem.*, *71*, 1652 (1967).

124. C. A. Kraus, *J. Am. Chem. Soc.*, *30*, 1323 (1908).

125. D. S. Berns, in "Solvated Electron," *Advan. Chem. Ser. 50*, 82 (1965).
126. G. Lepoutre and M. J. Sienko, Eds., *Metal–Ammonia Solutions*, Benjamin, New York, 1964.
127. J. Jortner and S. A. Rice, in "Solvated Electron," *Advan. Chem. Ser., 50*, 7 (1965).
128. L. C. Kenausis, E. C. Evers, and C. A. Kraus, *Proc. Natl. Acad. Sci. U.S., 48*, 121 (1962).
129. J. C. Thompson, in "Solvated Electron," *Advan. Chem. Ser., 50*, 96 (1965).
130. E. Arnold and A. Patterson, *J. Chem. Phys., 41*, 3098 (1964).
131. S. Golden, C. Guttman, and T. R. Tuttle, *J. Am. Chem. Soc., 87*, 135 (1965).
132. D. S. Berns, G. Lepoutre, E. A. Bockelman, and A. Patterson, *J. Chem. Phys., 35*, 1820 (1961).
133. R. A. Ferrel, *Phys. Rev., 108*, 167 (1957).
134. C. G. Kuper, *Phys. Rev., 122*, 1007 (1961).
135. W. T. Sommer, *Phys. Rev. Letters, 12*, 271 (1964).
136. N. R. Kestner, J. Jortner, M. H. Cohen, and S. A. Rice, *Phys. Rev., 140A*, 56 (1965).
137. H. Schnyders, L. Meyer, and S. A. Rice, *Phys. Rev., 150*, 127 (1966).
138. J. H. Baxendale, *Radiation Res. Suppl., 4*, 139 (1964).
139. C. A. Kraus and W. W. Lucasse, *J. Am. Chem. Soc., 43*, 2529 (1921).
140. M. C. R. Symons, M. J. Blandamer, R. Catterall, and L. Shields, *J. Chem. Soc., 1964*, 4357.
141. J. H. Baxendale, E. M. Fielden, and J. P. Keene, *Science, 148*, 637 (1965).
142. J. Eloranta and H. Linschitz, *J. Chem. Phys., 38*, 2214 (1963).
143. M. Anbar and E. J. Hart, *J. Phys. Chem., 69*, 1244 (1965).
144. L. M. Dorfman, in "Solvated Electron," *Advan. Chem. Ser., 50*, 36 (1965).
145. W. Weyl, *Pogg. Ann., 121*, 601 (1864).
146. M. Gold, W. L. Jolly, and K. S. Pitzer, *J. Am. Chem. Soc., 84*, 2264 (1962).
147. E. Arnold and A. Patterson, *J. Chem. Phys., 41*, 3089 (1964).
148. E. Becker, R. H. Lindquist, and B. J. Alder, *J. Chem. Phys., 25*, 971 (1956).
149. T. R. Hughes, *J. Chem. Phys., 38*, 202 (1963).
150. J. V. Acrivos and K. S. Pitzer, *J. Phys. Chem., 66*, 1693 (1962).
151. K. Bar-Eli and T. R. Tuttle, *J. Chem. Phys., 40*, 2508 (1964).
152. J. L. Dye and L. R. Dalton, *J. Phys. Chem., 71*, 184 (1967).
153. D. E. O'Reilly and T. Tsang, *J. Chem. Phys., 42*, 333 (1965).
154. T. R. Tuttle, C. Guttman, and S. Golden, *J. Chem. Phys., 45*, 2206 (1966).
155. J. L. Down, J. Lewis, B. Moore, and G. Wilkinson, *J. Chem. Soc., 1959*, 3767.
156. F. Cafasso and B. R. Sundheim, *J. Chem. Phys., 31*, 809 (1959).
157. F. S. Dainton, D. M. Wiles, and A. N. Wright, *J. Chem. Soc., 1960*, 4283.
158. F. S. Dainton, D. M. Wiles, and A. N. Wright, *J. Polymer Sci., 45*, 111 (1960).
159. G. Fraenkel, S. H. Ellis, and D. T. Dix, *J. Am. Chem. Soc., 87*, 1406 (1965).
160. H. Normant and M. Larchevegne, *Compt. rend. (France), 260*, 5062 (1965).
161. H. Normant, J. Normant, T. Cuvigny, and B. Angelo, *Bull. Soc. Chim. France, 1965*, 1561.
162. H. Normant, *Bull Soc. Chim. France, 1966*, 3362.
163. H. J. Chen and M. Bersohn, *J. Am. Chem. Soc., 88*, 2663 (1966).

164. C. A. Hutchison and R. C. Pastor, *J. Chem. Phys.*, *21*, 1959 (1953).

165. R. C. Douthit and J. L. Dye, *J. Am. Chem. Soc.*, *82*, 4472 (1960).

166. M. Ottolenghi, K. Bar-Eli, H. Linschitz, and T. R. Tuttle, *J. Chem. Phys.*, *40*, 3729 (1964).

167. H. Linschitz, M. G. Berry, and D. Schweitzer, *J. Am. Chem. Soc.*, *76*, 5833 (1954).

168. J. D. W. van Voorst and G. J. Hoijtink, *J. Chem. Phys.*, *42*, 3995 (1965).

169. G. N. Lewis and D. Lipkin, *J. Am. Chem. Soc.*, *64*, 2801 (1942).

170. G. N. Lewis and J. Bigeleisen, *J. Am. Chem. Soc.*, *65*, 520 (1943).

171. P. J. Zandstra and G. J. Hoijtink, *Mol. Phys.*, *3*, 371 (1960).

172. G. J. Hoijtink and J. D. W. van Voorst, *J. Chem. Phys.*, *45*, 3918 (1966).

173. J. H. Baxendale and G. Hughes, *Z. Phys. Chem.* (*Frankfurt*), *14*, 323 (1958).

174. J. Jortner and J. Rabani, *J. Am. Chem. Soc.*, *83*, 4868 (1961).

175. E. J. Kirschke and W. L. Jolly, *Science*, *147*, 45 (1965).

176. G. Dobson and L. I. Grossweiner, *Trans. Faraday Soc.*, *61*, 708 (1965).

177. M. S. Matheson, W. A. Mulac, and J. Rabani, *J. Phys. Chem.*, *67*, 2613 (1963).

178. M. Ottolenghi and H. Linschitz, in "Solvated Electron," *Advan. Chem. Ser.*, *50*, 149 (1965).

179. D. H. Geske and A. Maki, *J. Am. Chem. Soc.*, *82*, 2671 (1960).

180. E. Clementi, *Phys. Rev.*, *133A*, 1274 (1964).

181. G. Favini and M. Simonetta, *Theoret. Chim. Acta*, *1*, 294 (1963).

182. W. Reppe, O. Schlichting, K. Klager, and T. Toepel, *Ann. Chem.*, *560*, 1 (1948).

183. T. J. Katz, W. H. Reinmuth, and D. E. Smith, *J. Am. Chem. Soc.*, *84*, 802 (1962).

184. T. J. Katz and H. L. Strauss, *J. Chem. Phys.*, *32*, 1873 (1960).

185. H. L. Strauss and G. K. Fraenkel, *J. Chem. Phys.*, *35*, 1738 (1961).

186. H. M. McConnell, *J. Chem. Phys.*, *24*, 632, 764 (1956).

187. I. Bernal, P. H. Rieger, and G. K. Fraenkel, *J. Chem. Phys.*, *37*, 1489 (1962).

188. H. L. Strauss, T. J. Katz, and G. K. Fraenkel, *J. Am. Chem. Soc.*, *85*, 2360 (1963).

189. C. A. Coulson and W. E. Moffitt, *Phil. Mag.*, *40*, 1 (1949).

190. D. E. Wood and H. M. McConnell, *J. Chem. Phys.*, *37*, 1150 (1962).

191. J. dos Santos-Veiga, *Mol. Phys.*, *5*, 639 (1962).

192. (a) H. M. McConnell and A. D. McLachlan, *J. Chem. Phys.*, *34*, 1 (1961); (b) L. C. Snyder and A. D. McLachlan, *ibid.*, *36*, 1159 (1962).

193. (a) M. Szwarc, M. Levy, and R. Milkovich, *J. Am. Chem. Soc.*, *78*, 2656 (1956); (b) M. Szwarc, *Nature*, *178*, 1168 (1956).

194. J. L. R. Williams, T. M. Laakso, and W. J. Dulmage, *J. Org. Chem.*, *23*, 638 (1958).

195. J. Jagur-Grodzinski and M. Szwarc, *Proc. Roy. Soc.* (*London*), *Ser. A*, *288*, 224 (1965).

196. R. Asami and M. Szwarc, *J. Am. Chem. Soc.*, *84*, 2269 (1962).

197. G. Spach, H. Monteiro, M. Levy, and M. Szwarc, *Trans. Faraday Soc.*, *58*, 1809 (1962).

198. M. Matsuda, J. Jagur-Grodzinski, and M. Szwarc, *Proc. Roy. Soc.* (*London*), *Ser. A*, *288*, 212 (1965).

199. S. C. Chadha, J. Jagur-Grodzinski, and M. Szwarc, *Trans. Faraday Soc.*, *63*, 2994 (1967).

200. K. F. O'Driscoll, R. J. Boudreau, and A. V. Tobolsky, *J. Polymer Sci.*, *31*, 115 (1958).
201. K. F. O'Driscoll and A. V. Tobolsky, *J. Polymer Sci.*, *31*, 123 (1958).
202. C. Walling, E. R. Briggs, W. Cummings, and F. R. Mayo, *J. Am. Chem. Soc.*, *72*, 48 (1950).
203. A. V. Tobolsky and D. B. Hartley, *J. Polymer Sci. A*, *1*, 15 (1963).
204. C. Stretch and G. Allen, *Polymer*, *2*, 151 (1961).
205. K. F. O'Driscoll and A. V. Tobolsky, *J. Polymer Sci.*, *37*, 363 (1959).
206. C. G. Overberger and N. Yamamoto, *J. Polymer Sci. A-1*, *4*, 3101 (1966).
207. J. E. Mulvaney, C. G. Overberger, and A. Schiller, *Advan. Polymer Sci.*, *3*, 106 (1964).
208. D. B. George and A. V. Tobolsky, *J. Polymer Sci.*, *B2*, 1 (1964).
209. R. K. Graham, D. L. Dunkelberger, and W. E. Goode, *J. Am. Chem. Soc.*, *82*, 400 (1960).
210. J. F. Garst, J. G. Pacifici, and E. R. Zabolotny, *J. Am. Chem. Soc.*, *88*, 3872 (1966).
211. J. Saltiel and G. S. Hammond, *J. Am. Chem. Soc.*, *85*, 2515, 2516 (1963). See also, G. S. Hammond et al., *ibid.*, *86*, 3197 (1964).
212. W. Hückel, *Theoretische Grundlage der Organischen Chemie*, Akademie-Verlag, Berlin, 1934.
213. W. E. Bachman, *J. Am. Chem. Soc.*, *55*, 1179 (1933).
214. W. Schlenk and E. Bergmann, *Ann. Chem.*, *464*, 1 (1928).
215. H. V. Carter, B. J. McClelland, and E. Warhurst, *Trans. Faraday Soc.*, *56*, 455 (1960).
216. H. E. Bent and A. J. Harrison, *J. Am. Chem. Soc.*, *66*, 969 (1944).
217. N. Hirota and S. I. Weissman, *J. Am. Chem. Soc.*, *86*, 2537, 2538 (1964).
218. N. Hirota and S. I. Weissman, *J. Am. Chem. Soc.*, *82*, 4424 (1960).
219. A. G. Evans, J. C. Evans, and E. H. Godden, *Trans. Faraday Soc.*, *63*, 136 (1967).
220. B. J. McClelland, *Chem. Rev.*, *64*, 301 (1964).
221. I. C. Lewis and L. S. Singer, *J. Chem. Phys.*, *43*, 2712 (1965).
222. O. W. Howarth and G. K. Fraenkel, *J. Am. Chem. Soc.*, *88*, 4514 (1966).
223. T. Förster and K. Kasper, *Z. Phys. Chem.*, *1*, 275 (1954), and *Z. Elektrochem.*, *59*, 976 (1955).
224. B. Stevens and E. Hutton, *Nature*, *186*, 1045 (1960).
225. G. J. Hoijtink, *Z. Electrochem.*, *64*, 156 (1960).
226. A. Ron and O. Schnepp, *J. Chem. Phys.*, *44*, 19 (1966).
227. E. Beckman and T. Paul, *Ann. Chem.*, *266*, 1 (1891).
228. E. R. Zabolotny and J. F. Garst, *J. Am. Chem. Soc.*, *86*, 1645 (1964).
229. A. Carrington, H. C. Longuet-Higgins, and P. F. Todd, *Mol. Phys.*, *8*, 45 (1964).
230. T. J. Katz, M. Yoshida, and L. C. Siew, *J. Am. Chem. Soc.*, *87*, 4516 (1965).
231. H. J. Dauben and M. R. Rifi, *J. Am. Chem. Soc.*, *85*, 3041 (1963).
232. N. L. Bauld and M. S. Brown, *J. Am. Chem. Soc.*, *87*, 4390 (1965).
233. R. L. Ward and S. I. Weissman, *J. Am. Chem. Soc.*, *79*, 2086 (1957).
234. F. C. Adam and S. I. Weissman, *J. Am. Chem. Soc.*, *80*, 1518 (1958).
235. S. I. Weissman, *Z. Elektrochem.*, *64*, 47 (1960).
236. M. T. Jones and S. I. Weissman, *J. Am. Chem. Soc.*, *84*, 4269 (1962).

237. P. J. Zandstra and S. I. Weissman, *J. Chem. Phys.*, *35*, 757 (1961).
238. P. J. Zandstra and S. I. Weissman, *J. Am. Chem. Soc.*, *84*, 4408 (1962).
239. N. M. Atherton and S. I. Weissman, *J. Am. Chem. Soc.*, *83*, 1330 (1961).
240. R. Chang and C. S. Johnson, *J. Am. Chem. Soc.*, *88*, 2338 (1966).
241. L. H. Piette and W. A. Anderson, *J. Chem. Phys.*, *30*, 899 (1959).
242. J. M. Fritsch, T. P. Layloff, and R. N. Adams, *J. Am. Chem. Soc.*, *87*, 1724 (1965).
243. S. Arai, D. A. Grev, and L. M. Dorfman, *J. Chem. Phys.*, *46*, 2572 (1967).
244. M. Levy and M. Szwarc, *J. Am. Chem. Soc.*, *82*, 521 (1960).
245. A. G. Evans and J. C. Evans, *J. Chem. Soc.*, *1963*, 6036.
246. A. G. Evans and B. J. Tabner, *J. Chem. Soc.*, *1963*, 5560.
247. A. G. Evans and J. C. Evans, *Trans. Faraday Soc.*, *61*, 1202 (1965).
248. A. G. Evans and D. B. George, *J. Chem. Soc.*, *1961*, 4653.
249. H. M. McConnell, *J. Chem. Phys.*, *35*, 508 (1961).
250. S. I. Weissman, *J. Am. Chem. Soc.*, *80*, 6462 (1958).
251. V. V. Voevodskii, S. P. Solodovnikov, and V. M. Chibrikin, *Dokl. Akad. Nauk SSSR*, *129*, 1082 (1959).
252. W. L. Reynold, *J. Phys. Chem.*, *67*, 2866 (1963).
253. J. Jagur, M. Levy, M. Feld, and M. Szwarc, *Trans. Faraday Soc.*, *58*, 2168 (1962).
254. D. Gill, J. Jagur-Grodzinski, and M. Szwarc, *Trans. Faraday Soc.*, *60*, 1424 (1964).
255. J. Jagur-Grodzinski and M. Szwarc, *Trans. Faraday Soc.*, *59*, 2305 (1963).
256. D. Dadley and A. G. Evans, *J. Chem. Soc.* (*B*), *1967*, 418.
257. (a) R. Lipman, J. Jagur-Grodzinski, and M. Szwarc, *J. Am. Chem. Soc.*, *87*, 3005 (1965); (b) J. Stearne, J. Smid, and M. Szwarc, *Trans. Faraday Soc.*, *62*, 672 (1966).
258. S. N. Khanna, M. Levy, and M. Szwarc, *Trans. Faraday Soc.*, *58*, 747 (1962).
259. A. A. Arest-Yakubovich, A. R. Gantmakher, and S. S. Medvedev, *Dokl. Akad. Nauk SSSR*, *139*, 1331 (1961).
260. D. Nicholls and M. Szwarc, *Proc. Roy. Soc.* (*London*), *Ser. A.*, *301*, 223, 231 (1967).
261. S. Pluymers and G. Smets, *Makromol. Chem.*, *88*, 29 (1965).
262. W. Schlenk and R. Ochs, *Chem. Ber.*, *49*, 608 (1916).
263. G. Wittig and D. Wittenberg, *Ann. Chem.*, *606*, 1 (1957).
264. G. A. Russell and E. G. Janzen, *J. Am. Chem. Soc.*, *84*, 4153 (1962).
265. F. J. Smentowski, *J. Am. Chem. Soc.*, *85*, 3036 (1963).
266. A. G. Evans, J. C. Evans, and B. J. Tabner, *Proc. Chem. Soc.*, *1962*, 338.
267. R. Waack and M. A. Doran, *J. Organometal. Chem.*, *3*, 92 (1965).
268. E. Franta, J. Chaudhuri, A. Cserhegyi, J. Jagur-Grodzinski, and M. Szwarc, *J. Am. Chem. Soc.*, *89*, 7129 (1967).
269. A. Prock, M. Djibelian, and S. Sullivan, *J. Phys. Chem.*, *71*, 3378 (1967).
270. M. A. Doran and R. Waack, *J. Organometal. Chem.*, *3*, 94 (1965).
271. D. H. Richards and M. Szwarc, *Trans. Faraday Soc.*, *55*, 1644 (1959).
272. M. Morton, A. Rembaum, and E. E. Bostick, *J. Polymer Sci.*, *32*, 530 (1958).
273. A. V. Tobolsky and D. B. Hartley, *J. Am. Chem. Soc.*, *84*, 1391 (1962).
274. F. Bahsteter, J. Smid, and M. Szwarc, *J. Am. Chem. Soc.*, *85*, 3909 (1963).
275. J. H. Binks and M. Szwarc, *J. Chem. Phys.*, *30*, 1494 (1959).

276. A. H. Samuel and J. L. Magee, *J. Chem. Phys.*, *21*, 1080 (1953).
277. R. L. Platzman, report No. 305 to AEC, 1953, p. 34.
278. J. W. Hunt and J. K. Thomas, *Radiation Res.*, in press.
279. M. Eigen, W. Kruse, G. Maass, and L. de Maeyer, *Progress in Reaction Kinetics*, Vol. 2, Pergamon Press, New York, 1967, p. 285.
280. A. Badmani, J. Jagur-Grodzinski, and M. Szwarc, to be published.
281. R. N. Adams, *J. Electroanal. Chem.*, *8*, 151 (1964).
282. P. A. Malachesky, T. A. Miller, T. Layloff, and R. N. Adams, *Exchange Reactions* (Proc. Symp., Upton, N.Y.) 1965, p. 157.
283. R. A. Marcus, *J. Phys. Chem.*, *67*, 853 (1963).
284. B. E. Conway, *Theory and Principles of Electrode Processes*, The Ronald Press, New York, 1965.
285. J. Jordan and R. A. Javick, *Electrochim. Acta*, *6*, 23 (1962).
286. G. J. Hoijtink, unpublished results.
287. D. M. Hercules, *Science*, *145*, 808 (1964).
288. R. E. Visco and E. A. Chandross, *J. Am. Chem. Soc.*, *86*, 5350 (1964).
289. K. S. V. Santhanam and A. J. Bard, *J. Am. Chem. Soc.*, *87*, 139 (1965).
290. E. A. Chandross, J. W. Longworth, and R. E. Visco, *J. Am. Chem. Soc.*, *87*, 3259 (1965).
291. S. W. Feldberg, *J. Am. Chem. Soc.*, *88*, 390 (1966).
292. D. M. Hercules, R. C. Lansbury, and D. K. Roe, *J. Am. Chem. Soc.*, *88*, 4578 (1966).
293. E. A. Chandross, to be published.
294. W. Bruning and S. I. Weissman, *J. Am. Chem. Soc.*, *88*, 373 (1966).
295. R. A. Marcus, *J. Phys. Chem.*, *67*, 853 (1963); *Ann. Rev. Phys. Chem.*, *15*, 155 (1964).
296. D. L. Maricle and A. H. Maurer, *J. Am. Chem. Soc.*, *89*, 188 (1967).
297. A. Zweig, D. L. Maricle, J. S. Brinen, and A. H. Maurer, *J. Am. Chem. Soc.*, *89*, 473 (1967).
298. B. L. Funt and K. C. Yu, *J. Polymer Sci.*, *62*, 359 (1962).
299. B. L. Funt and F. D. Williams, *J. Polymer Sci. A*, *2*, 865 (1964).
300. (a) B. L. Funt and S. W. Laurent, *Can. J. Chem.*, *42*, 2728 (1964); (b) B. L. Funt and S. N. Bhadani, *ibid.*, *42*, 2733 (1964); *J. Polymer Sci. A*, *3*, 4191 (1965).
301. N. Yamazaki, S. Nakahama, and S. Kambara, *J. Polymer Sci. B*, *3*, 57 (1965).
302. B. L. Funt, D. Richardson, and S. N. Bhadani, *Can. J. Chem.*, *44*, 711 (1966).
303. R. A. Marcus, *J. Chem. Phys.*, *43*, 2654 (1965).
304. E. A. Chandross and R. E. Visco; to be published.
305. R. S. Mulliken, *J. Phys. Chem.*, *56*, 801 (1952).
306. H. Kainer and A. Uberle, *Chem. Ber.*, *88*, 1147 (1955).
307. R. Foster and T. J. Thomson, *Trans. Faraday Soc.*, *58*, 860 (1962).
308. W. Liptay, G. Briegleb, and K. Schindler, *Z. Elektrochem.*, *66*, 331 (1962).
309. D. Bryce-Smith and J. E. Lodge, *J. Chem. Soc.* (*London*), *1962*, 2675.
310. C. Lagercrantz and M. Yhland, *Act. Chem. Scand.*, *16*, 508, 1043, 1799 (1962).
311. R. L. Ward, *J. Chem. Phys.*, *38*, 2588; *39*, 852 (1963).
312. D. Bijl, H. Kainer, and A. C. Rose-Innes, *J. Chem. Phys.*, *30*, 765 (1959).
313. L. S. Singer and J. Kommandeur, *J. Chem. Phys.*, *34*, 133 (1961).

314. M. Bose and M. M. Labes, *J. Am. Chem. Soc.*, *83*, 4505 (1961).
315. D. B. Chesnut and W. D. Phillips, *J. Chem. Phys.*, *35*, 1002 (1961).
316. A. Ottenberg, C. J. Hoffman, and J. Osiecki, *J. Chem. Phys.*, *38*, 1898 (1963).
317. (a) H. Scott, G. A. Miller, and M. M. Labes, *Tetrahedron Letters, 1963*, 1073; (b) H. Scott and M. M. Labes, *J. Polymer Sci. B*, *1*, 413 (1963).
318. L. P. Ellinger, *Polymer*, *5*, 559 (1964).
319. J. W. Beitenbach and O. F. Olaj, *J. Polymer Sci. B*, *2*, 685 (1964).
320. L. P. Ellinger, *Polymer*, *6*, 549 (1965).
321. K. Takakura, K. Hayashi, and S. Okamura, *J. Polymer Sci. B*, *3*, 565 (1965).
322. H. Nomori, M. Hatano, and S. Kambara, *J. Polymer Sci. B*, *4*, 261 (1966).
323. D. Laurin and G. Paravano, *J. Polymer Sci. B*, *4*, 797 (1966).
324. J. Pac and P. H. Plesch, *Polymer*, *8* (5), 237, 252 (1967).
325. B. L. Funt, in *Macromolecular Reviews*, A. Peterlin, M. Goodman, S. Okamura, B. H. Zimm, and H. F. Mark, Eds, Interscience, New York, 1967.
326. L. S. Marcoux, J. M. Fritsch, and R. N. Adams, *J. Am. Chem. Soc.*, *89*, 5766 (1967).
327. S. A. Rice, *Account of Chem. Res.*, *1*, 81 (1968).
328. S. Penczek, J. Jagur-Grodzinski, and M. Szwarc, *J. Am. Chem. Soc.*, *90*, 2174 (1968).

Author Index

Numbers in parentheses are reference numbers and show that an author's work is referred to although his name is not mentioned in the text. Numbers in *italics* indicate the pages on which the full references appear.

A

Aalbersberg, W. I., 350(46), 351, 353 (46), 375(90), *430*, *431*
Abe, K., 52(331), *75*
Abraham, R. J., 14(432), 49(412), 52 (333), *75*, *78*
Acrivos, J. V., 383, *433*
Adam, F. C., 413(234), *435*
Adam, W., 171, *186*
Adams, G. E., 377(102), *432*
Adams, R. N., 414, 416(242), 425, *436* *437*
Adcock, W., 149, 153, 161, 164, *186*, 192, 244, 254(1), *255*
Adrian, F. J., 20(51), *68*
Ahmad, M., 57(439), 58(439), *78*
Ainsworth, J., 47(288), *74*
Albrecht, A. C., 372(88), *431*
Alder, B. J., 256(6), *291*, 382, 383, *433*
Allen, G., 406, *435*
Allen, L. C., 19(46), 22(46), 32, *68*
Allen, T., 18(31), 20, 21(31), *67*
Allerhand, A., 13(23), *67*
Allred, A. L., 122, 130, 131(36), *144*
Amako, Y., 21(34), *67*
Ames, D. P., 83(12), *107*
Anbar, M., 382(143), *433*
Andersen, K. K., 148(3), 155(15), 164 (15), 169(3,15), 173(3), *186*, 192, 255, 293, 294(6), 299(6), 301(6), 303(6), 308(6), 309(6), 314–316(6), 320(6), *321*
Anderson, J. K., 61(380), 64(380), *77*
Anderson, J. M., 98(66), *108*
Anderson, W. A., 62(384), 65(384), *77*, 412, *436*
Andresen, H. G., 8(13), 45(13a), *67*

Andrewskii, D. N., 41(309), *75*
Anet, F. A. L., 57(439), 58(439), *78*
Angelo, B., 384(161), *433*
Appenrodt, J., 325(2), 388(2), 400(2), *429*
Arai, S., 352(101), 376(101), 377, 417 (243), *432*, *436*
Arest-Yakubovich, A. A., 422(259), *436*
Armstrong, R. J., 246(10), 254(10), *255*
Arnold, E., 380(130,147), *433*
Aronson, J. A., 42(225), *72*
Asami, R., 400, *434*
Aston, J. G., 18(32), 21(32), 38(149,150, 160), 39(169), 42(221), 43(235,244), 44(244,255), 55(354), 59(360), 62 (149), 63(150), 64(403,404), 67, *70–73*, *76*, *77*, 216
Aten, A. C., 352(54,55), *430*
Atherton, N. M., 410, *436*
Axelrod, M., 94(48), *108*
Axford, D. W. E., 47(290), 48(290), *74*

B

Bachman, W. E., 397, *435*
Bachmann, W. E., 357(62), *431*
Baciocchi, E., 312(36), *321*
Baddeley, G., 87(25), *107*
Badder, F. A., 258
Badger, R., 24(68), 65(68), *68*
Badmani, A., 392(280), *437*
Bahsteter, F., *436*
Baird, N. C., 121, 128, 137(51), *144*, *145*
Bak, B., 17(29), 24(69), 37(138), 40(195, 196), 62(385), *67*, *68*, *70*, *72*, 77
Balch, C., 312(35), *321*

439

Subject Index

A

B

C

D

E

Progress in Physical Organic Chemistry

CUMULATIVE INDEX, VOLUMES 1–6